CMS/CAIMS Books in Mathematics

Volume 16

CMS/CAIMS Books in Mathematics is a collection of monographs and graduate-level textbooks published in cooperation jointly with the Canadian Mathematical Society- Societé mathématique du Canada and the Canadian Applied and Industrial Mathematics Society-Societé Canadienne de Mathématiques Appliquées et Industrielles. This series offers authors the joint advantage of publishing with two major mathematical societies and with a leading academic publishing company. The series is edited by Karl Dilcher, Frithjof Lutscher, Nilima Nigam, and Keith Taylor. The series publishes high-impact works across the breadth of mathematics and its applications. Books in this series will appeal to all mathematicians, students and established researchers. The series replaces the CMS Books in Mathematics series that successfully published over 45 volumes in 20 years.

CMS
SMC

CAIMS
SCMAI

Eric L. F. Roettger · Hugh C. Williams

The Enchantment of Numbers

Springer

Eric L. F. Roettger
Department of Mathematics and Computing
Mount Royal University
Calgary, AB, Canada

Hugh C. Williams
Department of Mathematics and Statistics
University of Calgary
Calgary, AB, Canada

ISSN 2730-650X ISSN 2730-6518 (electronic)
CMS/CAIMS Books in Mathematics
ISBN 978-3-031-76567-4 ISBN 978-3-031-76568-1 (eBook)
https://doi.org/10.1007/978-3-031-76568-1

Mathematics Subject Classification: 00 General and overarching topic; collections, 01 History and biography, 11 number theory, 68 computer science

This Springer imprint is published by the registered company Springer Nature Switzerland AG
The registered company address is: Gewerbestrasse 11, 6330 Cham, Switzerland

To Shaunessy and Lynn.

Preface

Seek out distant horizons, and cherish those you attain.
—Māori Proverb

Human beings do not usually live in isolation from one another; what we find most interesting about ourselves derives from our interconnection with others. The same is true about numbers. This is a book about numbers and their remarkable relationships, and it is intended for people who simply like numbers. We are always surprised by how many individuals fall into this category. In the case of this volume, it is easy for us to list: students, number enthusiasts, amateur and professional number theorists, but there are many others who may not even be aware of a latent passion for numbers. In this case, it is our hope that this book might enable or at least encourage that realization.

Although there are many categories of numbers, when most people think about numbers they think about the whole numbers, or integers–the numbers that we use for counting. These, of course, are just abstract ideas; they cannot be perceived by our usual senses of taste, touch, sight, smell or hearing. They are constructs of the mind only, and as such are pristine and cannot be sullied by use. They are perfect, beautiful and eternal. The German mathematician Leopold Kronecker (1823–1891) is reported to have put it this way: "God made the integers, all else is the work of man". Perhaps this is why many of us find them so attractive. In general, however, it is impossible to predict why numbers might appeal to particular individuals, but for many it is the patterns that they make. Certainly, our own fascination with numbers developed from examining certain progressions or sequences of them, for example, the primes.

No matter how many initial terms of any sequence of numbers you are given, there are infinitely many successions of numbers with these same initial terms. If you are given 1, 3, 5, 7 and 13 as initial terms and asked to provide the next term in the sequence, you could select any number that you please because there are infinitely many rules by which the next term could be found. Thus, it is important in characterizing a particular sequence to give the first few terms and the rule by which all subsequent terms can be determined. Consider the sequence 4, 14, 194, 37634, . . . , where the next term is found by squaring the immediately previous term and subtracting 2. Notice that prime 7 divides the second term of the sequence and prime 31 divides the fourth. Indeed, if $2^{n+1} - 1$ is a prime, it must divide the nth term of the sequence. Furthermore, if n is even and $2^{n+1} - 1$ divides the nth term of the sequence, then it must be a prime. Isn't that amazing? Why is this true? It was this kind of result and the questions concerning it that instilled in us a life-long passion for the integers. They are at once charming and bewitching. In *Antony and Cleopatra*, Shakespeare wrote the following sublime tribute to Cleopatra, "Age cannot wither her, nor custom stale/Her infinite variety: other women cloy/The appetites they feed, but she makes hungry/Where most she satisfies". For those of us who have been smitten by the integers, an only slightly rewritten accolade could be applied to them.

As mathematicians, we appreciate the ineffable beauty of numbers. Paul Erdös, the "man who only loved numbers", has been quoted as saying:

> Why are numbers beautiful? It's like asking why is Beethoven's Ninth Symphony beautiful. If you don't see why, someone can't tell you. I *know* numbers are beautiful. If they aren't beautiful, nothing is.

Unfortunately, in order to appreciate this beauty, one must have some familiarity with mathematics, something that either through circumstance or inclination, many people do not possess. It is easy to understand that many individuals might be put off by the abstraction and abstruse language of mathematics. As Barbie said, "math class is tough". However, since mathematics by its very nature is abstraction, there is no way we can avoid it, nor would we wish to. We will try in this book to introduce ideas slowly at first and to bring in notation only as it is needed, but as Euclid is supposed to have said to King Ptolemy I, "there is no royal road" or shortcut to learning mathematics. However, the rewards for doing so are enormous. Once you understand the fundamental tenets of mathematics (logical thinking), there is nothing you cannot learn, and just as in learning to read, this opens many doors which would otherwise remain closed, to the wonders of our marvelous world.

Most of this work should be accessible to anyone with a high school education, but we include some more advanced material from time to time. It is our hope that this might challenge the reader to learn more, and to that end, we include a list of sources for further reading at the end of each chapter. The reader should be advised that while some of these sources are less approachable than others, they all contain significant portions of easily understandable information, which are relevant to our theme, particularly in their introductory sections and bibliographies. These lists consist mainly of items with which we are familiar or which sit on our personal bookshelves and should not, therefore, be regarded as anywhere near complete. We apologize here to the many authors whose work we have not cited. We also encourage readers to consult the Internet, but with some caution as some sites are less accurate than they might be. In addition, we warn the reader that the Internet is plagued by frustrating and all-too-frequent instances of seemingly arbitrary document disappearance. Nevertheless, we have occasionally included some URLs in this work, and we highly recommend Neil Sloane's *Online Encyclopedia of Integer Sequences* (OEIS) and Chris Caldwell's *PrimePages*.

We have included a number of theorems but, although mathematics is based upon the concept of proof, we have nevertheless for the most part elected to omit proofs from this work. We have included a few where they seemed important or illustrative of a certain important idea, but it is our experience in trying to encourage students to develop an interest in mathematics that it is best first to discuss results, rather than to indulge in the very technical, esoteric and often formidable process of devising proofs. As mathematicians, we bypass these arguments very reluctantly, but we hope that the resulting economy and accessibility of the remaining material makes up for this sacrifice. Readers interested in demonstrations of any of the results mentioned in what follows can find relevant material in the sources mentioned at the end of each chapter.

In order to assist the beginner in getting acquainted with the topics discussed we provide some practice problems at the end of most chapters. These problems focus on application or calculation which we hope will strengthen one's understanding of the material. Often they encourage the reader to replicate methods that are discussed in the text and thereby give the reader the opportunity to enjoy working with the integers. We have also provided solutions to some of these problems at the back of the book. These problems are headed by the notation [**BB**]. The principal objective of this mainly didactic book is a description of certain entertaining aspects of the integers and of their connections to one another, but we also try to provide to the reader some idea of why these rather surprising phenomena occur. Thus, this volume should be considered at a level intermediate between a collection of disjoint mini-essays and an elementary number theory textbook.

It has long been recognized that play is a vital component of human activity, and when mathematical practitioners play with mathematics, the result is often called *recreational mathematics*, a study in which more professional mathematicians delight than are perhaps willing to admit. A significant component of this book is devoted to recreational number theory. Although it goes well beyond this, we will consider Number Theory to be that branch of mathematics dedicated to the study of whole numbers (the integers); recreational number theory is number theory practised as recreation or entertainment rather than a strictly research and/or application-driven professional activity. In short, it's just supposed to be fun. What constitutes recreational number theory is not easily described, but the material in this book straddles several of the areas, which are generally regarded as forming the major part of this subject, namely:

1. solution of problems involving integers,
2. certain number sequences,
3. primes,
4. construction of algorithms,
5. special stratagems and devices for solving problems,

6. numbers of special forms,
7. digital properties of integers,
8. solution of selected Diophantine equations,
9. history of number theory.

We have not made any attempt at a comprehensive approach to any of these topics. Instead, we have organized this book into 21 chapters, each of which discusses some aspects of a particular number-theoretic theme. We have selected these in deference to Mark Twain's advice to, "write about what you know;" thus, we have elected to present this material as viewed through the prism of our personal research experience. Over the years, our acquisition of this knowledge has added not only to our understanding of certain properties of the integers but also contributed greatly to our enjoyment of life. There is something very satisfying about losing oneself in the study of a particular problem involving the integers—it's a delightful form of meditation and highly recommended.

Any one of these chapters might involve one or several of the above-listed nine recreational areas. In the first five chapters, we introduce the ingredients upon which the remaining chapters will largely depend. We discuss the concepts, some possible recreational aspects and introduce all the necessary notation for a novice to follow along; furthermore, none of these chapters is intended to be overly recondite or abstruse. Again, much more information can be found in the sources cited at the end of each chapter.

While there are many excellent books dedicated to various facets of recreational number theory, most of them do not deal with number theory itself in much depth. We will not delve into number theory here with the insight provided by a textbook, but we will from time to time, where appropriate, include some ancillary information about number theory and its history. Naturally, a work such as this represents a compilation of the contributions of many individuals, past and present. We will usually, on introducing these researchers, include birth and death dates for those who are now deceased. We will also provide some brief indication of their profession and nationality, but as so many of these contributors were or are American mathematicians, in the interest of brevity, we omit stating their citizenship.

An inspiration for this work is the book by the electrical engineer Albert H. Beiler (1898–1988) entitled, *Recreations in the Theory of Numbers—The Queen of Mathematics Entertains*. This volume, published by Dover Publications in 1964 and still in print, is composed of 26 chapters. Each chapter is devoted to a particular aspect of number theory, but from the viewpoint of enjoyment rather than scholarship. Many parts of it are, as might be expected after 60 years, now out of date, but the author's passion for the material is still infectious. We will discuss several, but not all, of the topics in this book both from the perspective of modern number theory and its computational component. This latter study is devoted to finding and implementing efficient computer algorithms for solving various problems arising in the theory of numbers. We will also update several of Beiler's statements; however, we cannot match his rather rich writing style. Many years of survival in academia tend to edit exhilaration from one's mode of expression, but we hope that a fascination that has survived for almost 60 years provides some testimony to our enthusiasm for this material.

In this book, we discuss properties of the integers that have charmed, captivated and delighted us, beginning in our teens and extending to the present. Indeed, we must confess that even in our professional lives we have often practised number theory as a recreational pursuit. The reader should bear this in mind when scrutinizing this work. The book is admittedly idiosyncratic, and a somewhat bittersweet aspect of writing it is our realization that as computer technology continues to advance, much the numerical material will soon be out of date. Nevertheless, it is our hope that our readers will enjoy reading this volume as much as we did writing it.

Calgary, Canada

Eric L. F. Roettger
Hugh C. Williams

References

Chris K. Caldwell, editor, The PrimePages: Prime Number Research and Records, published electronically at https://t5k.org

N. J. A. Sloane, editor, The On-Line Encyclopedia of Integer Sequences, published electronically at https://oeis.org

Acknowledgements

We are most grateful for the efforts of two proofreaders who volunteered to review an earlier version of this book. Rose Rong, a grade 12 high school student, and Aaron Christie, a mathematician and Canadian federal public servant, were able to eliminate a number of misprints, errors and grammatical howlers. Nevertheless, the reader is warned that in spite of these efforts and our own, many blunders and solecisms undoubtedly remain. These, of course, are the sole responsibility of the authors and we apologize in advance for any inconvenience caused by our inattention.

One author (ELFR) owes thanks to Karim Dharamsi, Vice-Provost Academic at Mount Royal University, and Michael Quinn, Provost and Vice-President Academic at Vancouver Island University, for providing both research and computational support.

We also want to thank the editors of the Canadian Mathematical Society's CMS/CAIMS Books in Mathematics series for accepting our rather unusual contribution to their sequence of titles.

Donna Chernyk, Senior Editor (mathematics) for Springer Nature, was most helpful in guiding us through the publication process, and we wish to express our gratitude to her for her assistance.

We want further to acknowledge the patience and forbearance of our families for both encouraging and tolerating us during the production of this book.

Contents

Symbols

We list below, in the order they appear in the book, several of the symbols we frequently use. However, the reader is cautioned that occasionally the same symbol may be used to represent two or more different objects. For instance, F_n is used for both the nth Fibonacci number and Fermat number. In these cases, the exact meaning of the symbol should be clear through context.

$a > b$	a is greater than b	7
$a \geq b$	a is greater than or equal to b	7
$a < b$	a is less than b	7
$a \leq b$	a is less than or equal to b	7
$\lvert x \rvert$	The absolute value of x	7
$a \mid b$	a divides b	7
$a \nmid b$	a does not divide b	7
$s(n)$	The sum of all aliquot divisors of n	7
$\neq$	Not equal to	8
$\#p_n$	The primorial of p_n	9
$\sqrt{n}$	The square root of n	10
$\pm$	The choice of addition or subtraction	11
$\gcd(a, b)$	The greatest common divisor of a and b	11
$\mathbb{Q}$	The rational numbers	11
$\mathbb{Z}$	The integers	11
$a \cdot b$	a times b	11
$\rightarrow$	Implies	12
$\text{lcm}[a, b]$	The least common multiple of a and b	13
$[a] \pmod{m}$	Residue class of a modulo m	14
$a \equiv b \pmod{m}$	m divides $a - b$	14
a^{-1}	Multiplicative inverse of a	15

CSR (mod m)	Complete set of residues modulo m	15
$g(a_1, a_2, \ldots, a_n)$	Frobenius number	17
$(r_0 r_1 \ldots r_k)_b$	An integer written in base-b	23
$\log_b$	The power to which b needs to be raised to yield a given number	24
$\leftarrow$	Replace by	32
ϵ	Is an element of	33
P	$(10^8)^{(10^8)}$	35
L	$P^{(10^8)}$	35
$\lceil x \rceil$	The ceiling function	47
$\lfloor x \rfloor$	The floor function	47
P	Computational procedure	48
$O(f(x))$	Big O-notation	49
$o(f(x))$	Little O-notation	49
$\Omega(f(x))$	Big Ω-notation	49
$\lim_{x \to \infty}$	The limit as x approaches infinity	49
e	Euler's constant: $\lim_{n \to \infty} (1 + 1/n)^n = 2.71828...$	50
$\tilde{O}(f(x))$	Soft Big O-notation	50
RSR (mod m)	Reduced set of residues modulo m	50
$\phi(m)$	The Euler phi function or totient of m	50
$\text{ord}_m\, a$	The order of a modulo m	53
$a \not\equiv b \pmod{m}$	m does not divide $a - b$	55
$\text{ind}_g\, y$	Index of y to the base g modulo m	55
$(a\|q)_l$	lth power residuacity symbol	61
$(a\|q)$	Legendre symbol	61
$(P\|Q)$	Jacobi symbol	62
$\sum$	Values being summed	69
$\prod$	Values being multiplied	69
$n!$	n factorial: $\prod_{i=1}^{n} i$	70
$a\|\|b$	a exactly divides b	70
$v_p(a)$	$v_p(a) = m$ where $p^m \|\| a$	70
$\sigma_p(n)$	The sum of the p-ary digits of n	71
$\binom{n}{k}$	n choose k	72
C_n	The nth Catalan number	76
$E(m)$	Encryption function	95
$D(c)$	Decryption function	95
ϕ	$\frac{1+\sqrt{5}}{2}$	104
F_n	The nth Fibonacci number	105

(X_n)	An ordered set of integers or sequence	105
L_n	The nth Lucas number	109
$\epsilon(n)$	The Legendre symbol $(n\|5)$	112
$\sigma(n)$	The sum of all divisors of n	119
$s_k(n)$	The kth element of the aliquot sequence for n	121
M_n	The nth Mersenne number, $2^n - 1$	123
$(U_n),\ (V_n)$	Lucas sequences	134
$r(p)$	Rank of appearance of p	136
$\Phi(m)$	Lucas sequences analogue of $\phi(m)$	136
$(\bar{U}_n),\ (\bar{V}_n)$	Lehmer sequences	138
$\sim$	Asymptotically equivalent	140
$\pi(x)$	The number of primes less than x	146
$\mathrm{li}(x)$	Logarithmic integral function	146
$\int_a^b$	Integral of a function evaluated between a and b	147
$\lim\limits_{\epsilon\to 0^+}$	The limit as ϵ approaches 0 from the right	147
$\mathrm{Li}(x)$	Euler's logarithmic integral	147
L_p	Pseudosquare	151
F_k	The kth Fermat number	155
i	$i^2 = -1$	173
LDS	Linear divisibility sequence	175
$[a_0; a_1, a_2, \ldots a_n]$	Simple continued fraction	186
$\frac{A_i}{B_i}$	The ith convergent	187
$[a_0; a_1, a_2, \ldots]$	Infinite simple continued fraction	187
$\mathbb{Q}(\sqrt{d})$	Set of all quadratic irrationals of radicand d	194
$\overline{\alpha}$	Conjugate of an element	194
$\frac{P+\sqrt{D}}{Q}$	Quadratic irrational	194
$p(D)$	The period length of the SICF of $\sqrt{D}$	236
$G(D)$	$G(D) = \frac{p(D)}{\sqrt{D}\log\log 4D}$	237
max	The maximal value	237
$Pr(a_{j+1} = k)$	The probability that the partial quotient a_{j+1} is some integer k	238
p_n	The nth prime	243
A	Mills' constant	245
α	Wright's α	246
$f_A(x)$	$x^2 + x + A$, where A is an integer	249
$\mathbb{Q}(\sqrt{D})$	All numbers of the form $(a + b\sqrt{D}/c)$, where $a,\ b,\ c\ (\neq 0)$ are integers	249

Chapter 1
Introduction

This is a book about interesting and entertaining numbers, but not just any numbers. We will restrict our attention to the *integers*, numbers which can be written without a fractional component, e.g. 3, -5 and 89. Of course, numbers like 3.75, $22/7$, $\sqrt{2}$ are not integers. Because counting is fundamental to almost all human activity, the integers were the only category of numbers of which our early ancestors were aware. Indeed, today we distinguish those integers used in counting, the positive integers, as the *natural numbers*. What makes the integers special is that while the sum, difference or product of any two integers is also an integer, their quotient is in general not an integer. We learned this in elementary school when performing long division. Consider $315/29$: we can divide 315 by 29 and find that $315 = 10 \times 29 + 25$. The quotient is 10 and the remainder is 25. More generally, it was known to Euclid that if a and b are integers and b is positive, then there exist integers q (the *quotient*) and r (the *remainder*), where r is nonnegative and less than b such that

$$a = q \times b + r.$$

This *Euclidean condition* is a cornerstone in the study of the properties of the integers.

Next to nothing is known about Euclid (fl. 300 BCE) which is remarkable, given the profound effect his work, especially the *Elements*, has exerted on Western thought. What is known has been recounted by Sir Thomas Heath (1861–1940) and more recently by David Fowler (1937–2004). We will have cause to mention Euclid's work again.

We observe that, on dividing by b, we have $a/b = q + r/b$; also, because r is an integer, it must belong to the set $\{0, 1, 2, \ldots, b-1\}$. If $r = 0$, then a/b is an integer and we say that *b divides a* or, alternatively, b is a *divisor* of a. In most cases, however, it is far more likely that b will not divide a and a/b is not an integer. While this might appear to be a complicating factor in sorting out the characteristics of the integers, it is also what makes them interesting. Indeed, we even have a "folk"

E. L. F. Roettger and H. C. Williams, *The Enchantment of Numbers*, CMS/CAIMS Books in Mathematics 16, https://doi.org/10.1007/978-3-031-76568-1_1

theorem, a theorem not attributable to any single individual, that all natural numbers are interesting. For, consider the set of all uninteresting natural numbers; this set of positive integers must have a least member, but that would make this integer, the least uninteresting natural number, interesting, a contradiction.

Consider the following example of a curious property of some special numbers written with repeated digits:

$$
\begin{aligned}
(111 \times 111) - 10 \times (11 \times 11) &= 11111 \\
(1111 \times 1111) - 10 \times (111 \times 111) &= 1111111 \\
(11111 \times 11111) - 10 \times (1111 \times 1111) &= 111111111 \\
(111111 \times 111111) - 10 \times (11111 \times 11111) &= 11111111111 \\
(1111111 \times 1111111) - 10 \times (111111 \times 111111) &= 1111111111111 \\
(11111111 \times 11111111) - 10 \times (1111111 \times 1111111) &= 111111111111111 \\
(111111111 \times 111111111) - 10 \times (11111111 \times 11111111) &= 11111111111111111.
\end{aligned}
$$

The question that might occur to you is, "does this peculiar property of these numbers persist?" At first glance, this might seem like a difficult question, but it is easy to verify that for any positive integer n

$$(10^n - 1)^2 - 10 \times (10^{n-1} - 1)^2 = 9 \times (10^{2n-1} - 1).$$

Since a number of the form

$$\underbrace{111\ldots1}_{n \text{ ones}}$$

can be written as $(10^n - 1)/9$, it is easy to see that because

$$\left(\frac{10^n - 1}{9}\right)^2 - 10 \times \left(\frac{10^{n-1} - 1}{9}\right)^2 = \frac{10^{2n-1} - 1}{9},$$

the above pattern will persist indefinitely. As we will see in Chap. 10, the above identity is a special case of a more general one involving the Lucas sequences.

As another example of an interesting number, consider the integer 142857 and observe that

$$
\begin{aligned}
1 \times 142857 &= 142857 \\
2 \times 142857 &= 285714 \\
3 \times 142857 &= 428571 \\
4 \times 142857 &= 571428 \\
5 \times 142857 &= 714285 \\
6 \times 142857 &= 857142.
\end{aligned}
$$

Each of these products contains the same 6 digits of the original number 142857 in the same cyclic order. Imagine these digits arranged clockwise on the circumference of a circle:

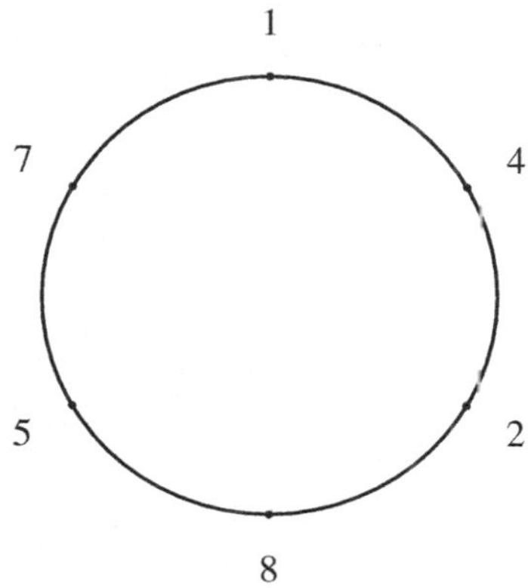

Each of the products corresponds to an array of 6 digits selected from those on the circle and in the same order, but starting at a different place. We call these arrays of digits *cyclic rearrangements* of 142857. Note also that $142 + 857 = 999$. This all seems most remarkable. Are there other numbers, besides 142857, which possess these same properties? What's going on here? If these are the sorts of questions that occur to you, then you should be warned that you are very likely to become a lover of recreational number theory.

As a further example, we mention that during his military service in the Franco-Prussian War (1870–1871), Édouard Lucas, a great mathematical recreationist, invented the problem of the "pile of cannonballs"; in this problem, the solver is required to find the conditions under which a pile of cannonballs with a square base and organized as a square pyramid contains a number of cannon balls which is the square of an integer. Lucas believed (rightly, it turned out) that there could only be 1 or 4900 cannonballs in such a pile, but could not prove it, even though he thought he had. This problem, which is equivalent to finding x, y satisfying

$$1^2 + 2^2 + 3^2 + \cdots + x^2 = (x(x+1)(2x+1))/6 = y^2,$$

where x and y are positive integers, is an example of what is called a Diophantine equation, and this one turned out to be much more difficult to solve than Lucas had anticipated. It was finally solved in 1919 by a very complicated argument of the English mathematician George Neville Watson (1886–1965).

While recreational mathematics is often regarded as the precinct of amateurs and enthusiasts, it can, as we have seen above, be a source of very interesting and often challenging problems. Because of this or just the simple desire for diversion, even serious theoreticians often indulge in this pastime. Indeed, it must be said that many professional number theorists began their careers as recreational number theorists, because a fascination with the integers began in their youth when they actually started to become curious about numbers. Often this process was stimulated by remarks made by individuals with more advanced knowledge, such as teachers or other authorities. For instance, the equation $x^2 - Dy^2 = 1$, where D is a given positive non-square integer and the solutions x and y are constrained to be integers, is called a *Pell equation*. For example, the Pell equation $x^2 - 7y^2 = 1$ has the solution $x = 8, y = 3$. One of us (HCW) can recall, on reading Carmichael's book *Diophantine Analysis* as a teenager, the statement (in a footnote on p. 33) that the least integer value of x such that $x^2 - 1621y^2 = 1$, where y must also be an integer, is a number of 76 digits. As this is an enormous number, how could he possibly know this? He explained that this x could be computed by developing $\sqrt{1621}$ into a continued fraction, whatever that is. These seemingly innocuous remarks motivated a life-long interest in the fascinating properties of continued fractions, particularly those of the square roots of positive integers, and of the Pell equation (see Chaps. 14, 17 and 19).

What we have discussed so far is just a tiny sample of the many number delights that are to follow. However, since mathematics is the ultimate example of reductionism, it is essential that we begin at the beginning. In as much as mathematics is also a language consisting of abstract symbols, we have first to learn some words, and this means that we must acquire knowledge of some notation. We then need to learn the grammar—how to manipulate the symbols and the accompanying notation. This is something that is best done, as in the case of any other language, by trying it and learning from mistakes. We start in the next chapter.

Further Reading

Recreational number theory problems

Aaron Bakst, *Mathematics: Its Mystery and Mastery*, 3rd Edition, D. Van Nostrand Co., NJ, 1967 pp. 31–50.

Albert H. Beiler, *Recreations in the Theory of Numbers—the Queen of Mathematics Entertains*, Dover Publications, NY, 1964.

Jerome S. Meyer, *Fun with Mathematics*, World Publishing Co., NY, 1952.

Pile of cannonballs problem

W. S. Anglin, The square pyramid puzzle, *American Mathematical Monthly*, Vol. 97 (1990), 120–194.

The Pell equation

Robert D. Carmichael, *The Theory of Numbers and Diophantine Analysis*, Dover Publications, NY, 1959.
Edward J. Barbeau, *Pell's Equation*, Springer, NY, 2003, pp. 64–74.
Harold Davenport, *The Higher Arithmetic*, Dover Publications, NY, 1983, pp. 79–114.

Euclid

David Fowler, *The Mathematics of Plato's Academy*, 2nd Edition, Clarendon Press, Oxford, 2003, pp. 204–208.
Sir Thomas Heath, *A History of Greek Mathematics*, Vol. 1, Dover Publications, NY, 1981, pp. 354–360.

Chapter 2
Division, Factors, Primes, Congruences, GCD, etc.

Before going any further, we need to review some basic notation that will be used in what follows. We usually employ Latin letters to denote unspecified integers. We denote by $a > b$ ($a \geq b$) that a is greater than b (a is greater than or equal to b). Also, $a < b$ ($a \leq b$) means that a is less than b (a is less than or equal to b). As examples, we note that $5 > 4, 5 \geq 4, 4 < 5, 4 \leq 5$. Notice that if $a \leq b$ and $a \geq b$, then $a = b$; also, if $a \geq b$, then $-a \leq -b$. We use $|x|$ (the *absolute value* of x) to denote x when $x \geq 0$ or $-x$ when $x < 0$. Thus, $|x| \geq 0$. For example, $|-5/4| = 5/4$, $|3/2| = 3/2$.

We recall that if a and b are integers and b is positive, then there exists q (the quotient) and r (the remainder), where r is in $\{0, 1, 2, \ldots, b-1\}$ such that

$$a = q \times b + r.$$

In the case that $r = 0$, we say that b *divides* a, b is a *divisor* of a, or b is a *factor* of a. If b divides a, we denote this by $b \mid a$; if b does not divide a, we write $b \nmid a$. Thus, $5 \mid 30$, but $7 \nmid 30$. Observe that if a and b are positive integers such that $a \mid b$ and $b \mid a$, then a must be equal to b. If a and b are positive and b is a divisor of a less than a, we say that b is an *aliquot* or *proper divisor* of a, e.g. 10 is an aliquot divisor of 30, but 30 is not. By using the above concept of divisors, we can define *prime numbers* or *primes* as positive integers that are only divisible by 1 and themselves. Alternately, we could say that a prime is an integer greater than 1 which has only one aliquot divisor.

Suppose we let the function $s(n)$ denote the sum of all the aliquot divisors of n. For example,

$$s(60) = 1 + 2 + 3 + 4 + 5 + 6 + 10 + 12 + 15 + 20 + 30 = 108$$

and

E. L. F. Roettger and H. C. Williams, *The Enchantment of Numbers*, CMS/CAIMS Books in Mathematics 16, https://doi.org/10.1007/978-3-031-76568-1_2

$$s(108) = 1 + 2 + 3 + 4 + 6 + 9 + 12 + 18 + 27 + 36 + 54 = 172.$$

If we define $s_0(n) = n$, $s_1(n) = s(n)$, and for any positive integer k

$$s_k(n) = s(s_{k-1}(n)),$$

we get

$$s_0(60) = 60, \quad s_1(60) = 108, \quad s_2(60) = s(s_1(60)) = s(108) = 172.$$

Notice that if p is any prime, then $s(p) = 1$. For a given n, we call the sequence

$$n\ (= s_0(n)), s_1(n), s_2(n), s_3(n), \ldots$$

the *aliquot sequence belonging to n*. For example, the aliquot sequence belonging to 60 is

$$60, 108, 172, 136, 134, 70, 74, 40, 50, 43, 1, \ldots.$$

Since $s(1) = 0$, this sequence cannot be extended any further. We mention that $s(0) = 0$ and that $s(n) \neq 0$ for any positive integer except $n = 1$.

In mathematics the term *conjecture* refers to a hypothesis that is put forward without proof. An old conjecture of Eugène Catalan (1814–1894) (later modified by L. E. Dickson (1874–1954)) states that for any n, the entries in the aliquot sequence belonging to n are bounded by a value that depends only on n, but this statement has vigorously resisted a proof. If you wish to investigate this further, you are cautioned that this is a more difficult problem than it seems. For example, the aliquot sequence belonging to 276 (the smallest number for which we have no indication of the truth of the *Catalan-Dickson conjecture*) has been computed to 2145 terms with $s_{2145}(276) = 2 \times 3 \times 7 \times 13 \times 101 \times C_{209}$, where C_{209} is an integer which is not a prime, consisting of 209 decimal digits, and for which no divisors have been found. Yet we still have no idea as to whether $s_k(276)$ is bounded as k gets large. Indeed, it appears to be increasing.

We can partition the positive integers greater than 1 (1 is not a prime; it is called a *unit*) in a very interesting way, as the integers that are prime and those that are not prime, i.e. *composite numbers*. The set of prime numbers begins as $\{2, 3, 5, 7, 11, 13, 17, 19, \ldots\}$. It was known to Euclid that this set is unbounded, or infinitely large. We will present Euclid's proof here as a theorem. In this demonstration, we make an assumption opposite to what we would like to prove and then show that this leads to a contradiction, and hence the assumption must be false; thus, what we want to prove must be true.

Theorem 2.1 *There are infinitely many prime numbers.*

Proof Assume that there are only finitely many prime numbers, say $p_1, p_2, \ldots, p_n$. We will now construct a number by taking the product of all the primes in this finite set and then adding 1. We call this number q, so we see $q = p_1 \times p_2 \times \cdots \times p_n + 1$.

Note that none of the primes p_i can divide q, thus q is a new prime or has a prime divisor that is not in our list. Either way we have a contradiction. □

To illustrate this proof, we take a small, finite list of primes and construct q as in the theorem; we will see that the number q will have a prime divisor that was not in our original finite list. Consider the set of primes $\{2, 3, 5, 7, 11, 13, 17, 19\}$. Here $q = 2 \times 3 \times 5 \times 7 \times 11 \times 13 \times 17 \times 19 + 1 = 9699691$. This number is not obviously composite, but one can verify that both 347 and 27953 are prime divisors of q. Hence, we have found not one but two new primes not from the original set, using the method of the theorem.

Suppose we consider the ordered set of all primes $\mathcal{P} = \{2, 3, 5, 7, \ldots\}$. That is $\mathcal{P} = \{p_1, p_2, p_3, \ldots\}$, where $p_i > p_{i-1}$. Today we often denote the product

$$p_1 \times p_2 \times p_3 \times \cdots \times p_n$$

by $\#p_n$, pronounced the *primorial of* p_n. For instance, $\#p_5 = 2 \times 3 \times 5 \times 7 \times 11 = 2310$. If we harken back to Euclid's proof, we see that $\#2 + 1$, $\#3 + 1$, $\#5 + 1$, $\#7 + 1$, $\#11 + 1$ are all primes, but $\#13 + 1 = 30031$ is not a prime; it is divisible by 59. According to the *PrimePages*, the largest known prime of the form $\#p_n + 1$ is the 169966 digit $\#392113 + 1$.

Often, for a composite number, we are curious about what primes divide it. When a composite number m is written as a product of primes, say $m = p_1 \times p_2 \times \cdots \times p_n$, we call this the *prime factorization* of m. Of course, these primes do not have to be distinct. For example, for the composite number 24, we have $24 = 2 \times 2 \times 2 \times 3$. It is customary to denote the product of n equal numbers a by a^n, where n is called the *exponent* and a the *base*. This expression is termed the nth *power of* a. Thus, we can write $24 = 2^3 \times 3$. Indeed any positive integer m greater than 1 can be written as

$$m = p_1^{n_1} \times p_2^{n_2} \times p_3^{n_3} \times \cdots \times p_k^{n_k},$$

where the p_i $(i = 1, 2, \ldots, k)$ are distinct primes. For example, $360 = 2^3 \times 3^2 \times 5$. It is important to note that it is a result going back to (yes) Euclid that the above expression for m, the *prime power decomposition of* m, is unique up to the order of the primes in it. Every positive integer greater than 1 has a unique prime decomposition (apart from rearrangements of ordering of the multiplication). This result is called the *fundamental theorem of arithmetic* or the *unique factorization theorem*. It is not uncommon for the number 1 to be mistakenly called a prime; this is an error that often occurs in newspaper articles about primes. However, in the walls of the multiplicative structure of the integers, the number 1 is simply not a brick; also, you can easily see that if 1 were a prime we would no longer have unique factorization, as $6 = 2 \times 3$ and $6 = 1 \times 2 \times 3$.

An easy-to-understand method of proving an arbitrary integer n is prime is to trial-divide by all positive integers less than its square root and greater than 1. If any of the values selected divides n then you have discovered an integer, greater than 1 and less than the number itself, which divides this number, and hence n cannot be prime. On the other hand, if none of the integers in this range divides n, then it must be prime. In this case, it is easier to work with the contrapositive of the previous statement: if our number is not prime, then it is divisible by some number less than its square root. To see this, let the number being considered be denoted by n. Since n is not prime we can write it as $a \times b$ such that $1 < a \leq b < n$, where a and b are integers. Imagine a is larger than $\sqrt{n}$, then $a \times b$ is greater than $\sqrt{n} \times \sqrt{n}$. This implies $a \times b$ is greater than n, which is impossible, as $a \times b$ exactly equals n. Thus, if we wish to establish that 1621 is a prime, we first observe that if it is composite, it must have a prime divisor $\leq \sqrt{1621} \approx 40.26$.

The above primality test can now be implemented as an algorithm, called the *Sieve of Eratosthenes*, to produce a list from 2 to some bound b of all primes less than b. Eratosthenes (c. 276–c. 195 BCE) was an ancient Greek polymath and head of the Library of Alexandria. He made contributions to mathematics, geography, philosophy, music and even poetry. He is best known today for his remarkably accurate computation of the circumference of the Earth.

Steps in the Sieve of Eratosthenes:

Step 1: Make a list of all integers from 2 to the preselected bound b.
Step 2: Circle the 2 and then cross out all multiples of 2 in the list.
Step 3: Circle the next value to the right that is not crossed out and cross out all multiples of this value.
Step 4: Repeat Step 3 until all values less than or equal to $\sqrt{b}$ have been circled or crossed out.
Step 5: The resulting circled values or the integers not crossed out are all the primes less than the selected bound.

For example, let our preselected bound be $b = 40$. We begin by Step 1 making our list as follows:

2 3 4 5 6 7 8 9 10 11 12 13 14 15 16 17 18 19 20
21 22 23 24 25 26 27 28 29 30 31 32 33 34 35 36 37 38 39 40.

Now by Step 2 we have

② 3 ~~4~~ 5 ~~6~~ 7 ~~8~~ 9 ~~10~~ 11 ~~12~~ 13 ~~14~~ 15 ~~16~~ 17 ~~18~~ 19 ~~20~~
21 ~~22~~ 23 ~~24~~ 25 ~~26~~ 27 ~~28~~ 29 ~~30~~ 31 ~~32~~ 33 ~~34~~ 35 ~~36~~ 37 ~~38~~ 39 ~~40~~.

Executing Step 3 twice we have

② ③ ~~4~~ 5 ~~6~~ 7 ~~8~~ ~~9~~ ~~10~~ 11 ~~12~~ 13 ~~14~~ ~~15~~ ~~16~~ 17 ~~18~~ 19 ~~20~~
~~21~~ ~~22~~ 23 ~~24~~ 25 ~~26~~ ~~27~~ ~~28~~ 29 ~~30~~ 31 ~~32~~ ~~33~~ ~~34~~ 35 ~~36~~ 37 ~~38~~ ~~39~~ ~~40~~.

② ③ ~~4~~ ⑤ ~~6~~ 7 ~~8~~ ~~9~~ ~~10~~ 11 ~~12~~ 13 ~~14~~ ~~15~~ ~~16~~ 17 ~~18~~ 19 ~~20~~
~~21~~ ~~22~~ 23 ~~24~~ ~~25~~ ~~26~~ ~~27~~ ~~28~~ 29 ~~30~~ 31 ~~32~~ ~~33~~ ~~34~~ ~~35~~ ~~36~~ 37 ~~38~~ ~~39~~ ~~40~~.

Ultimately we must arrive at the point where all the integers less than $\sqrt{40}$ have either been circled or crossed out, leaving our list of primes less than 40 as 2, 3, 5, 7, 11, 13, 17, 19, 23, 29, 31 and 37. As none of these primes is a divisor of 1621, we have proved that 1621 must be a prime.

We motivate the next part of this chapter with the *Die Hard Jug Problem*. In the movie *Die Hard 2*, the protagonists McClane and Zeus need to defuse a bomb by placing a jug with exactly 4 gallons of water on a scale. They are given a 3 gallon jug and a 5 gallon jug and they are placed in a fountain with as much water as they need. This problem was not invented for the movie; in fact, it is very old, perhaps going back to the fifteenth century, and is an example of what are now called *water jug problems*. How this problem relates to number theory should not be immediately clear. Can you solve this problem? Can you generalize the problem to ones that are solvable and ones that are not?

If for some integers a, b we have another integer c that divides both a and b, then we say c is a *common divisor* of a and b. Note that $c \mid (a \pm b)$ in this case. For example, 15 is a common divisor of 105 and 135. The largest of all these common divisors we call the *greatest common divisor* (gcd) of a and b, and we denote this value by $\gcd(a, b)$. Notice that if $d = \gcd(a, b)$, then $\gcd(a/d, b/d) = 1$. This means that if we have the fraction a/b ($b \neq 0$), then we can write it as $\bar{a}/\bar{b}$, where $\bar{a} = a/d$, $\bar{b} = b/d$ and $\gcd(\bar{a}, \bar{b}) = 1$. This is called writing the fraction a/b in *lowest terms*. Numbers which can be written as a/b where a and b are integers and $b \neq 0$ are called *rational numbers*. Number theorists commonly denote the set of all rational numbers by $\mathbb{Q}$, for quotients, and the set of all integers by $\mathbb{Z}$, which comes from the German word “Zahl” for number. Clearly $\mathbb{Z}$ is a subset of $\mathbb{Q}$.

The greatest common factor of two integers can be found in a naïve way by finding the prime power decompositions of both a and b and then taking their intersection. However, as you will see later, factoring is a hard problem. Fortunately, we can find the $\gcd(a, b)$ without factoring either of a or b, by use of the *Euclidean algorithm*. For simplicity we will restrict ourselves to positive integers a and b and without loss of generality we will assume $a > b$.

We begin by putting

$$a = q_1 \times b + r_1, \quad \text{where} \quad 0 \leq r_1 < b.$$

At this point, we mention that it is customary for mathematicians to write ab as the product of the symbols a and b instead of $a \times b$. Also, we will usually write $a \times b$ as $a \cdot b$ when a and b are explicit integers, rather than symbols. We then repeat the above as follows until we get a remainder term of zero:

$$\begin{aligned} b &= q_2 r_1 + r_2, \quad \text{where} \quad 0 \leq r_2 < r_1 \\ r_1 &= q_3 r_2 + r_3, \quad \text{where} \quad 0 \leq r_3 < r_2 \\ r_2 &= q_4 r_3 + r_4, \quad \text{where} \quad 0 \leq r_4 < r_3 \\ &\vdots \end{aligned}$$

$$r_{n-2} = q_n r_{n-1} + r_n, \quad \text{where} \quad 0 \leq r_n < r_{n-1}$$
$$r_{n-1} = q_{n+1} r_n + 0.$$

The value r_n is the greatest common divisor of a and b. Notice that since $r_k < r_{k-1}$ and each remainder is a nonnegative integer, the algorithm always terminates. We next observe that if $a = bq + r$, then $\gcd(a, b) = \gcd(b, r)$. For suppose $d = \gcd(a, b)$; since $d \mid a$ and $d \mid bq$, we see that since $r = a - bq$, we must have $d \mid r$ and therefore $d \leq \gcd(b, r)$. Similarly, we find that $\gcd(b, r) \mid a$; since $\gcd(b, r)$ also divides b, we get $\gcd(b, r) \leq d$, the greatest common divisor of both a and b. We can only have $\gcd(a, b) = \gcd(b, r)$. Thus, we can verify that the algorithm produces the desired greatest common divisor by noting from our previous observation that

$$\gcd(a, b) = \gcd(b, r_1) = \gcd(r_1, r_2) = \cdots = \gcd(r_n, r_{n+1}) = \gcd(r_n, 0) = r_n.$$

As an example of the Euclidean algorithm, let $a = 1183709$ and $b = 457843$, so then

$$a = q_1 b + r_1 \rightarrow 1183709 = 2 \cdot 457843 + 268023,$$
$$b = q_2 r_1 + r_2 \rightarrow 457843 = 1 \cdot 268023 + 189820,$$
$$r_1 = q_3 r_2 + r_3 \rightarrow 268023 = 1 \cdot 189820 + 78203,$$
$$r_2 = q_4 r_3 + r_4 \rightarrow 189820 = 2 \cdot 78203 + 33414,$$
$$r_3 = q_5 r_4 + r_5 \rightarrow 78203 = 2 \cdot 33414 + 11375,$$
$$r_4 = q_6 r_5 + r_6 \rightarrow 33414 = 2 \cdot 11375 + 10664,$$
$$r_5 = q_7 r_6 + r_7 \rightarrow 11375 = 1 \cdot 10664 + 711,$$
$$r_6 = q_8 r_7 + r_8 \rightarrow 10644 = 14 \cdot 711 + 710,$$
$$r_7 = q_9 r_8 + r_9 \rightarrow 711 = 1 \cdot 710 + 1,$$
$$r_8 = q_{10} r_8 + 0 \rightarrow 710 = 1 \cdot 710 + 0.$$

Hence, the $\gcd(1183709, 457843) = 1$. What is most interesting here is that we now know that a and b have no common prime factors without our having to determine any of the prime factors of either.

If we let $d = \gcd(a, b)$, it can be shown there exist integers s and t such that $as + bt = d$. This is often referred to as Bézout's theorem or identity after the French mathematician Étienne Bézout (1730–1783). It is easy to prove Bézout's theorem by an *existence proof*; this is a proof which simply shows that the desired objects (in this case the integers s and t) must exist without providing a means of obtaining the objects. To this end we let d' be the least positive value of $ax + by$ for all integer values x and y. Such a d' must exist, and we can put $d' = as + bt$ for some integers s and t. We next show that d' must be the $\gcd(a, b)$. Certainly, since $d \mid a$ and $d \mid b$, we must have $d \mid d'$. We observe that because $a = a \times 1 + b \times 0$, we have $d' \leq a$ and because $b = a \times 0 + b \times 1$, we have $d' \leq b$. Let $a = qd' + r$ where $0 \leq r < d' \leq a$; this is the Euclidean condition. We have

$$r = a - qd' = (1 - qs)a - qtb.$$

If r is nonnegative, then since $r < d'$, it can only be zero by the definition of d'. Thus, $d' \mid a$ and similarly $d' \mid b$. It follows that $d' = \gcd(a, b)$.

We can use the Euclidean algorithm to produce a *constructive proof* of the existence of s and t; such a proof provides a method for computing or creating the desired object(s). The constructive version of the proof of Bézout's theorem is often called the *extended Euclidean algorithm* (EEA). For example if $a = 544$ and $b = 85$ and we run the Euclidean algorithm, we get

$$\begin{aligned} 544 &= 6 \cdot 85 + 34 \\ 85 &= 2 \cdot 34 + 17 \\ 34 &= 2 \cdot 17 + 0; \end{aligned}$$

hence, $d = \gcd(544, 85) = 17$. Now, if we isolate each remainder we can work our way down, writing this remainder in terms of a and b as follows:

$$\begin{aligned} &544 = 6 \cdot 85 + 34 \rightarrow 34 = 544 - 6 \cdot 85 \rightarrow 34 = a - 6b \\ &85 = 2 \cdot 34 + 17 \rightarrow 17 = 85 - 2 \cdot 34 \rightarrow 17 = b - 2(a - 6b) = a(-2) + b(13). \end{aligned}$$

Thus, $544s + 85t = d$, for $s = -2$ and $t = 13$. A more formal description of the EEA can be found in Chap. 14.

More generally, it is easy to see that, given integers a, b, c, there exist integers x and y such that $ax - by = c$ if and only if $\gcd(a, b) \mid c$. Can you use this proposition to show that if we have integers a, b, c, such that $c \mid ab$ and $\gcd(c, a) = 1$, then c must divide b? This is the result that is needed to prove the unique factorization theorem. By the way, if two integers a and c have $\gcd(a, c) = 1$, we say that a and c are *relatively prime* or *coprime*.

In adding fractions, we often make use of the *least common denominator*. For example, if we wish to add $3/8$ and $5/12$, we first observe that both of these fractions can be written with the same least common denominator: 24; we then have

$$3/8 + 5/12 = 9/24 + 10/24 = 19/24.$$

In number theory the least common denominator is usually referred to as the *least common multiple* (lcm). We denote the least common multiple of two integers a and b as $\operatorname{lcm}[a, b]$. This lcm is related to the gcd by

$$\operatorname{lcm}[a, b] = \frac{ab}{\gcd(a, b)}.$$

Thus,

$$\text{lcm}[8, 12] = \frac{96}{4} = 24.$$

We now return to the Die Hard jug problem. We note that the $\gcd(5, 3) = 1$. And since we have a 3 gallon jug, really the problem reduces to finding a way of measuring out exactly 1 gallon using the two jugs, at which point we can then add 3 gallons to this 1 gallon to solve the riddle. By the extended Euclidean algorithm we can solve $5 \cdot (-1) + 3 \cdot 2 = 1$. Using this, we can see that by filling the small jug twice and emptying the large jug once we are left with exactly 1 gallon. We do this by filling the small jug and adding it to the empty large jug, at which point we have 3 gallons in the large jug. We can fill the small jug again and use it to top up the large. This will take 2 gallons leaving 1 gallon in the small jug. We can next empty the large jug and then add the 1 gallon to it. Now we note that since we have 1 gallon in the large jug, we can top up with exactly 3 gallons from the small jug. Note this is not the only way to solve this particular puzzle; can you think of another way?

Generalizing the problem, if we are given two jugs of size a gallons and b gallons and $\gcd(a, b) = 1$, we can use the extended Euclidean algorithm to solve $as + bt = 1$ and then in this case the final goal would be $b + 1$ gallons. Do you think this puzzle can be solved if the initial jugs given had been 6 gallons and 3 gallons with a final goal of measuring exactly 4 gallons?

Earlier we partitioned the integers into primes and composites. Another way to partition the integers is to consider all integers with the same remainder when divided by some fixed positive integer m, called the *modulus*. We call each set with the same shared remainder a *residue class*. We denote these m residue classes by $[0], [1], [2], \ldots, [m-1]$. For instance, if we divide the integers by 3, the possible remainders are 0, 1 or 2. We can then partition the integers into three residue classes,

$$\begin{aligned}[0] &= \{\ldots, -6, -3, 0, 3, 6, \ldots\},\\ [1] &= \{\ldots, -5, -2, 1, 4, 7, \ldots\},\\ [2] &= \{\ldots, -4, -1, 2, 5, 8, \ldots\}.\end{aligned}$$

These ideas lead us to the area of modular arithmetic. We say a is *equivalent* to b *modulo* m if a and b have the same remainder when divided by m, and we denote this by

$$a \equiv b \pmod{m}.$$

This is called a *congruence modulo* m and is one of the most important concepts in elementary number theory; it owes its origin to the great German mathematician Carl Friedrich Gauss (1777–1855) as early as 1798. Imagine a and b have the same remainder r when divided by m. We call this r the *residue* of a (or of b) modulo m. We have $a = q_a m + r$ and $b = q_b m + r$, and $a - b = q_a m + r - (q_b m + r) = q_a m + r - q_b m - r = (q_a - q_b)m$. Thus, equivalently, a and b are in the same residue class if $a - b$ is an integer multiple of m, i.e. $m \mid a - b$. For example, $43 \equiv 5 \pmod{19}$, as $43 - 5 = 38 = 2 \cdot 19$.

Many of the rules of arithmetic we are used to still apply here. For instance, if $a \equiv b \pmod{m}$ and $c \equiv d \pmod{m}$, then we have

$$\begin{aligned} a + c &\equiv b + d \pmod{m}, \\ a - c &\equiv b - d \pmod{m}, \\ ac &\equiv bd \pmod{m}. \end{aligned}$$

It follows that if $p(x)$ is any polynomial with integer coefficients, then

$$p(a) \equiv p(b) \pmod{m},$$

whenever $a \equiv b \pmod{m}$. The process of manipulating the integers through the use of congruence relations is called *modular arithmetic*. However, division here is not straightforward. For instance, notice that

$$15 \equiv 9 \pmod{6}$$

and we might notice both 15 and 9 are divisible by 3 and then want to divide both sides by 3, but we notice that if we do that we get

$$5 \equiv 3 \pmod{6},$$

a statement which is most certainly false. However, if we also divide the modulus by 3, we get

$$5 \equiv 3 \pmod{2},$$

which is correct. Furthermore, observe that both 30 and 60 are divisible by 5 and both of the following are true:

$$30 \equiv 60 \pmod{6} \quad \text{and} \quad 6 \equiv 12 \pmod{6}.$$

This time dividing both sides by 5 does not upset the equivalence. Unexpectedly, this also relates to the Die Hard problem or, more precisely, it relates to the extended Euclidean algorithm. What is happening here is we are able to divide by a if and only if the $\gcd(a, m) = 1$. Moreover, we are not really dividing by a so much as multiplying by what we denote as a^{-1} such that $aa^{-1} \equiv 1 \pmod{m}$. Since we can find integers s, t by using the extended Euclidean algorithm on a and m, we have $as + mt = 1 \rightarrow as - 1 = m(-t) \rightarrow as \equiv 1 \pmod{m}$; hence we can put $a^{-1} \equiv s \pmod{m}$.

We call the set $\{0, 1, 2, \ldots, m - 1\}$ a *complete set of residues* (CSR) modulo m. This is because if we have any integer k, the residue of k modulo m, or the remainder on dividing k by m, must be in the CSR (mod m). Now let a, b and m be any integers such that $\gcd(b, m) = 1$. We observe that the set S of residues of the m integers

$a + jb$ $(j = 0, 1, 2, \ldots, m-1)$ must be a CSR (mod m). For suppose

$$a + j_1 b \equiv a + j_2 b \pmod{m}.$$

We get

$$(j_1 - j_2)b \equiv 0 \pmod{m}.$$

Since $\gcd(b, m) = 1$, we see by an earlier result that $j_1 - j_2 \equiv 0 \pmod{m}$. Since $0 \le j_1, j_2 \le m - 1$, we have $-m + 1 \le j_1 - j_2 \le m - 1$; thus since $m \mid j_1 - j_2$, we can only have $j_1 = j_2$. It follows that S must contain m distinct residues modulo m, and is therefore a CSR (mod m).

We need the above finding to solve the *Chicken McNugget problem*. Originally, McDonald's sold its chicken nuggets in packs of 9 and 20. The problem is to determine the largest number of nuggets that could not be purchased with these pack sizes. A more general (mathematicians love to generalize) version of this problem is the following: given pack sizes of a and b (of course a and b are positive integers with $\gcd(a, b) = 1$), what is the largest integer that cannot be represented by $ax + by$, where x and y are nonnegative integers? To solve this problem we first notice that the integer $ab - a - b$ cannot be represented by the above expression. For if $ab - a - b = ax + by$, then $y \equiv -1 \pmod{a}$ and $x \equiv -1 \pmod{b}$. Since $x, y \ge 0$, we have $x \ge b - 1$ and $y \ge a - 1$. Thus,

$$ab - a - b = ax + by \ge a(b-1) + b(a-1) = 2ab - a - b > ab - a - b,$$

which is impossible.

We next show that if an integer $n > ab - a - b = (a-1)(b-1) - 1$, then there exist integers $x, y \ge 0$ such that

$$n = ax + by.$$

Since $\gcd(a, b) = 1$, the set of a residues of $n - jb$ modulo a, where $j = 0, 1, 2, \ldots, a - 1$, is a CSR (mod a). Since 0 is a member of this CSR, there must exist a unique k with $0 \le k \le a - 1$ such that $a \mid n - kb$, or for some integer h, we have $n - kb = ha$. We find that

$$ha = n - kb \ge (a-1)(b-1) - (a-1)b = -a + 1,$$

and therefore

$$h \ge (-a + 1)/a = -1 + 1/a > -1 \ge 0.$$

Since $h \ge 0$, $0 \le k \le a - 1$ and $n = ha + kb$, we see that the largest integer that cannot be represented by $ax + by$ $(x, y \ge 0)$ is $ab - a - b$. In the case of the Chicken McNuggets problem this is $9 \cdot 20 - 9 - 20 = 151$.

The above problem is an instance of the more general *coin problem*, often called the *Frobenius problem* after the German mathematician Ferdinand Georg Frobenius (1849–1917). This problem can be stated as follows: Given positive integers $a_1, a_2, a_3, \ldots, a_n$ with greatest common divisor 1, compute the largest integer not representable by the expression

$$a_1x_1 + a_2x_2 + \cdots + a_nx_n,$$

where $x_1, x_2, \ldots, x_n$ must be nonnegative integers. This largest integer, called the *Frobenius number*, is often denoted by $g(a_1, a_2, \ldots, a_n)$.

We have already seen that

$$g(a_1, a_2) = a_1a_2 - a_1 - a_2.$$

In general, however, no closed formula in $a_1, a_2, \ldots, a_n$ is known for

$$g(a_1, a_2, \ldots, a_n) \quad \text{when} \quad n \geq 3,$$

although there are such formulas for certain special forms of $a_1, a_2, \ldots, a_n$. In the United Kingdom, McDonald's sold McNuggets packs in quantities of 6, 9 and 20. Can you prove that $g(6, 9, 20) = 43$?

We have, to this point, simply assumed that we know how to perform arithmetic operations on the integers. Of course, we all learned how to do this in elementary school, but the techniques we were taught were dependent on how integers were represented. We will discuss the problem of how to represent the integers in the next chapter.

Problems

2.1 (Euclidean condition) For integer pairs (a, b) find integers q, r such that r is in the set $\{0, 1, 2, \ldots, b-1\}$ and $a = q \times b + r$:
(a) **[BB]** $(a, b) = (100, 21)$
(b) $(a, b) = (52, 4)$
(c) $(a, b) = (372, 17)$

2.2 (Aliquot divisors) List all aliquot divisors of:
(a) 6
(b) 17
(c) **[BB]** 30
(d) 494

2.3 (Sum of aliquot divisors) Let $s(n)$ denote the sum of aliquot divisors for n, and calculate:
(a) $s(6)$
(b) $s(17)$
(c) **[BB]** $s(30)$
(d) $s(494)$

2.4 (Aliquot sequence) Recall the aliquot sequence belonging to n is $s_0(n) = n$, $s_1(n) = s(n)$ and for any positive integer k

$$s_k(n) = s(s_{k-1}(n))$$

Find the first 3 terms of the aliquot sequence belonging to:
(a) $n = 4$
(b) $n = 10$
(c) **[BB]** $n = 8$

2.5 Consider the set of primes in the set S, given below, and show that the product of these primes plus one has a prime factor that is not in S:
(a) **[BB]** $S = \{2, 3, 5, 7\}$
(b) $S = \{2, 3, 5, 7, 11\}$
(c) $S = \{2, 3, 5, 7, 11, 13\}$

2.6 (Prime factorization) Find the prime factorization of m:
(a) $m = 70$
(b) **[BB]** $m = 5848$
(c) $m = 5929$

2.7 (Primorial) Find $\#p_n$ for:
(a) **[BB]** $n = 3$
(b) $n = 4$
(c) $n = 6$

2.8 (Primality) Prove that n is a prime by trial-dividing all the primes less than its square root. You may use that the set of the first ten primes is

$$\mathcal{P} = \{2, 3, 5, 7, 11, 13, 17, 19, 23, 29\} :$$

(a) **[BB]** $n = 211$
(b) $n = 557$
(c) $n = 601$

2.9 (Sieve of Eratosthenes) Using bound $b = 60$ find all the primes less than 60 using the Sieve of Eratosthenes.

2.10 (Euclidean algorithm) Use the Euclidean algorithm to find the greatest common divisor for the given integer pairs:
(a) **[BB]** $a = 321, b = 241$
(b) $a = 1001, b = 55$
(c) $a = 665831, b = 364139$

2.11 (Extended Euclidean algorithm) Use the extended Euclidean algorithm to find integers s, t, such that $as + tb = \gcd(a, b)$:
(a) **[BB]** $a = 435, b = 240$
(b) $a = 211, b = 50$
(c) $a = 431, b = 34$

2.12 (Water jug problem) Given two jugs, of irregular shape, that are 5 gallons and 7 gallons, explain how these could be used to measure out exactly 1 gallon of water.

2.13 (Congruence modulo m) Solve for x in the following such that x is in $\{0, 1, 2, \ldots, m-1\}$ where m is the modulus:
(a) **[BB]** $7 \equiv x \pmod 3$ (here $m = 3$)
(b) $332 \equiv x \pmod{17}$
(c) $478 \equiv x \pmod{13}$
(d) $1789245433223443223456 \equiv x \pmod 5$

2.14 (Frobenius problem) Imagine beer is sold in cases of 3 or 10. What is the largest number of beer that could not be purchased with these case sizes?

Further Reading

Aliquot sequences

K. Chum, R. K. Guy, M. J. Jacobson, Jr. and A. S. Mosunov, Numerical and statistical analysis of aliquot sequences, *Experimental Mathematics*, **29:4** (2018), 414–425.
Richard K. Guy, *Unsolved Problems in Number Theory*, 3rd Edition, Springer, NY, 2004, pp. 92–95. (This book is highly recommended, as it contains a wealth of information concerning interesting and unsolved number theoretic problems.)

Primes

David Wells, *Prime Numbers*, John Wiley and Sons, NJ, 2005.

Eratosthenes

Sir Thomas Heath, *A History of Greek Mathematics*, Vol. II, Dover Publications, NY, 1981, pp. 104–109.
Ivor Thomas, *Greek Mathematical Works*, Vol. II, Loeb Classical Library, Harvard University Press, MA, 1993, pp. 260–273.

Sieve of Eratosthenes

Hugh C. Williams, *Édouard Lucas and Primality Testing*, Wiley-Interscience, John Wiley and Sons, NY, 1998, pp. 28–29.

Water jug problems

Jay Bennett, *Riddle of the Week #8: The 'Die Hard' Jug Problem*, popularmechanics.com, 2016.
Elizabeth B. Cowley, Note on a linear Diophantine problem, *The American Mathematical Monthly*, **33**, No. 7 (Aug. – Sep., 1926), 379–381.
H. E. Dudney, *Amusements in Mathematics*, Dover Publications, NY, 1970, pp. 109–111.
T. H. O'Beirne, *Puzzles and Paradoxes*, Dover Publications, NY, 1965, pp. 49–75.
Shai Simonsen, *Public key cryptography, From Calculus to Computers: Using the Last 200 Years of Mathematical History in the Classroom*, editors Richard Jardine and Amy Shell, MAA Notes, Vol. 68, November 2005, pp. 109–124. (In spite of the title, this paper does discuss the water jug problem. It's also an interesting paper in its own right)

The Euclidean algorithm

Richard A. Mollin, *Fundamental Number Theory*, CRC Press, Boca Raton, FL, 1998, pp. 35–42.

The Frobenius problem

Richard K. Guy, *Unsolved Problems in Number Theory*, 3rd Edition, Springer, NY, 2004, pp. 171–174.
Jeffrey Shallit, The Frobenius problem and its generalizations, *Developments in Language Theory*, LNCS5257, Springer, NY, 2008, pp. 72–83.

Chapter 3
Representations of Integers

In this chapter, we will discuss several of the large variety of techniques that have been developed to represent the integers. All cultures have discovered ways to denote numbers and, as one might expect, these systems vary enormously. One of the earliest means used to depict an integer is the tally, where the number is represented as a sequence of strokes on a piece of wood or a bone. Many preliterate societies used some sort of tally marks to perform basic counting; there is evidence of this as early as 25000 to 35000 years ago. In Europe, the tally marks used to count worked as follows:

1	𝍷	6	𝍸𝍷	11	𝍸𝍸𝍷	16	𝍸𝍸𝍸𝍷
2	𝍷𝍷	7	𝍸𝍷𝍷	12	𝍸𝍸𝍷𝍷	17	𝍸𝍸𝍸𝍷𝍷
3	𝍷𝍷𝍷	8	𝍸𝍷𝍷𝍷	13	𝍸𝍸𝍷𝍷𝍷	18	𝍸𝍸𝍸𝍷𝍷𝍷
4	𝍷𝍷𝍷𝍷	9	𝍸𝍷𝍷𝍷𝍷	14	𝍸𝍸𝍷𝍷𝍷𝍷	19	𝍸𝍸𝍸𝍷𝍷𝍷𝍷
5	𝍸	10	𝍸𝍸	15	𝍸𝍸𝍸	20	𝍸𝍸𝍸𝍸

Many cultures use a tally that counts groups of fives; for example, France, Portugal and Spain share the following tally to count up groups of five:

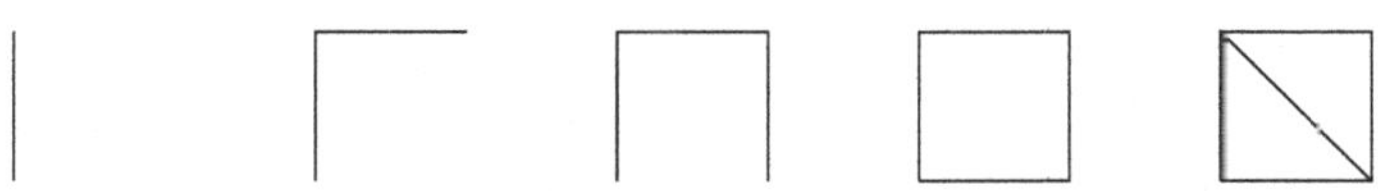

The Chinese use the character that means tally to tally, as it consists of five strokes:

Both of these last two can be used in the same manner as the European tally to count in groups of 5.

E. L. F. Roettger and H. C. Williams, *The Enchantment of Numbers*, CMS/CAIMS Books in Mathematics 16, https://doi.org/10.1007/978-3-031-76568-1_3

Of particular note, however, is an Inuit preliterate counting system that, rather than relying on a group of symbols or notches to count, relied instead on their grammar and vernacular to count in groups of 20, 15, 10, 5 and 1. Since this, at times, resembles something close to what we will call a base-20 system, a deeper discussion of it is more fitting later in this chapter.

The number system that is in common use today originated in India as long ago as the 1st century (CE). We call this system the *Hindu-Arabic numeration system*; it is a base-10 positional numeration system. Base-10 means that we use ten symbols which hold a numeric value, i.e. the set $\{0, 1, 2, 3, 4, 5, 6, 7, 8, 9\}$. The positional value of a number is a power of 10, and as children we learn that the positional values are

$$\ldots, 10^3, 10^2, 10^1, 10^0, \quad \text{or} \quad \ldots, \text{thousands, hundreds, tens, ones.}$$

Hence, a number such as 3245 has an expanded form that splits the number up into a sum of the products of the numeric value multiplied by its positional value, and so since childhood we have understood that

$$3245 = 3 \cdot 10^3 + 2 \cdot 10^2 + 4 \cdot 10^1 + 5 \cdot 10^0.$$

We mention this here so that when we next discuss some other early positional number systems, we can see how similar things are done and what might be missing. The novelty in this scheme is the symbol 0 for zero, which denotes the absence of quantity. Although this idea originated much earlier, it was through the publication of Fibonacci's *Liber Abbaci* in the thirteenth century that this system became known and widely used in the west.

In an arbitrary *base-b* system, we need b symbols and have positional values that are powers of b. In order to avoid confusion when we are writing down a base-b number, we will put a b in the subscript on the right of the number. For example, if we had the base-8 number $7 \cdot 8^2 + 3 \cdot 8^1 + 1 \cdot 8^0$, we would write this as 731_8. We will mention two frequently used choices for b here and show how we can convert between them and the Hindu-Arabic numeration system.

If we put $b = 2$, we call these representations the *binary numbers*. The symbols we use are simply $\{0, 1\}$, and in a computer we can imagine this as current off and current on. For binary, the positional values are just powers of two starting with exponent zero and going up by one as we go to the left. We can easily convert to Hindu-Arabic base-10 by expanding the number working in base-10. Suppose we are given the binary number 100010_2; if we wish to convert to base-10 we just note:

$$100010_2 = 1 \cdot 2^5 + 0 \cdot 2^4 + 0 \cdot 2^3 + 0 \cdot 2^2 + 1 \cdot 2^1 + 0 \cdot 2^0 = 32 + 2 = 34.$$

Converting a number n from Hindu-Arabic to any other base b can be done by a *making change method* where we imagine having coins that represent powers of the base b and we give out change that adds up to n using the largest coins possible. We can also perform a simple algorithm to do the conversion from Hindu-Arabic to base-b as follows: we repeatedly divide by the base b, while keeping track of the remainder (note in subsequent steps we only divide the whole part by the base-b). We terminate this process when we have zero whole parts after the division operation. At the end, we simply read off the remainders in reverse order to record our base-b number. Imagine we want to convert 178 to binary ($b = 2$), then we would perform the following operations:

$$\begin{aligned} 178 &= 2 \cdot 89 + 0 \\ 89 &= 2 \cdot 44 + 1 \\ 44 &= 2 \cdot 22 + 0 \\ 22 &= 2 \cdot 11 + 0 \\ 11 &= 2 \cdot 5 + 1 \\ 5 &= 2 \cdot 2 + 1 \\ 2 &= 2 \cdot 1 + 0 \\ 1 &= 2 \cdot 0 + 1. \end{aligned} \quad \uparrow$$

Now, as the arrow indicates, read the remainders from the bottom to the top yielding $10110010_2 = 178$. Based on this simple conversion process, we have the following theorem:

Theorem 3.1 *Given any positive integer n and base b, there exists a unique representation of n as*

$$n = r_0 b^k + r_1 b^{k-1} + \cdots + r_k,$$

where $0 \le r_i \le b - 1$ ($i = 0, 1, 2, \ldots, k$) and k is the unique integer such that $b^k \le n < b^{k+1}$.

We often write this representation of n as $(r_0 r_1 \ldots r_k)_b$. Notice that the number of bits needed in the binary representation of n is an integer bounded above by $\log_2 n + 1$, just as the number of digits required in the decimal representation of n is less than $\log_{10} n + 1$.

We remark here that if an integer n is represented in decimal notation as $(r_0 r_1 \ldots r_k)_{10}$, then since

$$n = r_0 10^k + r_1 10^{k-1} + \cdots + r_k,$$

we see that $2 \mid n$ if and only if $2 \mid r_k$ and $5 \mid n$ if and only if $5 \mid r_k$. Thus, it is easy to determine whether or not an integer n is even or divisible by 5 by simply examining its last digit. Also, since $10 \equiv 1 \pmod 9$, we have

$$n \equiv r_0 + r_1 + \cdots + r_k \pmod 9.$$

This is often referred to as the rule of casting out nines and can be used to check the accuracy of one's arithmetic. Since $3 \mid 9$, we see that $3 \mid n$ if and only if 3 divides the sum of the digits of n. For example, if $n = 397481521$, then

$$\begin{aligned} n &\equiv 3+9+7+4+8+1+5+2+1 \\ &\equiv 0+0+1+1+2+1+2+2+1 \\ &\equiv 1 \pmod{3}. \end{aligned}$$

Thus, $3 \nmid n$.

As long as the base b with which we are working has $b \leq 10$, we do not need any more symbols with higher numeric value than b. However, we can work in larger bases by simply assigning to new symbols to numeric values greater than 9. One commonly used higher-base system is the *hexadecimal* system, and it is simply base-16. Here we make use of the set of numbers $\{0, 1, 2, 3, 4, 5, 6, 7, 8, 9, A, B, C, D, E, F\}$, where $A = 10, B = 11, \ldots, F = 15$. This representation is commonly used by computers. The method to convert an arbitrary hexadecimal number to Hindu-Arabic is again just to expand and work in base-10. For example, it is easily verified that $AB9F_{16} = 43935$. To convert from base-10 to hexadecimal we can use the algorithm above but divide repeatedly by 16. It is fun to note here that hexadecimal is an unusual word, as it is a blend of the Greek *hex* for six and the Latin *decimal* or *decem* for ten. The all Latin wording would use the Latin *sex* for six, yielding the word *sexadecimal*, which was deemed inappropriate.

Now that we have a better idea of different base number systems we can return to the Inuit counting system mentioned earlier. Certain Inuit societies used a spoken number system that was based on the human body. The body represented 20, and was denoted by the word *yuinaq* and translates to *the whole body* or *person*. Here we will use Yupik Inuit words for the purpose of our discussion, but we note a similar description exists using Inupiaq. The Inuit would start counting with finger, *attauciq* meaning *appendage*, and then work their way down their body, this can be seen in the following table:

Value	Inuit (Yupik)	Translation
1	attauciq	appendage
2	malleghuk	one that follows
3	pingaayn	addendum
4	cetaaman	four objects
5	tahliiman	one arm
6	aghvinlegen	the first one across over
7	malleghunlegen	the second one across over
8	pingaayunlegen	the third one across over
9	cetaamanlegen	the fourth one across over
9	qullengunneghitaaxh	not being quite 10
10	qula	upper part of the body
11	qula attauciq-hwa	
12	qula malleghuk	
13	qula pingaayun	
14	qula cetaaman	
14	akiimiaghunneghitaaxh	not being quite 15
15	qula tahliiman	
15	akiimiaq	15 objects

16	qula aghvinlegen	
16	akiimiaq attauciq-hwa	
17	qula malleghunlegen	
17	akiimiaq malleghunek	
18	qula pingaayunlegen	
18	akiimiaq pingaayn	
19	qula cetaamanlegen	
19	yuinaunneghitaaxh	not being quite 20
20	yuinaq	the whole body or person

Thus, to express the number 22 we would make use of this pattern, and have yuinaq malleghuk. However, to indicate 40 we want to multiply 20 by 2 and this is done by yuinak malleghuk (here the appropriate suffixes -k, -t, -mek, -nek are used to indicate multiplication). Further, there is an interesting dominance of importance of 20, 15 and 10. Because of this 10 is only ever used in the singular, so the Inuit would not say 200 as 10×20, but rather 20×10, i.e. yuinaat qula. We should emphasize that 100 held special importance. This was expressed as five people, or in a more direct translation people five, yuinaat tahliiman (again, here the suffix -t directs us to multiply). Also, 100 is the number of pelts in a bundle (*naaneq*), so we could say either yuinaat tahliiman or nanneq to refer to 100.

The Inupiaq Inuit (Alaskan Inuit) language also has a numeration system based around the human body that is similar to base-20. This makes translation from Inupiaq (or Yupik) to Hindu-Arabic numerals somewhat difficult, and has been remedied by the creation of *Kaktovik numerals* in the early 1990s. With the help of their teacher, William Bartley, the students of the middle school *Harold Kaveolcok* (all 9 of them) in Kaktovik, Alaska, invented the following 20 symbols:

0	𝋀	5	𝋅	10	𝋊	15	𝋏
1	𝋁	6	𝋆	11	𝋋	16	𝋐
2	𝋂	7	𝋇	12	𝋌	17	𝋑
3	𝋃	8	𝋈	13	𝋍	18	𝋒
4	𝋄	9	𝋉	14	𝋎	19	𝋓

Using these symbols and positional values that are powers of 20 which are arranged as

$$\ldots 20^4, 20^3, 20^2, 20^1, 20^0,$$

we can write any integer. For instance, if we wanted to write

$$1607 = 4 \times 20^2 + 0 \times 20^1 + 7 \times 20^0 = \text{𝋄𝋀𝋇}.$$

To go from Kaktovik to Hindu-Arabic, simply expand in base 10, for example:

$$\text{𝋊𝋏𝋀} = 10 \times 20^2 + 15 \times 20^1 + 0 \times 20^0 = 4300.$$

Since the Inupiaq language did not have a word for zero, the students selected an image that looks like crossed arms, indicating nothing was being counted. The numerals the students selected also have the added feature that for some of the additions the new number looks like the the sum of the others, as $5 + 1 = 6$ is

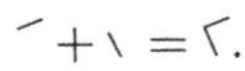

However, this is not always the case as $4 + 1 = 5$ yields

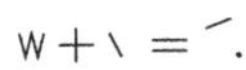

The ancient Babylonians inherited from the Sumerians a numeration system that used only two symbols. For illustrative purposes, we will use the symbol $\vee$ for 1 and $<$ for 10. These two symbols are examples of *proto-cuneiform*, which is what we call the earliest form of writing done on clay tablets. These two symbols would be "drawn" on the clay by pushing a wedge shaped tool into the clay. There is evidence that this system originated as early as 2000 BCE. Denoting numbers between 1 and 59 was done additively, meaning they would write down enough of each symbol such that the sum of all the symbols added up to the desired number, for instance,

$$47 = <<<<\vee\vee\vee\vee\vee\vee\vee.$$

When written on clay the symbols used for 10 and 1 would be stacked in a particular pattern; this is illustrated in the table below. If they wanted to write down any number greater than 59 they used positional values representing powers of 60, similar to the Hindu-Arabic system that uses powers of 10. Also, like the Hindu-Arabic system, the power on the farthest to the right is 0 and each power over to the left increases by one, i.e.

$$\ldots 60^4, 60^3, 60^2, 60^1, 60^0$$

are the positional values. Thus, if we wanted to write a number like 1979, we would notice that we have 32 groups of 60 and 59 left over; hence,

$$1979 = 32 \times 60^1 + 59 \times 60^0$$

or

$$1979 = <<<\vee\vee \quad <<<<<\vee\vee\vee\vee\vee\vee\vee\vee\vee.$$

From this we can see that writing any Hindu-Arabic integer in the Babylonian notation follows by writing the number as groups of powers of 60 and expressing those groups using Babylonian numbers between 1 and 59. This method of conversion is just the making change method described earlier.

Listed below are the first 80 Babylonian numbers drawn as they would have been on the clay tablets:

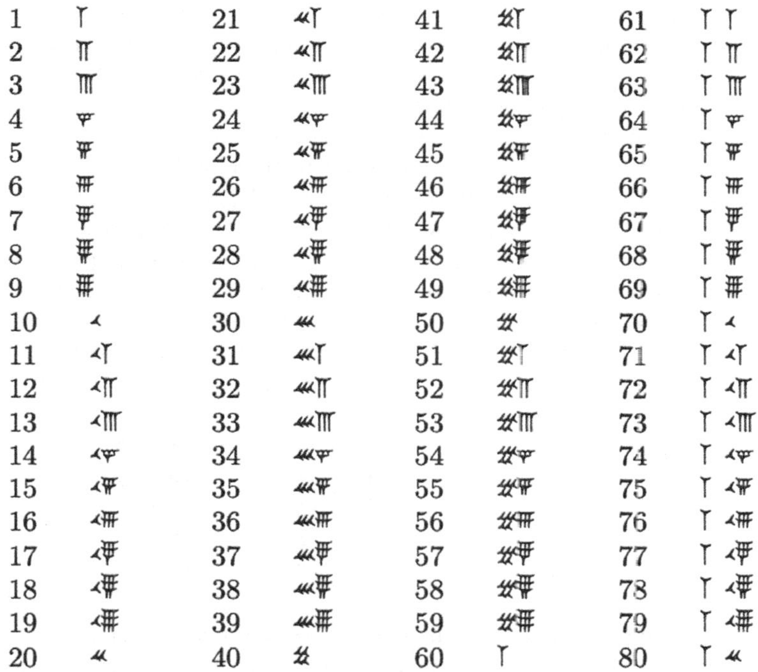

We note that the symbols for 60 and 1 are the same, and the only way to tell them apart is due to the way the numbers are written in columns while being pressed into the clay. If there is an empty space to the right of the symbol for 1, then it is understood to be 60, the empty space indicating that there is nothing in the 60^0 position and then 1 group of 60^1. This is certainly a weak point for this numeration system, and it could have been remedied by the inclusion of a third symbol denoting nothing, or zero.

It should also be mentioned that converting a Babylonian number to Hindu-Arabic is as easy as expanding the number by multiplying each positional value by the amount in it and adding them all up. For example, if we have

$$\vee\vee\vee\vee \quad <<<\vee\vee\vee\vee\vee\vee\vee \quad <<<<\vee\vee\vee\vee\vee\vee \quad <<<<$$

and we want to know what this is in the system we are most familiar with, then we would first notice that there are 4 positional values and we can then note:

$$\begin{aligned} &\vee\vee\vee\vee \quad <<<\vee\vee\vee\vee\vee\vee\vee \quad <<<<\vee\vee\vee\vee\vee\vee \quad <<<< \\ &= 4 \times 60^3 + 37 \times 60^2 + 46 \times 60^1 + 40 \times 60^0 \\ &= 4 \times 216000 + 37 \times 3600 + 46 \times 60 + 40 \times 1 \\ &= 1000000. \end{aligned}$$

Finally, we should say that although the Babylonian numeration system, often referred to as a *sexagesimal* system, uses positional powers of 60, it is not strictly speaking a base-60 system as it uses only two symbols, and a true base-60 system would need a hefty 60 symbols. Some vestiges of this scheme survive today in the form of degrees in a circle and in the measurement of time.

Another ancient number system is the Mayan numeration scheme. One advantage of the Mayan system over the Babylonian is that the Mayans had a symbol, 𝋠, that was used as a place holder to denote that there are none of a particular grouping. Some argue that it was not a true zero by the way it was, or more accurately, was not used in other areas of their mathematics. The Mayan system uses three symbols a shell, a pebble and a stick, denoted 𝋠, · and −, respectively. The pebble is the basic unit, and a stick is worth 5 pebbles, and hence we can use these three symbols to count up to 19 as follows:

0	𝋠	5	𝋥	10	𝋪	15	𝋯
1	𝋡	6	𝋦	11	𝋫	16	𝋰
2	𝋢	7	𝋧	12	𝋬	17	𝋱
3	𝋣	8	𝋨	13	𝋭	18	𝋲
4	𝋤	9	𝋩	14	𝋮	19	𝋳

To represent numbers 20 or larger they used a system that is nearly base-20. The lowest positional value is 20^0, and then the next one up is 20^1, however the next group of numbers is not the expected 20^2, but rather $18 \times 20 = 360$. From this point onward the next positional value is always 20 times the previous base. Another departure from our Hindu-Arabic system is that the positional values are not written right to left, but bottom to top, with 20^0 being on the bottom, thus our positional values are

$$\begin{array}{l} \vdots \\ 18 \times 20^3 \\ 18 \times 20^2 \\ 18 \times 20^1 \\ 20^1 \\ 20^0. \end{array}$$

For example, if we wanted to write 90363 as a Mayan number we would note that

$$90363 = 12 \times 7200 + 11 \times 360 + 0 \times 20 + 3 \times 1,$$

and so,

$$90373 =$$ 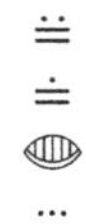 .

As with the Babylonian numbers, going from Mayan to Hindu-Arabic is as easy as expanding the number in our base-10 system. Take for example:

 $= 6 \times 144000 + 18 \times 7200 + 17 \times 360 + 14 \times 20 + 0 \times 1 = 1000000.$

It may seem peculiar that the third base from the bottom, which we might expect to be 20^2, is replaced instead by 18×20; however, we observe that $18 \times 20 = 360$. It is well known that the Mayans were excellent astronomers and could accurately measure a year's passing in days. Although the Mayans knew the year consisted of at least 365 days, they broke up the civil calendar, the *Haab*, as consisting of 18 *uinal* ("months") that were 20 *kin* (days) each. They then filled the year with an additional 5 days called *Uayeb*. Again, like the Babylonian numeration system, the Mayan numeration system is not exactly base-20.

The ancient Greeks had a numeration system that relied on letters in their alphabet corresponding to numeric values. For the purpose of our discussion here we will use modern notation for Greek letters; however, other symbols were used in ancient times. We begin by giving the table for the corresponding symbols and numeric values:

Symbol	Value	Symbol	Value	Symbol	Value
A'	1	I'	10	P'	100
B'	2	Λ'	20	Σ'	200
Γ'	3	K'	30	T'	300
Δ'	4	M'	40	Y'	400
E'	5	N'	50	Φ'	500
F'	6	Ξ'	60	X'	600
Z'	7	O'	70	Ψ'	700
H'	8	Π'	80	Ω'	800
Θ'	9	Q'	90	T'	900
$,A$	1000	$,I$	10000	$,P$	100000
$,B$	2000	$,\Lambda$	20000	$,\Sigma$	200000
$,\Gamma$	3000	$,K$	30000	$,T$	300000
$,\Delta$	4000	$,M$	40000	$,Y$	400000
$,E$	5000	$,N$	50000	$,\Phi$	500000
$,F$	6000	$,\Xi$	60000	$,X$	600000
$,Z$	7000	$,O$	70000	$,\Psi$	700000
$,H$	8000	$,\Pi$	80000	$,\Omega$	800000
$,\Theta$	9000	$,Q$	90000	$,T$	900000

Using the above and expressing numbers from left to right we can then write $4279 = {,}\Delta\Sigma'O'\Theta'$. Although much larger numbers can be written with Greek numerals than with the Roman numerals (see below), it is evident that this system has the same drawbacks, in that the scheme is rather cumbersome and there is a maximum size integer that can be represented.

As mentioned earlier, another ancient number system for the integers was the Roman numerals which were developed in ancient Rome and used throughout Europe into the Middle Ages. They are still used today for Super Bowl sequencing, page numbers of book prefaces, film production dates, building years of constructions, etc. The Roman numerals are a mostly additive numeration system that relies on the symbols:

Symbol:	I	V	X	L	C	D	M
Value:	1	5	10	50	100	500	1000

The first 40 Roman numerals in standard form are:

1	I	11	XI	21	XXI	31	XXXI
2	II	12	XII	22	XXII	32	XXXII
3	III	13	XIII	23	XXIII	33	XXXIII
4	IV	14	XIV	24	XXIV	34	XXXIV
5	V	15	XV	25	XXV	35	XXXV
6	VI	16	XVI	26	XXVI	36	XXXVI
7	VII	17	XVII	27	XXVII	37	XXXVII
8	VIII	18	XVIII	28	XXVIII	38	XXXVIII
9	IX	19	XIX	29	XXIX	39	XXXIX
10	X	20	XX	30	XXX	40	XL

Note in the above that if a smaller value is written to the left of a larger numeral then this represents a subtractive notation; for instance, we write 9 as IX, not VIIII. The only allowable subtractive notations are:

IV	IX	XL	XC	CD	CM
4	9	40	90	400	900

Hence, we cannot write 999 as IM. Beyond the subtraction rules noted above, Roman numerals are written with the largest values on the left and are recorded for groups of thousands, hundreds, tens using the following standard form:

	Thousands	Hundreds	Tens	Units
1	M	C	X	I
2	MM	CC	XX	II
3	MMM	CCC	XXX	II
4		CD	XL	IV
5		D	L	V
6		DC	LX	VI
7		DCC	LXX	VII
8		DCCC	LXXX	VIII
9		CM	XC	IX

The largest number that can be represented using this scheme is

$$3999 = MMMCMXCIX,$$

which is a rather unfortunate characteristic of this system. We would also note that the standard form above was not always adhered to and it is not uncommon for 4 to be written strictly additively as IIII or 40 as XXXX. It is easy to see why this cumbersome system was largely abandoned after the Hindu-Arabic scheme had been introduced to the western world.

To illustrate how intricate some number systems could become, we mention the mediaeval Cistercian numerals which were created by the Cistercian monks and date back to the thirteenth century. Initially developed to write numbers in the range 1–99 with a single glyph, they were later generalized (early sixteenth century) to write any integer from 1 to 9999. The Cistercians made use of 9 basic symbols:

Symbol: 
Value: 1 2 3 4 5 6 7 8 9

If these symbols are then reversed or reflected in the y-axis, we have

Symbol: 
Value: 10 20 30 40 50 60 70 80 90

Also, inverting or reflecting in the x-axis yields

Symbol: 
Value: 100 200 300 400 500 600 700 800 900

Finally, if we both reflect in the x-axis and the y-axis, or both invert and reverse, we get

Symbol: 

Value: 1000 2000 3000 4000 5000 6000 7000 8000 9000

Now if we wish to write a number, say 147, we simply stack the digits for 100, 40 and 7 on top of one another to form a single glyph, namely

147 = .

To read a glyph we look at the individual parts and recognize that each quadrant gives the positional values for 10^0, 10^1, 10^2 and 10^3, i.e.

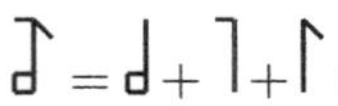 $= 9000 + 10 + 3 = 9013.$

Notice that since there were no groups of 100, simply nothing was put in the bottom right quadrant, and hence no need for a symbol denoting zero.

If we return our attention to binary numbers and we now also allow a signed value to each positional value, we can then express integers in multiple ways. For example:

$$(01101)_2 = 0 \times 2^4 + 1 \times 2^3 + 1 \times 2^2 + 0 \times 2^1 + 1 \times 2^0 = 13,$$
$$(100-1-1)_2 = 1 \times 2^4 + 0 \times 2^3 + 0 \times 2^2 - 1 \times 2^1 - 1 \times 2^0 = 13,$$
$$(10-101)_2 = 1 \times 2^4 + 0 \times 2^3 - 1 \times 2^2 + 0 \times 2^1 + 1 \times 2^0 = 13.$$

We can make the representation unique by insisting that all nonzero values not be adjacent; this is called a *non-adjacent form* or NAF. For example, $(10-101)_2$ is in NAF, while $(100-1-1)_2$ is not. With regular binary numbers, on average we can expect about half of the binary digits to be nonzero; however, with NAF we can expect only a third of the binary digits to be nonzero. In some algorithms, steps pertaining to a particular integer n which depend on the positional values in the representation of n will often execute more quickly when many of the positional values are zero. This can be useful in computing because it reduces the run-time of certain algorithms.

This of course leaves the need to convert a given binary representation to NAF. Given the input $B = (b_{m-1}b_{m-2}\dots b_1b_0)_2$, the following algorithm will produce output $C = (c_m c_{m-1}\dots c_1c_0)_{\text{NAF}}$. In the following we use the symbol $\leftarrow$ to mean "replace by".

Algorithm 1 An algorithm to convert a binary number to NAF with input $B = (b_{m-1}b_{m-2}\dots b_1b_0)_2$ and output $C = (c_m c_{m-1}\dots c_1c_0)_{\text{NAF}}$.

```
i ← 0
while B > 0 do
    if B is odd then
        c_i ← 2 − (B (mod 4))
        B ← B − c_i
    else
        c_i ← 0
    end if
    B ← B/2
    i ← i + 1
end while
```

For example if our input were $B = (110101)_2$ we can run the algorithm and see it has 7 iterations:

Step 1: $i = 0$
$B = 53 = (110101)_2$ is odd and 53 (mod 4) is 1, thus $c_0 = 1$ and our new value for B is $(53 - 1)/2 = 26 = (11010)_2$.

Step 2: $i = 1$
$B = 26 = (11010)_2$ is even, thus $c_1 = 0$ and our new value for B is $26/2 = 13 = (1101)_2$.

Step 3: $i = 2$
$B = 13 = (1101)_2$ is odd and 13 (mod 4) is 1, thus $c_2 = 1$ and our new value for B is $(13 - 1)/2 = 6 = (110)_2$.

Step 4: $i = 3$
$B = 6 = (110)_2$ is even, thus $c_3 = 0$ and our new value for B is $6/2 = 3 = (11)_2$.

Step 5: $i = 4$
$B = 3 = (11)_2$ is odd and 3 (mod 4) is 3, thus $c_4 = -1$ and our new value for B is $(3 + 1)/2 = 2 = (10)_2$.

Step 6: $i = 5$
$B = 2 = (10)_2$ is even, thus $c_5 = 0$ and our new value for B is $2/2 = 1 = (1)_2$.

Step 7: $i = 6$
$B = 1 = (1)_2$ is odd and 1 (mod 4) is 1, thus $c_6 = 1$ and our new value for B is $(1 - 1)/2 = 0 = (0)_2$. So we stop.

Thus, the NAF representation of 53 is $(10-10101)_{\text{NAF}}$.

There are other modern ways to represent integers as well. For instance, if we have two positive integers p, q such that $\gcd(p, q) = 1$, then every positive integer n can be represented as the sum or difference of numbers of the form $p^a q^b$, i.e.

$$n = \sum_{i=1}^{l} s_i p^{a_i} q^{b_i}, \quad \text{with } s_i \in \{-1, 1\} \text{ and } a_i, b_i \geq 0.$$

We call this a *double-base number system* (DBNS). We note that in the above we denote that i is a member of the set I by $i \in I$. In the above we have a *signed* DBNS since we allow s_i to be either 1 or -1; however, if we restrict $s_i = 1$ then we have an *unsigned* DBNS. For example, if we consider the unsigned case with $p = 2$ and $q = 3$, then we see that we can represent 7 in the following ways:

$$2^2 \cdot 3^0 + 2^0 \cdot 3^1 = 7,$$
$$2^1 \cdot 3^1 + 2^0 \cdot 3^0 = 7,$$
$$2^2 \cdot 3^0 + 2^1 \cdot 3^0 + 2^0 \cdot 3^0 = 7.$$

Hence, this representation scheme can be highly redundant. Even more so if we allow signed values for s_i, as now we have

$$2^3 \cdot 3^0 - 2^0 \cdot 3^0 = 7,$$
$$2^0 \cdot 3^2 - 2^1 \cdot 3^0 = 7,$$
$$2^1 \cdot 3^3 - 2^4 \cdot 3^1 + 2^0 \cdot 3^0 = 7,$$

as well as many others!

Indeed, we can find an unsigned DBNS representation for a positive integer n by first finding the largest s of the form $p^i q^j$ less than or equal to n. We then repeat this process for n replaced by $n - s$, etc. As an example, consider $n = 127$, $p = 2$, $q = 3$. We can find s by an exhaustive search and get $s = 108 = 2^2 \times 3^3$ and $n - s = 127 - 108 = 19$. Repeating this process with $n = 19$ yields $s = 18 = 2 \times 3^2$. Thus

$$127 = 2^2 \times 3^4 + 2 \times 3^2 + 2^0 \times 3^0.$$

Of course, there are other DBNS representations of 127. More efficient ways of finding s, given n, are known, but we will conclude our discussion of DBNS systems here.

One desirable quality in a number system is the ability to denote large numbers. For some systems with a maximal possible number, like Roman, Greek or Cistercian numerals, this is of course impossible. For others, like European tally, there is no limit to what can be expressed, but it is not really efficient nor meaningful.

According to Jewish tradition the biblical prophet Jeremiah (c. 650–c. 570 BCE) wrote, "As the host of heaven cannot be numbered, neither the sand of the sea measured,. . ." (Jeremiah 33:22, KJV). While this is a very early expression of the seeming boundlessness of grains of sand, it was likely not the first. Much later Archimedes (c.287–212 BCE) stated at the beginning of *The Sand Reckoner*, "There are some, King Gelon, who think that the number of the sand is infinite in multitude." He then went on to determine an upper bound for how many grains of sand would fit into the universe. In this he was trying to convince the reader that the number of grains of sand needed was not infinite, but could be bounded by a large expressible integer. What is interesting here is not the estimate of the size of the universe according to models of his era, but rather that Archimedes succeeded in inventing a way to display extremely large numbers.

At the time of Archimedes the Greeks could only express numbers up to 10,000, and this was called a *myriad*. In order to represent larger numbers Archimedes first considered a *myriad of myriads* (myriad-myriad), in today's notation this would be $10,000 \times 10,000 = 10^8$, he named all numbers up to 10^8 *first order*. He then considered a myriad myriads (10^8) the *unit of the second order*. He next considered multiples of the unit of the second order, namely: *second order* numbers up to $(10^8)^2$, and called this the *unit of the third order* and then *third order* numbers are those up to $(10^8)^3$. He repeats this until he reaches the (10^8)th *order*, again in today's notion that is $(10^8)^{(10^8)}$. Suppose we denote this quantity by P. He lets numbers up to P, be described as the *first period*. Continuing in a similar fashion, he lets P be the unit of the *first order of the second period*, and this accounts for numbers greater than the first period up to 10^8P. Next, 10^8P is the unit of the *second order of the second period*, and this goes up to $(10^8)^2P$. This is repeated until we are at (10^8)th *order of the second period*, the period ending with $(10^8)^{(10^8)}P$, which is P^2. We now let P^2, be the unit of *first order of the third period*, and we continue analogously to before until we have (10^8)th *order of the third period*, which ends in P^3. This process is then repeated until we have the (10^8)th *order of the* (10^8)th *period*. This is exactly

$$L = P^{10^8},$$

in today's notation, a number of 8×10^{16} digits and a very large number indeed. Notice that in our modern exponential notation we write a^{b^c} for $a^{(b^c)}$. In Archimedes' number scheme L is simply "a myriad-myriad units of the myriad-myriad-th-order of the myriad-myriad-th-period". Archimedes goes on to show that under the best available model of the universe at his time, the number of grains of sand needed to fill the universe is bounded above by 10^7 units of the 8th order of numbers or by $10^7 \times 10^{56} = 10^{63}$.

We can update this problem by using the current value for the radius of the observable universe: 4.40×10^{28} centimetres and a smaller object than a grain of sand. There is some controversy about whether space is discrete or continuous, but many physicists think that there is a nonzero absolute minimum volume of space of approximately 10^{-99} cubic centimetres. This latter quantity is roughly the cube of the Planck length, which physicists regard as the shortest physically measurable distance. Since the observable universe is a sphere, it has a volume of $\frac{4\pi}{3}(4.40 \times 10^{28})^3 < 10^{87}$ cubic centimetres. Thus, the total number of grains of space that could fit into the observable universe at this time (the universe keeps on expanding) is $< 10^{186}$. This is a very large number, but is still much less than Archimedes' P. The reader should be warned that the physics in the above is very naïve, but even so it is very difficult to produce a number larger than P which has a possible physical meaning.

A number such as L is inconceivably large, well beyond the human power of comprehension, but it is truly dwarfed by numbers which have been discussed recently to explain phenomena which arise in mathematical logic, computer science, combinatorics and graph theory. We can describe or produce representations of these huge numbers like Graham's number or the unimaginably larger TREE(3), but words fail completely to give any real impression of their extraordinary magnitude. Fortunately, in our excursion into number theory no such behemoths will be required; nevertheless, we will have occasion in what follows to mention some very large integers, several much larger than L.

Another modern problem is how to represent integers on computers. Since integers are stored in binary, the easiest way to represent them is with a fixed number of bits or bytes. We mention here that a *bit* is the smallest unit of storage and is simply a 0 or a 1, while a *byte* is collection of 8 bits. For example, IBM has three classes of fixed *word* sizes that will store an integer exactly, these classes are *small integer*, *large integer* (or just *integer*) and *big integer*. Small integer has a fixed 2 bytes to represent a signed integer. One bit goes to the sign and the remaining 15 bits are used for powers of 2 between 0 and 14. Hence, the largest magnitude of integer we can represent using small integer is

$$2^{14} + 2^{13} + 2^{12} + 2^{11} + 2^{10} + 2^{9} + 2^{8} + 2^{7} + 2^{6} + 2^{5} + 2^{4} + 2^{3} + 2^{2} + 2^{1} + 2^{0} = 32767.$$

Next, large integer dedicates a fixed 4 bytes to represent an integer, where again one bit is used for sign and the remaining 31 bits are used for powers of 2 from 0 to 30. Thus, large integer can represent integers in the range

$$-2147483647 \text{ to } 2147483647.$$

While big integer has a fixed 8 bytes, in a similar fashion as the others and can thus represent integers in the range

$$-9223372036854775807 \text{ to } 9223372036854775807.$$

Moreover, computers are able to work with integers outside these ranges if *arbitrary-precision integers* or *multiple-precision integers* are employed. In this case the number is still stored in binary but now in variable length arrays of words limited only by the size of the memory in the computer. However, this does come with complications, as the time required to perform arithmetic on such numbers can be much greater than that needed for integers of fixed word length.

We conclude this chapter with a brief look at the very powerful *residue numeral system* (RNS), where integers are represented by their value modulo a set of pairwise coprime integers. A RNS is set up by choosing a set of k integers to be our moduli, denoted as

$$\{m_1, m_2, \ldots, m_k\}.$$

We choose these so that the greatest common divisor of any two moduli is 1. Then we can represent any integer x ($0 \leq x < M$, where $M = m_1 m_2 \cdots m_k$) by the set

$$\{a_1, a_2, \ldots, a_k\},$$

where

$$\begin{aligned} x &\equiv a_1 \pmod{m_1}, \\ x &\equiv a_2 \pmod{m_2}, \\ &\vdots \\ x &\equiv a_k \pmod{m_k}, \end{aligned}$$

and for each a_i we insist $0 \leq a_i \leq m_i$.

For example, if we want to represent $x = 1000000$ using RNS, then we might use a set of moduli

$$\{m_1, m_2, m_3, m_4, m_5\} = \{11, 13, 17, 19, 23\}.$$

In this case $M = m_1 m_2 m_3 m_4 m_5 = 1062347 > 1000000$ and we can represent $x = 1000000$ by

$$\{a_1, a_2, a_3, a_4, a_5\} = \{1, 1, 9, 11, 6\},$$

since

$$\begin{aligned} 1000000 &\equiv 1 \pmod{11}, \\ 1000000 &\equiv 1 \pmod{13}, \\ 1000000 &\equiv 9 \pmod{17}, \\ 1000000 &\equiv 11 \pmod{19}, \\ 1000000 &\equiv 6 \pmod{23}. \end{aligned}$$

Notice that performing arithmetic using the RNS scheme is very simple. For example, if we use the above set of moduli ($\{m_1, m_2, m_3, m_4, m_5\} = \{11, 13, 17, 19, 23\}$),

$$a = \{1, 1, 9, 11, 6\} \quad \text{and} \quad b = \{2, 5, 1, 7, 19\},$$

we get the RNS for ab as the product of the corresponding entries modulo the corresponding modulus. Thus,

$$ab = \{2, 5, 9, 1, 22\},$$

as

$$\begin{aligned} 1 \cdot 2 &\equiv 2 \pmod{11}, \\ 1 \cdot 5 &\equiv 5 \pmod{13}, \\ 9 \cdot 1 &\equiv 9 \pmod{17}, \\ 11 \cdot 7 &\equiv 1 \pmod{19}, \\ 6 \cdot 19 &\equiv 22 \pmod{23}. \end{aligned}$$

Finally, if we are given a set of moduli and a set of remainders for some x value, it is possible to apply the below *Chinese remainder theorem* (CRT) to recover that value. The term *Chinese* here refers to the Chinese mathematician Sun-Tzu (fl. c. 250 CE) who dealt with a special case of the CRT.

Theorem 3.2 (The Chinese Remainder Theorem) *Let m_1, m_2,..., m_k be integers such that* $\gcd(m_i, m_j) = 1$ *for $i \neq j$ and $1 \leq i, j \leq k$. Then the set of simultaneous congruences*

$$\begin{aligned} x &\equiv a_1 \pmod{m_1}, \\ x &\equiv a_2 \pmod{m_2}, \\ &\vdots \\ x &\equiv a_k \pmod{m_k}, \end{aligned}$$

has a unique solution modulo $M = m_1 m_2 \cdots m_k$.

Proof We begin by defining $M_i = M/m_i$. By our assumptions, we then have $\gcd(M_i, m_i) = 1$, and hence (as discussed in Chap. 2) we can use the extended Euclidean algorithm to solve

$$y_i M_i \equiv 1 \pmod{m_i} \quad (i = 1, 2, \ldots, k)$$

for integers y_1, y_2, ..., y_k. We now claim

$$x = y_1 a_1 M_1 + y_2 a_2 M_2 + \cdots + y_k a_k M_k$$

is a solution to the stated simultaneous congruences. This can be seen by noting that for $x \equiv a_j \pmod{m_j}$ for $j = 1, 2, \ldots, k$, since $m_i \mid M_j$ (so $M_j \equiv 0 \pmod{m_i}$) whenever $i \neq j$ and $y_j M_j \equiv 1 \pmod{m_j}$. We must also show that this solution is unique, we will do so by contradiction. Let us assume we have two solutions, say x_0 and x_1, such that $x_0 \not\equiv x_1 \pmod{M}$. First we have $x_0 \equiv x_1 \equiv a_j \pmod{m_j}$ and thus by definition of congruence we have $m_j \mid x_0 - x_1$ $(j = 1, 2, \ldots k)$. Hence $\operatorname{lcm}[m_1, m_2, \ldots, m_k] \mid x_0 - x_1$ and since $\gcd(m_i, m_j) = 1$ for $i \neq j$ and $1 \leq i, j \leq k$ this yields $\operatorname{lcm}[m_1, m_2, \ldots, m_k] = M$. Combining these last two observations gives us $M \mid x_0 - x_1$, which implies $x_0 \equiv x_1 \pmod{M}$, which is a contradiction to our assumption. □

For clarity, let us walk through an example of the Chinese remainder theorem, by finding a solution to

$$\begin{aligned} x &\equiv 7 \pmod{11}, \\ x &\equiv 2 \pmod{13}, \\ x &\equiv 5 \pmod{15}. \end{aligned}$$

In this case we have $a_1 = 7, a_2 = 2, a_3 = 5$ and $m_1 = 11, m_2 = 13, m_3 = 15$, using the above notation. Thus

$$\begin{aligned} M &= m_1 m_2 m_3 = 11 \times 13 \times 15 = 2145, \\ M_1 &= m_2 m_3 = 13 \times 15 = 195, \\ M_2 &= m_1 m_3 = 11 \times 15 = 165, \\ M_3 &= m_1 m_2 = 11 \times 13 = 143. \end{aligned}$$

We can now use the extended Euclidean algorithm three times to solve

$$y_i M_i \equiv 1 \pmod{m_i} \quad \text{for } i = 1, 2, 3, \text{ i.e.}$$

$$\begin{aligned} y_1 195 &\equiv 1 \pmod{11} & & & y_1 8 &\equiv 1 \pmod{11} \\ y_2 165 &\equiv 1 \pmod{13} & &\text{or equivalently} & y_2 9 &\equiv 1 \pmod{13} \\ y_3 143 &\equiv 1 \pmod{15} & & & y_3 8 &\equiv 1 \pmod{15}. \end{aligned}$$

Thus, EEA gives us $y_1 = 7$, $y_2 = 3$ and $y_3 = 2$. Hence,

$$\begin{aligned} x &= y_1 a_1 M_1 + y_2 a_2 M_2 + y_3 a_3 M_3 \\ &= 7 \times 7 \times 195 + 3 \times 2 \times 165 + 2 \times 5 \times 143 \\ &= 11975. \end{aligned}$$

Finally, since $M = 2145$, $x \equiv 11975 \equiv 1250 \pmod{M}$.

Problems

3.1 (European tally) Convert the following to European tally:
(a) 3
(b) **[BB]** 22
(c) 34

3.2 (France, Portugal and Spain tally) Convert the following to the tally system used in France, Portugal and Spain:
(a) 4
(b) **[BB]** 13
(c) 22

3.3 (Chinese tally) Convert the following to the Chinese tally:
(a) 2
(b) 14
(c) 33

3.4 (Base-b to Hindu-arabic) Convert the following numbers to Hindu-Arabic:
(a) 10101_2
(b) **[BB]** 745_8
(c) $A9F_{16}$

3.5 (Hindu-Arabic to base-b) Convert the following numbers to the indicated bases:
(a) **[BB]** 176 to base-2
(b) 375 to base-5
(c) 82376 to hexadecimal

3.6 (Hindu-Arabic numerals to Yupik) How would the Yupik say each of the following quantities (explain):
(a) 23
(b) 60
(c) 40
(d) 400

3.7 (Kaktovik numerals to Hindu-Arabic numerals) Convert the following Kaktovik numerals to Hindu-Arabic numerals:
(a) **[BB]** 𝋅𝋀𝋊
(b) 𝋄𝋁𝋂
(c) 𝋈𝋎𝋅

3.8 (Hindu-Arabic numerals to Kaktovik numerals) Convert the following Hindu-Arabic numerals to Kaktovik numerals:
(a) 73
(b) **[BB]** 4352
(c) 10365

3.9 (Babylonian numerals to Hindu-Arabic numerals) Convert the following Babylonian numerals to Hindu-Arabic numerals:
(a) **[BB]** <∨ <<∨∨∨∨∨∨∨
(b) <<∨∨ <<<<∨∨ <<∨∨∨∨
(c) ∨ <<<<∨ ∨∨ <<<<∨∨∨∨∨∨∨

3.10 (Hindu-Arabic numerals to Babylonian numerals) Convert the following Hindu-Arabic numerals to Babylonian numerals:
(a) 201
(b) **[BB]** 10623
(c) 789999

3.11 (Mayan numerals to Hindu-Arabic numerals) Convert the following Mayan numerals to Hindu-Arabic numerals:

(a) **[BB]**  (b)  (c) 

3.12 (Hindu-Arabic numerals to Mayan numerals) Convert the following Hindu-Arabic numerals to Mayan numerals:
(a) **[BB]** 201
(b) 10623
(c) 789999

3.13 (Greek numerals to Hindu-Arabic numerals) Convert the following Greek numerals to Hindu-Arabic numerals:
(a) **[BB]** $T'\Pi'B'$
(b) , $F\Psi'M'Z'$
(c) , $\Omega Y'\Lambda'$

3.14 (Hindu-Arabic numerals to Greek numerals) Convert the following Hindu-Arabic numerals to Greek numerals:
(a) 43
(b) **[BB]** 6789
(c) 89001

3.15 (Roman numerals to Hindu-Arabic numerals) Convert the following Roman numerals to Hindu-Arabic numerals:
(a) XVII
(b) **[BB]** $DCXXIX$
(c) $CDXLIX$

3.16 (Hindu-Arabic numerals to Roman numerals) Convert the following Hindu-Arabic numerals to Roman numerals (use subtractive notations):
(a) 84
(b) **[BB]** 955
(c) 3243

3.17 (Cistercian numerals to Hindu-Arabic numerals) Convert the following Cistercian numerals to Hindu-Arabic numerals:
(a) 
(b) **[BB]** 
(c) 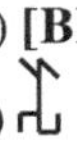

3.18 (Hindu-Arabic numerals to Cistercian numerals) Convert the following Hindu-Arabic numerals to Cistercian numerals:
(a) 72
(b) **[BB]** 851
(c) 5673

3.19 (NAF to Hindu-Arabic numerals) Convert the following numbers in NAF to Hindu-Arabic numerals:
(a) **[BB]** $(10-1)_{\text{NAF}}$
(b) $(10-10)_{\text{NAF}}$
(c) $(1000-101)_{\text{NAF}}$

3.20 (Hindu-Arabic numerals to NAF) Convert the following Hindu-Arabic numerals to NAF:
(a) **[BB]** 7
(b) 124
(c) 251

3.21 (DBNS to Hindu-Arabic numerals) Convert the following DBNS (with $p = 2$ and $q = 3$) numbers to Hindu-Arabic numerals:
(a) **[BB]** $2^0 \cdot 3^3 + 2^0 \cdot 3^0$
(b) $2^1 \cdot 3^2 + 2^1 \cdot 3^0$
(c) $2^2 \cdot 3^3 + 2^1 \cdot 3^1$

3.22 (Hindu-Arabic numerals to DBNS) Convert the following Hindu-Arabic numerals to DBNS (with $p = 2$ and $q = 3$) using 2 sums or fewer (warning: answers are not unique):
(a) 10
(b) **[BB]** 100
(c) 1000

3.23 **[BB]** Express the number $x = 165000$ as a RNS using the moduli

$$\{m_1, m_2, m_3, m_4\} = \{7, 17, 37, 67\}.$$

3.24 Express the number $x = 2000000$ as a RNS using the moduli

$$\{m_1, m_2, m_3, m_4\} = \{71, 73, 79, 83\}.$$

3.25 **[BB]** Find the number x that has been expressed using as a RNS where

$$\{m_1, m_2\} = \{13, 17\} \quad \text{and} \quad \{a_1, a_2\} = \{3, 7\}.$$

3.26 Find the number x that has been expressed using as a RNS where

$$\{m_1, m_2, m_3\} = \{5, 7, 11\} \quad \text{and} \quad \{a_1, a_2, a_3\} = \{1, 2, 3\}.$$

Further Reading

Writing numbers

Graham Flegg, *Numbers: Their History and Meaning*, Dover Publications, NY, 2002, pp. 39–77.

Fibonacci

Keith Devlin, *Finding Fibonacci*, Princeton University Press, Princeton, NJ, 2017.

The Inuit number system

John H. Koo and Gust Bartman, The number system in Yupik Eskimo, *American Philosophical Society*, **124**, No. 1 (1980), 48–51.

The Mayan calendar

The Mayan Calendar, http://webexhibits.org/calendars/calendar-mayan.html

NAF

Douglas R. Stinson, *Cryptography Theory and Practice*, 2nd Edition, Chapman & Hall/CRC, NY, 2002, pp. 258–259.

D. Hankerson, A. Menezes, and S. A. Vanstone, *Guide to Elliptic Curve Cryptography*, Springer-Verlag, 2004, pp. 98.

Double-base number systems

Vassil Dimitrov, Laurent Imbert, and Pradeep K. Mishra, The double-base number system and its application to elliptic curve cryptography, *Mathematics of Computation*, **77** (2008), 1075–1104.

Valerié Berthé and Laurent Imbert, On converting to the double-base number system, *Proc. SPIE '04, Advanced Signal Processing Algorithms, Architectures, and Implementations XIV*, 2004, pp. 70–78.

The Sand Reckoner

Sir Thomas L. Heath, *The Works of Archimedes*, Dover Publications, NY, 2002, pp. 221–232.

Very large integers

Antonio Padilla, *Fantastic Numbers and Where to Find Them: A Cosmic Quest from Zero to Infinity*, Farrar, Straus and Giroux, NY, 2022.

Chapter 4
Integer Powers

We will devote this chapter to the problem of computing $a^n \pmod{m}$ and its application to certain number-theoretic problems. Here n and m may be quite large integers. Before embarking on this, we remind the reader of the following rules regarding exponential arithmetic. For $a \neq 0$, $b \neq 0$, we have the following exponent rules:

$$\begin{aligned}
&\text{Product Rule:} \quad a^m a^n = a^{m+n}\\
&\text{Quotient Rule:} \quad a^m/a^n = a^{m-n}\\
&\text{Power Rule:} \quad (a^m)^n = a^{mn}\\
&\text{Power of a Product Rule:} \quad (ab)^n = a^n b^n\\
&\text{Power of a Fraction Rule:} \quad (a/b)^n = a^n/b^n\\
&\text{Zero Exponent:} \quad a^0 = 1\\
&\text{Negative Exponent:} \quad a^{-n} = 1/a^n.
\end{aligned}$$

We will use the product rule, power rule and power of product rule in what follows.

There are certain numerical procedures in which we need to compute $a^n \pmod{m}$ quickly. Of course we know how to determine $a^n \pmod{m}$ in a slow way, as we could simply multiply $a \times a \times \cdots \times a$, performing the $n-1$ multiplications and then finding the remainder modulo m at the end. This is wasteful in two ways. First, we would need a lot of memory to store the powers of a along the way as they would grow quickly for large values of a and n; we can solve this by replacing each multiplication with a multiplication followed by a reduction modulo m, called a *modular multiplication*. Second, if we exploit what we have learned in the previous section about binary numbers, we can perform this procedure in far fewer than $n-1$ modular multiplications, which are the most expensive of the elementary arithmetic operations. For example, if we want to calculate $7^{49} \pmod{67}$, we first note $49 = 110001_2$, so $49 = 32 + 16 + 1$ and hence $7^{49} = 7^{32}7^{16}7^1$ by the product rule above. We can then calculate $7^2 \pmod{67}$, $7^4 \equiv (7^2)^2 \pmod{67}$, $7^8 \equiv (7^4)^2 \pmod{67}$,

E. L. F. Roettger and H. C. Williams, *The Enchantment of Numbers*, CMS/CAIMS Books in Mathematics 16, https://doi.org/10.1007/978-3-031-76568-1_4

$7^{16} \equiv (7^8)^2 \pmod{67}$, $7^{32} \equiv (7^{16})^2 \pmod{67}$, with only 5 modular multiplications. These modular multiplications yield

$$\begin{aligned} 7^2 &\equiv 49 \pmod{67}, \\ 7^4 &\equiv 49^2 \equiv 56 \pmod{67}, \\ 7^8 &\equiv 56^2 \equiv 54 \pmod{67}, \\ 7^{16} &\equiv 54^2 \equiv 35 \pmod{67}, \\ 7^{32} &\equiv 35^2 \equiv 19 \pmod{67}. \end{aligned}$$

We finish with two more modular multiplications to find

$$7^{49} \equiv 7^{32}7^{16}7^1 \equiv 19 \times 35 \times 7 \equiv 32 \pmod{67}.$$

Hence, we only need to perform 7 modular multiplications rather than the 48 required by the naïve way, which is a significant saving. We note that in this algorithm calculating $a^2 \pmod{m}$, $a^4 \pmod{m}$, $a^8 \pmod{m}$ and so on requires only that the current power of a be stored.

We can implement this as an algorithm as follows.

Algorithm 1 An algorithm to compute $a^n \pmod{m}$ with input integers a, m and $n = (b_k b_{k-1} \ldots b_1 b_0)_2$ and output $a^n \pmod{m}$.

```
solution ← 1
for i = 0 to k do
    if b_i = 1 then
        solution ← solution × a (mod m)
    else
    end if
    a ← a × a (mod m)
end for
return solution
```

We can also write this as follows.

Algorithm 2 An algorithm to compute $a^n \pmod{m}$ with input integers a, m and $n = (b_0 b_1 \ldots b_{k-1} b_k)_2$ and output $a^n \pmod{m}$.

```
r_0 ← a^{b_0} = a
for i = 1 to k do
    if b_i = 1 then
        r_i ← r_{i-1}^2 × a (mod m)
    else
        r_i ← r_{i-1}^2 (mod m)
    end if
end for
return r_k = a^n (mod m)
```

The proof that this algorithm works follows from defining

$$s_j \quad (j = 0, 1, 2, \ldots, k)$$

by $s_0 = b_0 = 1$ and

$$s_{i-1} = 2s_i + b_{i+1} \quad (i = 0, 1, 2, \ldots, k-1),$$

then $n = s_k$. Thus, if

$$r_j \equiv a^{s_j} \pmod{m},$$

we have

$$r_{j+1} = r_j^2 a^{b_{j+1}} \pmod{m};$$

hence, $r_k \equiv a^{s_k} = a^n \pmod{m}$.

Also, we note $k = \lceil \log_2 n \rceil$. Here the *ceiling function* $\lceil x \rceil$ maps real values of x to the least integer greater than or equal to x. For example, $\lceil 3.8 \rceil = 4$ and $\lceil -2.2 \rceil = -2$. Later on in this book we will also make use of the *floor function* $\lfloor x \rfloor$, which denotes the greatest integer less than or equal to x. Note that $\lfloor x \rfloor = \lceil x \rceil$ only when x is an integer; otherwise $\lceil x \rceil = \lfloor x \rfloor + 1$. In considering the floor function, we point out that $x - 1 < \lfloor x \rfloor \le x$, $\lfloor x + m \rfloor = \lfloor x \rfloor + m$ when m is an integer, and

$$\left\lfloor \frac{x}{m} \right\rfloor = \left\lfloor \frac{\lfloor x \rfloor}{m} \right\rfloor$$

when m is a positive integer. We also observe that

$$\lfloor x \rfloor + \lfloor y \rfloor \le \lfloor x + y \rfloor \le \lfloor x \rfloor + \lfloor y \rfloor + 1.$$

If we were to calculate $871^{504} \pmod{1009}$ by using Algorithm 2, we would first convert 504 to binary, i.e. $504 = (111111000)_2$. We would then set $r_0 = 871$ and execute the steps below

- $b_1 = 1 \rightarrow r_1 = r_0^2 a = 871^2 \times 871 \equiv 373 \pmod{1009}$
- $b_2 = 1 \rightarrow r_2 = r_1^2 a = 459^2 \times 871 \equiv 459 \pmod{1009}$
- $b_3 = 1 \rightarrow r_3 = r_2^2 a = 459^2 \times 871 \equiv 357 \pmod{1009}$
- $b_4 = 1 \rightarrow r_4 = r_3^2 a = 357^2 \times 871 \equiv 926 \pmod{1009}$
- $b_5 = 1 \rightarrow r_5 = r_4^2 a = 926^2 \times 871 \equiv 805 \pmod{1009}$
- $b_6 = 0 \rightarrow r_6 = r_5^2 = 805^2 \equiv 247 \pmod{1009}$
- $b_7 = 0 \rightarrow r_7 = r_6^2 = 247^2 \equiv 469 \pmod{1009}$
- $b_8 = 0 \rightarrow r_8 = r_7^2 = 469^2 \equiv 1008 \pmod{1009}$.

Thus, $871^{504} \equiv 1008 \pmod{1009}$.

In a similar way we can use NAF to calculate $a^n \pmod{m}$. Using the same example as earlier, $7^{49} \pmod{67}$, we note this time that

$$49 = (10-10001)_{\text{NAF}}.$$

Thus,

$$7^{49} \equiv 7^{64}(7^{16})^{-1}7^1 \equiv 26 \times 35^{-1} \times 7 \equiv 26 \times 23 \times 6 \equiv 32 \pmod{67}.$$

Notice, though, that we do not divide by 35; rather, we use the EEA to find $35^{-1} \pmod{67}$. This time we used 5 modular multiplications to calculate powers of 7 up to 64, and then another 2 modular multiplications to complete the calculation. However, some extra work was needed to perform the EEA. In general this can be done using the following algorithm.

Algorithm 3 An algorithm to calculate $a^n \pmod{m}$ with input integers a, m and $n = (b_k b_{k-1} \ldots b_1 b_0)_{\text{NAF}}$ and output $a^n \pmod{m}$.

```
solution ← 1
for i = 0 to k do
   if b_i = 1 then
      solution ← solution × a (mod m)
   end if
   if b_i = −1 then
      solution ← solution × a^{-1} (mod m)
   end if
   a ← a × a
end for
return solution
```

Since we are using the NAF representation, we note $k = \lceil \log_2 n \rceil + 1$ here. Unfortunately, because of the need to precompute $a^{-1} \pmod{m}$, this is probably not a very good way to proceed, unless we wish to compute $a^n \pmod{m}$ for many different values of n and fixed values of a and m.

We can also compute $a^n \pmod{m}$ in an efficient manner if we have a representation for n as a DBNS. Returning to the same example, $7^{49} \pmod{67}$, one more time we note that $49 = 2^4 3^1 + 2^0 3^0$. From this we see that

$$7^{49} \equiv 7^{2^4 3^1 + 2^0 3^0} \equiv (7^{16})^3 7^1 \equiv 7^{16} 7^{16} 7^{16} 7 \equiv 35 \times 35 \times 35 \times 7 \equiv 32 \pmod{67}.$$

Here the total number of modular multiplications is a mere 7: 4 to get to $7^{16} \pmod{67}$ and then 3 more to finish the problem.

It is now clear that computing $a^n \pmod{m}$ naïvely with $n-1$ multiplications is slower than any of the more savvy methods mentioned here. It is now customary to think about the efficiency of a computational procedure P, often called an *algorithm*—we note that we have already used the term algorithm, but we are now formalizing it. We describe the efficiency in terms of the number of bit operations that are needed to carry out, or *execute*, P. This is usually expressed as a function, called the *time complexity*, of P. This is a function of the *input length* of P, the total number of bits needed to describe the input to P. For example, consider the problem of adding two binary integers a and b. The input length is bounded above by $k = \lceil \log_2 a \rceil + \lceil \log_2 b \rceil$, and the time complexity of the addition process is less than k because in order to perform this procedure we require fewer than k bit additions.

In analysing the complexity of an algorithm, it is helpful to make use of something called *Big O notation*. We begin by defining the symbol O as follows. Let $f(x)$ and $g(x) > 0$ be functions which are defined for real values of x. We will say

$$f(x) = O(g(x)),$$

if there exists some fixed real value $c > 0$ and x_0 such that $|f(x)| < cg(x)$ for all $x \geq x_0$. For instance, if $f(x) = 7x^4 - 2x^3 + 4$ and $g(x) = x^4$, then $f(x) = O(g(x))$. This is true since we can find a $c > 0$ and x_0 such that $|f(x)| < cg(x)$ for all $x \geq x_0$, as follows by letting $x_0 = 1$ and noting

$$\begin{aligned} |f(x)| &= |7x^4 - 2x^3 + 4| < 7x^4 + |2x^3| + 1 \leq 7x^4 + 2x^4 + 1x^4 \leq 10x^4 \\ &= 10g(x) \end{aligned}$$

and thus,

$$|f(x)| < 10g(x).$$

In addition to big O notation, we will have occasion later in this book to use *little o notation* and *big Ω notation*. We write

$$f(x) = o(g(x))$$

when

$$\lim_{x \to \infty} \frac{f(x)}{g(x)} = 0.$$

Hence,

$$3x^2 + 2x + 1 = o(x^3).$$

Note that $f(x) = o(1)$ means that $f(x) \to 0$ as $x \to \infty$. We say that

$$f(x) = \Omega(g(x))$$

if there exists a positive constant c such that

$$f(x) > c|g(x)|$$

for all $x > x_0$. For example,

$$3x^2 + 2x + 1 = \Omega(x^2).$$

The way big O notation applies to algorithms is as follows: we count the number of arithmetic bit operations, i.e. addition, subtraction, multiplication and division as a function of the input length, and express this in big O notation. In Algorithm 2 above, the type of arithmetic operation we are interested in is modular multiplication. Since the number of these multiplications at each step is at most two and there are $\lceil \log_2(n) \rceil$ steps in our algorithm, the number of modular multiplications required is $O(\log n)$. Here and elsewhere we use $\log x$ to denote the *natural* or *Napierian logarithm*, sometimes denoted by $\log_e x$ or $\ln x$, where

$$e = \lim_{n \to \infty} (1 + 1/n)^n = 2.71828\ldots$$

is *Euler's constant*, named after the Swiss mathematician, physicist, astronomer, geographer, logician and engineer, Leonhard Euler (1707–1783). Notice that since

$$\log_b x = \frac{\log_e x}{\log_e b},$$

we have $\log_b x = O(\log x)$ for any fixed base b. As an example, in computing $a^n \pmod m$ by Algorithm 2, the input length is bounded above by $k = \lceil \log_2 a \rceil + \lceil \log_2 m \rceil + \lceil \log_2 n \rceil$ and, as we have seen, the time complexity of the exponentiation process is $\leq 2k$ modular multiplications. The optimal time complexity $M(h)$ of the modular multiplication process of two integers of size (number of bits) h is still unknown, but the best current result is

$$M(h) = O(h \log h).$$

Thus, the time complexity of Algorithm 2 is $O(kM(k)) = O(k^2 \log k)$. We often write $f(x) = \tilde{O}(g(x))$ (soft O of $g(x)$) if for some fixed h independent of the value of x,

$$f(x) = O(g(x)(\log x)^h).$$

Thus, we can also write the time complexity of Algorithm 2 as $\tilde{O}(k^2)$.

In the case that k is the input length of an algorithm, $g(k)$ is a polynomial in k and the algorithm executes in $O(g(k))$ steps or bit operations, we say the algorithm executes in *polynomial time*. Similarly if $g(k)$ is an exponential function of k, we say that the algorithm executes in *exponential time*. Suppose k is the input length for Algorithm 2; since $k^2 \log k = O(k^3)$, we see that Algorithm 2 executes in polynomial time, whereas the naïve way of computing $a^n \pmod m$ executes in $O(n)$ operations, and since $n = 2^{\log_2 n} = O(2^k)$, this algorithm executes in exponential time.

Let m be any positive integer and consider the set of all positive integers less than m and relatively prime to m. We call such a set a *reduced set of residues* modulo m or $RSR \pmod m$. We define $\phi(m)$ to be the number of elements in the $RSR \pmod m$. For example, consider $m = 60$. We have

$$RSR \text{ (mod 60)} = \{1, 7, 11, 13, 17, 19, 23, 29, 31, 37, 41, 43, 47, 49, 53, 59\},$$

and therefore $\phi(60) = 16$. The $\phi(m)$ function is usually called *Euler's phi function* and is sometimes called the *totient* of m. It is obvious that for any prime number p, we have $\phi(p) = p - 1$. Indeed, it is easy to show that $\phi(m) = m - 1$, if and only if m is a prime. We also have a simple result for the Euler phi function of prime powers below.

Theorem 4.1 *Let p be a prime and let r be a positive integer. Then*

$$\phi(p^r) = p^{r-1}(p-1).$$

Proof We want to examine how many integers from 1 to p^r are relatively prime to p^r. Note that any integer not relatively prime to p^r must be an integer multiple of p, i.e. jp where $j = 1, 2, 3, \ldots, p^{r-1}$. Consequently, exactly p^{r-1} integers are not relatively prime to p^r, thus $p^r - p^{r-1}$ integers are relatively prime to p^r. Hence, we have our conclusion, $\phi(p^r) = p^{r-1}(p-1)$. □

However, we will often be interested in $\phi(m)$, where m is not a prime power. Consider $\phi(15)$. If we write out all the positive integers less than 15 and strike out all that are not relatively prime to 15, i.e. integers either divisible by 3 or 5 we have

$$RSR \text{ (mod 15)} = \{1, 2, \cancel{3}, 4, \cancel{5}, \cancel{6}, 7, 8, \cancel{9}, \cancel{10}, 11, \cancel{12}, 13, 14\}.$$

It is clear, then, that $\phi(15) = 8$. Notice also that since $\phi(p) = p - 1$, we have $\phi(3) = 2$ and $\phi(5) = 4$, and that $\phi(15) = \phi(3)\phi(5)$. This may seem like something of a coincidence; however, it is not because the Euler phi function is a multiplicative function. We say a function $f(x)$, defined over the integers, is a *multiplicative function* if $f(ab) = f(a)f(b)$ whenever a and b are relatively prime. To show the Euler phi function possesses this property, it is enough to show that the two sets RSR (mod mn) and RSR (mod m) $\times$ RSR (mod n) are the same size, that is, there is a *bijection* between them, namely a function that pairs each element from one set with exactly one element from another set. The function that has a bijection between RSR (mod mn) and RSR (mod m) $\times$ RSR (mod n) is

$$f(a) = (a \text{ (mod } m), a \text{ (mod } n)).$$

To show our selected function is a bijection, we make use of the CRT from Chap. 3.

Theorem 4.2 *Let m, n be positive integers such that* $\gcd(m, n) = 1$. *Then*

$$\phi(mn) = \phi(m)\phi(n).$$

We now have all the ingredients needed to prove the following theorem.

Theorem 4.3 *If m is a positive integer and*

$$m = p_1^{r_1} p_2^{r_2} \cdots p_k^{r_k},$$

where p_i $(i = 1, 2, \ldots, k)$ are distinct primes and $r_i \geq 1$ $(i = 1, 2, \ldots, k)$, then

$$\phi(m) = p_1^{r_1-1}(p_1 - 1)p_2^{r_2-1}(p_2 - 1) \cdots p_k^{r_k-1}(p_k - 1).$$

It is often useful to rewrite the above formula as

$$\phi(m) = m(1 - 1/p_1)(1 - 1/p_2) \cdots (1 - 1/p_k).$$

Notice that $\phi(m) < m$ for all integers $m > 1$. Also, if $m > 2$, then $\phi(m)$ is always even. There are a number of unsolved problems concerning ϕ, not the least of which is a problem posed in 1932 by Derrick Henry Lehmer (1905-1991): does $\phi(m) \mid m - 1$ when m is composite? This problem is still unsolved, but we do know that such an m must be odd, cannot have a square divisor (this is easy to prove), must be the product of at least 14 primes and therefore be bigger than 10^{20} (this has been refined recently to 10^{22}). More is known if we insist that m be divisible by 3. In this case we must have $k > 40000000$ and $m > 10^{360000000}$. Another interesting problem is whether given some positive even integer k, we have any values of m such that $\phi(m) = k$. For example, if $k = 12$, then $\phi(m) = 12$ for $m = 13$ only, but if $k = 14$, there are no values of m which satisfy the equation. However, for $k = 6$, we have the solutions $m = 7, 9, 14, 18$, and these are all.

A very important application of $\phi(m)$ occurs as *Euler's theorem.*

Theorem 4.4 (Euler's theorem.) *If the* $\gcd(a, m) = 1$, *then*

$$a^{\phi(m)} \equiv 1 \pmod{m}.$$

This theorem follows easily from the observation that if we take each element of the RSR (mod m), multiply by a and reduce the product modulo m, then the resulting set is just a reordering of the original RSR (mod m). Thus, if we put $R \equiv$ product of all the elements of the RSR (mod m), we must have

$$a^{\phi(m)} R \equiv R \pmod{m}.$$

Since $\gcd(R, m) = 1$, we get Euler's theorem.

By Euler's theorem, we can write

$$a^{-1} \equiv a^{\phi(m)-1} \pmod{m}.$$

Thus, provided we are given $\phi(m)$, we see by our earlier remarks that the time complexity of evaluating a^{-1} (mod m) is $O(k^2 \log k)$, where $k = \log_2 m$. However, a careful analysis of the complexity of using the EEA to evaluate a^{-1} reveals that this complexity measure can be reduced to $O(k^2)$. Furthermore, the EEA does not require that we know $\phi(m)$ in advance.

We also observe that if $\gcd(a, m) = 1$, there must exist a least positive integer x such that $a^x \equiv 1 \pmod{m}$; this x is called *the order of a modulo m*, and we denote it by $\text{ord}_m a$. We are now able to define a *primitive root modulo m*. We will say a is a primitive root modulo m, if a and m are relatively prime and $\text{ord}_m a = \phi(m)$. For example, we can see that 2 is a primitive root of 19 as follows: since 19 is prime it is clear all positive integers less than 19 are relatively prime to 19, thus $\phi(19) = 18$. Also, we can verify

$$\begin{array}{lll} 2^1 \equiv 2 \pmod{19}, & 2^2 \equiv 4 \pmod{19}, & 2^3 \equiv 8 \pmod{19}, \\ 2^4 \equiv 16 \pmod{19}, & 2^5 \equiv 13 \pmod{19}, & 2^6 \equiv 7 \pmod{19}, \\ 2^7 \equiv 14 \pmod{19}, & 2^8 \equiv 9 \pmod{19}, & 2^9 \equiv 18 \pmod{19}, \\ 2^{10} \equiv 17 \pmod{19}, & 2^{11} \equiv 15 \pmod{19}, & 2^{12} \equiv 11 \pmod{19}, \\ 2^{13} \equiv 3 \pmod{19}, & 2^{14} \equiv 6 \pmod{19}, & 2^{15} \equiv 12 \pmod{19}, \\ 2^{16} \equiv 5 \pmod{19}, & 2^{17} \equiv 10 \pmod{19}, & 2^{18} \equiv 1 \pmod{19}. \end{array}$$

Hence, $\text{ord}_{19} 2 = \phi(19)$. We point out that this also serves as an example of *Fermat's little theorem*, which is a corollary to Euler's theorem and is stated below.

Corollary 4.1 *If p is a prime, then*

$$a^p \equiv a \pmod{p}.$$

We next have the following important theorem concerning which positive integers m have primitive roots. We will need this result shortly in order to look at solutions of certain binomial congruences.

Theorem 4.5 *A positive integer m has a primitive root if and only if*

$$m = 2, 4, q^k, 2q^k,$$

where $k \geq 1$ and q is an odd prime.

We now return to the second example in Chap. 1. Numbers with the same properties as 142857 are called *cyclic numbers* or *carousel numbers*. Recall that for the integer 142857 we have

$$\begin{aligned} 1 \cdot 142857 &= 142857 \\ 2 \cdot 142857 &= 285714 \\ 3 \cdot 142857 &= 428571 \\ 4 \cdot 142857 &= 571428 \\ 5 \cdot 142857 &= 714285 \\ 6 \cdot 142857 &= 857142. \end{aligned}$$

Each of these products contains the same 6 digits of the original number 142857 in the same cyclic order. We want to know if there are other such numbers.

It is well known that the decimal part of any rational number consists of a single block of digits followed by a repeating block of some other set of digits, often called the *period* or *repetend*. For example,

$$46/57 = .807070707\ldots;$$

thus 07 is the repetend of $46/57$. In fact, if q is an odd prime and 10 is a primitive root of q, then a cyclic number is equivalent to the repetend of $1/q$.

For example, consider $q = 17$. It is easy to verify that 10 is a primitive root of 17 and the repetend of $1/17$ is the 16 digit block

$$0588235294117647.$$

If we regard this block as a number (including the leading 0) and multiply it by any integer between 1 and 16, we get a cyclic rearrangement of the digits of the original block. Also, we have

$$05882352 + 94117647 = 99999999.$$

That this latter property holds for all such q, for which 10 is a primitive root, follows from an old (1836) result of the French mathematician Étienne Midy (c. 1775–1850).

In order to find a primitive root of m ($\neq 2, 4$), we must first produce a primitive root of the prime q. For small values of q, this can be done quite simply by trial, but when q is large this can be a difficult problem unless the prime power decomposition of $q - 1$ is known. This follows from the easily verified result below.

Theorem 4.6 *If* $g^{(q-1)/p} \not\equiv 1 \pmod{q}$ *for each of the distinct prime divisors* p *of* $q - 1$, *then* g *is a primitive root of* q.

Of course, in order to use Theorem 4.6, we have to know a suitable candidate for g in advance. In practice, it turns out that after only a few informed guesses at a value for g, applications of Theorem 4.6 will usually result in a primitive root of q. This is a consequence of a still unproved, but widely believed, conjecture of the Austrian mathematician Emil Artin (1898–1962) concerning the distribution of primes for which a given non-square integer a ($\neq -1$) is a primitive root. For example, under this conjecture the set of primes for which $a = 2$ or 10 is a primitive root has a fixed density given by *Artin's constant*, which is the product of $1 - \frac{1}{p(p-1)}$ over all the prime values of p and is approximately .374. This means that for any prime q, the probability that 10 is a primitive root of q is roughly .374. A list of the smallest such q can be found in A001913 of the OEIS.

If we let g be a primitive root of m and we have some y such that y and m are relatively prime, we know there must exist some i such that

$$y \equiv g^i \pmod{m}.$$

We call this i the *index of y to base g modulo m*; we denote it as $i = \mathrm{ind}_g\, y$. For example, 11 is the index of 15 to the base 2 modulo 19, i.e. $11 = \mathrm{ind}_2\, 15$. Some useful results about indices are in the theorem below.

Theorem 4.7 *Let g be a primitive root of m, then we have*

$$\begin{aligned}
&\mathrm{ind}_g\, 1 \equiv 0 \pmod{\phi(m)},\\
&\mathrm{ind}_g(ab) \equiv \mathrm{ind}_g\, a + \mathrm{ind}_g\, b \pmod{\phi(m)},\\
&\mathrm{ind}_g(a^n) \equiv n\, \mathrm{ind}_g\, a \pmod{\phi(m)}.
\end{aligned}$$

Notice that indices behave like the more familiar logarithms and can be used to shorten computations modulo m for those values of m specified by Theorem 4.5. However, it is first necessary to have tables of indices and power residues modulo m. For example, if we let $m = 50$, then by Theorem 4.5 m has a primitive root $g = 3$. Since $m = 50 = 2 \times 5^2$, we can then see that $\phi(50) = 20$ and that

$$RSR \pmod{50} = \{1, 3, 7, 9, 11, 13, 17, 19, 21, 23, 27, 29, 31, 33, 37,\\ 39, 41, 43, 47, 49\}.$$

As in the previous chapter, we denote that i is a member of a set I by $i \in I$. We can now calculate the following small table for each value $y \in RSR \pmod{50}$ and $i = \mathrm{ind}_g\, y$:

Table 4.1 $i = \mathrm{ind}_g\, y$ for each $y \in RSR \pmod{50}$

y	i	y	i	y	i	y	i	y	i	y	i	y	i	y	i	y	i	y	i
1	20	7	15	11	8	17	19	21	16	27	3	31	4	37	7	41	12	47	11
3	1	9	2	13	17	19	14	23	13	29	6	33	9	39	18	43	5	49	10

Suppose, for example, we wish to find some integer x such that $x^3 \equiv 11 \pmod{50}$. By taking indices on both sides, we can change this congruence to $3\,\mathrm{ind}_g\, x \equiv \mathrm{ind}_g\, 11 \pmod{20}$. Thus, $\mathrm{ind}_g\, x \equiv 3^{-1} 8 \equiv 7 \times 8 \equiv 16 \pmod{20}$ and $x \equiv 3^{16} \equiv 21 \pmod{50}$. It is easy to compute such tables when m is small, but for large m, it is a computational challenge, given g, to find the index of some $a \pmod{m}$. This is called the *discrete logarithm problem* and its presumed difficulty ensures the security of the Diffie-Hellman key exchange protocol, the basis for the security of much Internet commerce (see Chap. 7).

Problems

4.1 (Binary exponentiation) Use Algorithm 1 to compute $a^n \pmod m$ for the following:
(a) **[BB]** $5^{17} \pmod{10}$
(b) $4^{66} \pmod{13}$
(c) $11^{80} \pmod{15}$

4.2 (Binary exponentiation) Use Algorithm 2 to compute $a^n \pmod m$ for the following:
(a) $3^{19} \pmod{10}$
(b) **[BB]** $2^{129} \pmod 5$
(c) $76^{38} \pmod{11}$

4.3 (Binary exponentiation) Use Algorithm 3 to compute $a^n \pmod m$ for the following:
(a) $7^7 \pmod{10}$
(b) $4^{64} \pmod 5$
(c) **[BB]** $18^{15} \pmod{11}$

4.4 (Big O notation) Show that $f(x) = O(g(x))$ showing there exists some fixed real value $c > 0$ and x_0 such that $|f(x)| < cg(x)$ for all $x \geq x_0$:
(a) **[BB]** $f(x) = 9x^2 + 2x + 6$ and $g(x) = x^2$
(b) $f(x) = 5x^4 - x^3 + 9x - 16$ and $g(x) = x^4$
(c) $f(x) = 1x^6 - x^4 - 9x^2 + 6$ and $g(x) = x^6$

4.5 (Reduced set of residues) Find the reduced set of residues modulo m (RSR $(\text{mod } m)$) below:
(a) **[BB]** RSR (mod 12)
(b) RSR (mod 20)
(c) RSR (mod 35)

4.6 (Euler's phi function) Calculate the Euler phi function, $\phi(m)$, for each m below:
(a) $m = 35$
(b) **[BB]** $m = 12$
(c) $m = 2205$

4.7 (Primitive root) Verify the primitive root, g, has the order $\phi(m)$ for each given pair:
(a) $g = 2, m = 5$
(b) **[BB]** $g = 2, m = 11$
(c) $g = 3, m = 17$

4.8 (Indices) Use indices to solve for x in each of the following using the values found in Table 4.1:
(a) **[BB]** $x^7 \equiv 9 \pmod{50}$

(b) $x^{11} \equiv 41 \pmod{50}$
(c) $x^{17} \equiv 41 \pmod{50}$

Further Reading

Exponentiation methods

Dan Gordon, A survey of fast exponentiation methods, *Journal of Algorithms*, **27** (1998), 129-146.

The floor function

Ivan Niven, Herbert S. Zucherman and Hugh L. Montgomery, *An Introduction to the Theory of Numbers*, Fifth Edition, John Wiley and Sons, NY, 1991, pp. 180–184.

DBNS exponentition

V. S. Dimitrov, G.A. Jullien and W.C. Miller, An algorithm for modular exponentiation, *Information Processing Letters*, Vol. 66, issue. 3, (1998), 155–159.

Elementary computational complexity

H. C. Williams, *Édouard Lucas and Primality Testing*, Wiley-Interscience, John Wiley and Sons, NY, 1998, pp. 7–12.

Euler's totient function

Richard K. Guy, *Unsolved Problems in Number Theory*, 3rd Edition, Springer, NY, 2004, pp. 138–149.

Complexity of the Euclidean algorithm

E. Bach and J. Shallit, *Algorithmic Number Theory*, MIT Press, Cambridge MA, 1996, pp. 67–99, 102.

Primitive roots and indices

Richard A. Mollin, *Fundamental Number Theory with Applications*, CRC Press, Boca Raton FL, 1998, pp. 145–159.

Cyclic numbers

Martin Gardner, *Mathematical Circus*, Alfred A. Knopf, NY, 1979, pp. 111–122.
Katsuya Mori, On a certain inverse problem for carousel numbers, *Integers*, **20** (2020), #A77.

Artin's primitive root conjecture

D. H. Lehmer and Emma Lehmer, Heuristics Anyone?, *Studies in Mathematical Analysis and Related Topics: Essays in Honor of George Polya*, Stanford University Press, Stanford CA, 1962, pp. 202–210.
Daniel Shanks, *Solved and Unsolved Problems in Number Theory*, 4th edition, Chelsea Publishing Co., NY, 1993, pp. 80–83.
Pieter Moree, Artin's primitive root conjecture-a survey, *Integers*, **12** (2012), 1305–1416.

Discrete logarithm problem

Joshua Holden, *The Mathematics of Secrets: Cryptography from Caesar Ciphers to Digital Encryption*, Princeton University Press, NJ, 2017, pp. 182–200.

Chapter 5
The Binomial Congruence

Our next goal is to explore the solvability of the general binomial congruence

$$x^n \equiv a \pmod{m}$$

for given a, n and m, where a and m are relatively prime. Let $m = q_1^{r_1} q_2^{r_2} \cdots q_k^{r_k}$ be the prime decomposition of m and let x be some integer that satisfies

$$x^n \equiv a \pmod{m}$$

(provided such an x exists), then for any i, $1 \leq i \leq k$

$$x^n \equiv a \pmod{q_i^{r_i}}.$$

Thus, if we can solve

$$x^n \equiv a \pmod{q_i^{r_i}}.$$

for all i, we can then also recover a solution to

$$x^n \equiv a \pmod{m}$$

by using the CRT from Chap. 3. Consequently, moving forward, we will only need to show how to solve

$$x^n \equiv a \pmod{q^r},$$

where q^r is any $q_i^{r_i}$ for $1 \leq i \leq k$. We begin by letting $h = q^r$, where q is an odd prime. By Theorem 4.5, h has a primitive root; we will denote it by g. Now if we take indices of both sides of

$$x^n \equiv a \pmod{h},$$

E. L. F. Roettger and H. C. Williams, *The Enchantment of Numbers*, CMS/CAIMS Books in Mathematics 16, https://doi.org/10.1007/978-3-031-76568-1_5

by Theorem 4.7, we have

$$n \operatorname{ind}_g x \equiv \operatorname{ind}_g a \pmod{\phi(h)}.$$

From this we deduce that

$$x^n \equiv a \pmod{h}$$

has solutions if and only if $d \mid \operatorname{ind}_g a$ where $d = \gcd(n, \phi(h))$. These ideas can be used to devise the following theorem.

Theorem 5.1 *Let m be a positive integer with a primitive root. If n is a positive integer and a is an integer such that $\gcd(a, m) = 1$ and $d = \gcd(n, \phi(m))$, then*

$$x^n \equiv a \pmod{m}$$

has a solution if and only if

$$a^{\phi(m)/d} \equiv 1 \pmod{m}.$$

In the case that solutions exist, there are exactly d incongruent solutions modulo m.

Suppose we apply the above theorem to the case where $m = 50 = 2 \times 5^2$. Notice by Theorem 4.5 that $m = 50$ has a primitive root and $\phi(50) = 20$. Now if we were to let $a = 3$ and $n = 4$, then we could use Theorem 5.1 to check for the existence of solutions to $x^4 \equiv 3 \pmod{50}$ by calculating $a^{\phi(m)/d} \pmod{m}$ where $d = \gcd(n, \phi(m)) = \gcd(4, 20) = 4$ to see if $a^{\phi(m)/d} \equiv 1 \pmod{m}$. In this case $3^{20/4} \equiv 43 \not\equiv 1 \pmod{50}$, and hence no solutions exist for $x^4 \equiv 3 \pmod{50}$. However, if $a = 11$ and n remains 4, we can similarly check whether $x^4 \equiv 11 \pmod{50}$ has solutions. Note here $11^5 \equiv 1 \pmod{5}$, and hence $x^4 \equiv 11 \pmod{50}$ has solutions; indeed, it has 4 incongruent solutions. The table below shows $x^4 \pmod{50}$ for all values of x such that $\gcd(x, 50) = 1$.

x	$x^4 \pmod{50}$	x	$x^4 \pmod{50}$	x	$x^4 \pmod{50}$	x	$x^4 \pmod{50}$
1	1	13	11	27	41	39	41
3	31	17	21	29	31	41	11
7	1	19	21	31	21	43	1
9	11	21	31	33	21	47	31
11	41	23	41	37	11	49	1

From the table we can see that our 4 incongruent solutions are $x = 9, 13, 37$ and 41.

If we restrict ourselves to the case of $m = q$ a prime, then $\phi(m) = \phi(q) = q - 1$. Let l be any integer such that $l \mid q - 1$. We can define the symbol

$$(a|q)_l$$

by

$$(a|q)_l \equiv a^{(q-1)/l} \pmod{q},$$

where $(a|q)_l \in \{0, 1, 2, \ldots, q-1\}$. Now, if we have $n \mid q-1$, the equation

$$x^n \equiv a \pmod{q}$$

has solutions by Theorem 5.1 if and only if

$$a^{(q-1)/n} \equiv 1 \pmod{q},$$

which is exactly when $(a|q)_n = 1$. Furthermore, when $n \mid q-1$ we see that we have

$$(a|q)_n = 1$$

for exactly $(q-1)/n$ values of a in the RSR $\pmod{q}$ and

$$(a|q)_n \neq 1$$

for the remaining $(n-1)(q-1)/n$ values of a in the RSR $\pmod{q}$. Thus, for $n > 2$ the number of values a in the RSR $\pmod{q}$ such that $(a|q)_n \neq 1$ outnumbers those for which $(a|q)_n = 1$ by a factor of $n-1$.

We are often interested in the case where $n = 2$ in

$$x^n \equiv a \pmod{m}.$$

If for a positive integer m and integer a relatively prime to m we have solutions to

$$x^2 \equiv a \pmod{m},$$

then we say a is a *quadratic residue* of m. If no solutions exist, we say a is a *quadratic nonresidue* of m.

By Fermat's little theorem we have $a^{q-1} \equiv 1 \pmod{q}$, when q is an odd prime and q does not divide a. We can then see here that q divides

$$a^{q-1} - 1 = (a^{(q-1)/2} - 1)(a^{(q-1)/2} + 1).$$

If q divides $(a^{(q-1)/2} - 1)$, then $x^2 \equiv a \pmod{q}$ has a solution. However, if q does not divide $(a^{(q-1)/2} - 1)$, then q divides $(a^{(q-1)/2} + 1)$ and $x^2 \equiv a \pmod{q}$ does not have a solution. Hence, a is a quadratic residue of q when

$$a^{(q-1)/2} \equiv 1 \pmod{q},$$

and a is a quadratic nonresidue of q when

$$a^{(q-1)/2} \equiv -1 \pmod{q}.$$

The above is called *Euler's criterion* for quadratic residuocity.

Suppose we consider the set of all quadratic residues modulo q. This must be included in the set of integers of the form $k^2 \pmod{q}$, where $k = 1, 2, 3, \ldots, q-1$; however, if $k^2 \equiv h^2 \pmod{q}$, then $k \equiv h$ or $k \equiv -h \equiv q - h \pmod{q}$. Thus, there must be exactly $(q-1)/2$ distinct quadratic residues of q, and therefore $(q-1)/2$ distinct quadratic nonresidues of q. For example, consider $q = 11$ and the integers

$$1^2 \equiv 1,\quad 2^2 \equiv 4,\quad 3^2 \equiv 9,\quad 4^2 \equiv 5,\quad 5^2 \equiv 3,\quad 6^2 \equiv 3,\quad 7^2 \equiv 5, 8^2 \equiv 9,\quad 9^2 \equiv 4,\quad 10^2 \equiv 1 \pmod{11}.$$

Thus, the quadratic residues of 11 are 1, 3, 4, 5, 9, and the quadratic nonresidues of 11 are 2, 6, 7, 8, 10.

In the case of $l = 2$ we call $(a|q)_2$ the *Legendre symbol* and we simply write $(a|q)$. We can see that $(a|q) \in \{0, -1, 1\}$, as

$$(a|q) \equiv a^{(q-1)/2} \pmod{q}.$$

The following theorem contains some well-known results about the Legendre symbol.

Theorem 5.2 *If q is a odd prime, then*

(i) $(t^2|q) = (1|q) = 1$ *when* $\gcd(t, q) = 1$,

(ii) $(a|q) = (b|q)$ *when* $a \equiv b \pmod{q}$,

(iii) $(ab|q) = (a|q)(b|q)$,

(iv) $(-1|q) = (-1)^{(q-1)/2} = \begin{cases} 1, & \text{if } q \equiv 1 \pmod{4}; \\ -1, & \text{if } q \equiv -1 \pmod{4}, \end{cases}$

(v) $(2|q) = (-1)^{(q^2-1)/8} = \begin{cases} 1, & \text{if } q \equiv \pm 1 \pmod{8}; \\ -1, & \text{if } q \equiv \pm 3 \pmod{8}. \end{cases}$

The most remarkable result about the Legendre symbol is the *law of quadratic reciprocity*.

Theorem 5.3 *If p and q are distinct odd primes, then*

$$(p|q) = (-1)^{(p-1)(q-1)/4}(q|p).$$

Since p and q are distinct primes, it is simply amazing that the existence of a solution to the congruence

$$x^2 \equiv p \pmod{q}$$

should have anything to do with the existence of a solution to

$$x^2 \equiv q \pmod{p};$$

nevertheless, Theorem 5.3 provides a simple connection between these two different congruences.

Let us now consider the following example: suppose we want to know if

$$x^2 \equiv 871 \pmod{1009}$$

has a solution, or equivalently if 871 is a quadratic residue of 1009. We could simply use one of the algorithms discussed at the beginning of Chap. 4 to compute

$$871^{504} \pmod{1009}$$

and then apply Euler's criterion. Or we could proceed using the Legendre symbol as follows:

$$\begin{aligned} 871^{504} \pmod{1009} &\equiv (871|1009) = (67|1009)(13|1009) \\ &= (-1)^{(67-1)(1009-1)/4}(1009|67)(-1)^{(13-1)(1009-1)/4}(1009|13) \\ &= (1009|67)(1009|13) \\ &= (4|67)(8|13) = (4|67)(4|13)(2|13) = -1. \end{aligned}$$

Thus,

$$x^2 \equiv 871 \pmod{1009}$$

has no solutions, that is, 871 is a quadratic nonresidue of 1009.

Although the law of quadratic reciprocity was conjectured by Euler in 1749 and subsequently by the French mathematician and astronomer Adrien-Marie Legendre (1752–1833) in 1788, it was first proved by Gauss in 1801. Since that time many investigators, including Gauss, have provided additional proofs distinct from Gauss' original proof. Indeed, there are now over 300 proofs of this result in the literature. A particularly simple proof was provided by D. H. Lehmer in (1957). Also, in the many years since Gauss, this law has been generalized in a number of directions.

In the above example we needed to factor 871 in order to use Legendre symbols. This need to factor can be removed if we were to use the *Jacobi symbol*, a generalization of the Legendre symbol by the German mathematician Carl Gustav Jacob Jacobi (1804–1851). The Jacobi symbol is defined as follows (here we will use capital letters rather than lower case in an attempt to remove any confusion with the Legendre symbol): let Q be an odd integer such that

$$Q = \prod_{i=1}^{k} q_i^{r_i},$$

where q_i $(i = 1, 2, \ldots, k)$ are distinct primes, then the Jacobi symbol is

$$(P|Q) = \prod_{i=1}^{k} (P|q_i)^{r_i},$$

where the $(P|q_i)$ $(i = 1, 2, \ldots, k)$ are Legendre Symbols. We have a set of results for the Jacobi symbols, similar to those in Theorems 5.2 and 5.3, in the following theorem.

Theorem 5.4 *If P, R are integers and Q is an odd integer, then*

(i) $(P^2|Q) = (1|Q) = 1$ *when* $\gcd(P, Q) = 1$,
(ii) $(P|Q) = (R|Q)$ *when* $P \equiv R \pmod{Q}$,
(iii) $(PR|Q) = (P|Q)(R|Q)$,
(iv) $(-1|Q) = (-1)^{(Q-1)/2} = \begin{cases} 1, & \text{if } Q \equiv 1 \pmod 4; \\ -1, & \text{if } Q \equiv -1 \pmod 4, \end{cases}$
(v) $(2|Q) = (-1)^{(Q^2-1)/8} = \begin{cases} 1, & \text{if } Q \equiv \pm 1 \pmod 8; \\ -1, & \text{if } Q \equiv \pm 3 \pmod 8, \end{cases}$
(vi) $(P|Q) = (-1)^{(P-1)(Q-1)/4}(Q|P)$ *when P is odd.*

We remark here that if $(P|Q) = -1$, then P is a quadratic nonresidue of Q, but if $(P|Q) = 1$, it is not necessarily the case that P is a quadratic residue of Q. For example $(2|15) = 1$, but $(2|5) = (2|3) = -1$; hence, we cannot solve

$$x^2 \equiv 2 \pmod{15}.$$

We can now compute $(871|1009)$ again, this time using the Jacobi symbol and the rules in the previous theorem. We have

$$\begin{aligned}(871|1009) &= (-1)^{(871-1)(1009-1)/4}(1009|871) = (1009|871) = (138|871) \\ &= (2|871)(69|871) = (1)(69|871) = (-1)^{(69-1)(871-1)/4}(871|69) \\ &= (43|69) = (-1)^{(43-1)(69-1)/4}(69|43) = (26|43) = (2|43)(13|43) \\ &= (-1)(13|43) = (-1)(-1)^{(13-1)(43-1)/4}(43|13) = (-1)(43|13) \\ &= (-1)(4|13) = (-1)(1) = -1.\end{aligned}$$

Although factoring is not entirely removed, the only factoring we are doing is removing factors of 2 which is easier than factoring in general. We can use the above results to devise an algorithm of time complexity $O(k^2)$, where $k = \log_2 Q$, for evaluating $(P|Q)$ when $1 \le P < Q$.

As we see from the example following Theorem 5.1, it is one thing to know that a binomial congruence has a solution, but quite another to find the solution(s), especially when m is large. There are, however, certain circumstances when this problem is not too difficult. We know that we can use the EEA to find integers s and t such that

$$sn + t\phi(m) = d.$$

Thus,

$$s\frac{n}{d}+t\frac{\phi(m)}{d}=1,$$

and therefore

$$a^{sn/d}\equiv a^{1-t\phi(m)/d}\pmod{m}.$$

Since by Theorem 5.1 we may assume that

$$a^{\phi(m)/d}\equiv 1\pmod{m},$$

we get

$$a^{sn/d}\equiv a\pmod{m}.$$

It follows that in the case of $d=1$, we have $x\equiv a^s\pmod{m}$ as a solution of

$$x^n\equiv a\pmod{m}.$$

Thus, if $\gcd(n,\phi(m))=1$, we can readily find a solution to the binomial congruence. However, when $d>1$, the problem can become, at least theoretically, quite difficult. Consider the simple case of $n=2$ and suppose 4 does not divide $\phi(m)$. If we put $s=(\phi(m)/2+1)/2$, we see that s is an integer and

$$a^{2s}\equiv aa^{\phi(m)/2}\equiv a\pmod{m}.$$

Thus, a^s is a solution of

$$x^2\equiv a\pmod{m}.$$

But if $4\mid\phi(m)$, then the solution to this simple quadratic congruence can be quite difficult, even when m is a prime. In this circumstance, we see that

$$a^{(p+1)/2}\pmod{p}$$

is a solution of

$$x^2\equiv a\pmod{p}$$

when $p\equiv 3\pmod 4$, but when $p\not\equiv 3\pmod 4$, the problem is somewhat more challenging. Legendre noticed that if $p=8k+5\equiv 5\pmod 8$, then

$$x\equiv\pm a^{k+1}\quad\text{when}\quad a^{2k+1}\equiv 1\pmod{p}$$

or

$$x\equiv\pm(1/2)(4a)^{k+1}\quad\text{when}\quad a^{2k+1}\equiv -1\pmod{p}.$$

Legendre's observation was extended to other values of p by Alberto Tonelli (1849–1920) in 1891. His idea was rediscovered and extended by Daniel Shanks (1917–1996) in 1972. A generalized version of these ideas can be used to solve

$$x^r \equiv a \pmod{p}$$

when r is a prime divisor of $p-1$. Given some h such that

$$(h|p)_r \neq 1,$$

this algorithm is of complexity $O(r(\log p)^4)$. The main difficulty in this process resides in finding h. In practice this is not usually a problem because the probability that a randomly selected h (modulo p) satisfies $(h|p)_r \neq 1$ is $1 - 1/r$. Thus we expect that h can be found after only a few trials. Of course, if we already know a primitive root g of p, then we must have $(g|p)_r \neq 1$.

From a theoretic standpoint, however, it would be useful to have some idea of an upper bound on the least value of h such that $(h|p)_r \neq 1$. Unfortunately, the best available unconditional result is that if we use $g(p,k)$ to denote this h, then for any fixed k

$$g(p,k) < cp^{1/4} \log p,$$

where c is a constant and $c \leq 1.1$. It seems, though, that this is very likely far from the best possible result. When $r = 2$, a result conditional on the truth of the widely believed Generalized Riemann Hypothesis (GRH) gives

$$g(p,2) < (\log p)^2, \quad g(p,k) < 2(\log p)^2 \quad (k \text{ fixed}).$$

Here, as mentioned in Chap. 4, log denotes the Napierian logarithm or $\log_e$. In fact, many number theorists think that there exists some constant c such that

$$g(p,2) < c \log p \log\log p.$$

Results such as these are very difficult to derive and remain at the forefront of research in what is called analytic number theory.

Problems

5.1 (Binomial congruence solutions) Use Theorem 5.1 to determine which of the binomial congruences below are solvable (no need to find the actual solution):
(a) **[BB]** $x^{10} \equiv 6 \pmod{25}$
(b) $x^4 \equiv 9 \pmod{98}$
(c) $x^{15} \equiv 10 \pmod{121}$
(d) $x^{14} \equiv 5 \pmod{34}$

5.2 (Euler's criterion) For q values below, list all quadratic residues/nonresidues modulo q verified using Euler's criterion:
(a) **[BB]** $q = 5$
(b) $q = 7$
(c) $q = 13$

5.3 (Quadratic residues) For q values below, list all quadratic residues modulo q found by solving

$$k^2 \pmod{q}, \quad \text{where } k = 1, 2, 3, \ldots, q - 1:$$

(a) $q = 5$
(b) **[BB]** $q = 7$
(c) $q = 13$

5.4 (Legendre symbol) Use Theorems 5.2, 5.3 and the Legendre symbol to determine if the following congruences have solutions:
(a) **[BB]** $x^2 \equiv 35 \pmod{431}$m
(b) $x^2 \equiv 15 \pmod{191}$
(c) $x^2 \equiv 45 \pmod{101}$
(d) $x^2 \equiv 17 \pmod{419}$

5.5 (Jacobi symbol) Use Theorem 5.4 and the Jacobi symbol to determine if the following congruences have solutions:
(a) **[BB]** $x^2 \equiv 13 \pmod{223}$
(b) $x^2 \equiv 33 \pmod{97}$
(c) $x^2 \equiv 5 \pmod{199}$
(d) $x^2 \equiv 55 \pmod{607}$

Further Reading

Jacobi symbol algorithm

E. Bach and J. Shallit, *Algorithmic Number Theory*, MIT Press, Cambridge MA, 1996, pp. 111–114.

Reciprocity laws

D. H. Lehmer, A low energy proof of the reciprocity law, *American Mathematical Monthly*, **64** (1957), 103–106.
Franz Lemmermeyer, *Reciprocity Laws: From Euler to Eisenstein*, Springer-Verlag, Berlin, 2000.

Solution of binomial congruences

Eric Bach and Jeffrey Shallit, *Algorithmic Number Theory*, Vol. 1, MIT Press, 1996, pp. 155–163.

Least k-th power nonresidue

H. C. Williams, *Édouard Lucas and Primality Testing*, Wiley-Interscience, John Wiley and Sons, NY, 1998, pp. 223–227.

Enrique Treviño, The least k-th power non-residue, *Journal of Number Theory*, **149** (2015), 201–224.

Kevin J. McGown and Enrique Treviño, The least quadratic non-residue, *Contemporary Mathematics*, Vol. 775, Mexican Mathematics in the World, Amer. Math. Soc., 2021, pp. 205–232.

Chapter 6
The Binomial Coefficients

Before proceeding any further, we introduce some useful notation. In mathematics, we often use the upper case Greek letter Σ (sigma) to denote a sum and the Greek letter $\prod$ (pi) to denote a product. We put some sort of conditions above and/or below the Σ (resp. $\prod$) to indicate what values are being summed (resp. multiplied). For instance, to sum the first 5 integers we could write

$$\sum_{i=1}^{5} i = 1 + 2 + 3 + 4 + 5,$$

and to multiply the first 5 integers we could write

$$\prod_{i=1}^{5} i = 1 \cdot 2 \cdot 3 \cdot 4 \cdot 5.$$

Suppose we have three different objects a, b, c, and we wish to arrange them in a row. We see that we can do this as

$$abc, acb, bac, bca, cab, cba.$$

The first object in any particular arrangement can be any one of the three, but the second can only be one of the two left over, and the last is the one left over. When looked at in this way, we see that there can only be

$$3 \cdot 2 \cdot 1 = 6$$

distinct arrangements of the original three objects. It is easy to generalize this to n objects, in that if we begin with n different objects, then there can only be

E. L. F. Roettger and H. C. Williams, *The Enchantment of Numbers*, CMS/CAIMS Books in Mathematics 16, https://doi.org/10.1007/978-3-031-76568-1_6

$$n \cdot (n-1) \cdot (n-2) \cdot \cdots \cdot 1 = \prod_{i=1}^{n} i$$

distinct arrangements of them.

We call $\prod_{i=1}^{n} i$ the *factorial* of n or n *factorial*, written $n!$. We define $0! = 1$, as there can only be one way to arrange 0 objects. The factorials grow very quickly; for example, we have $1! = 1$, $2! = 2$, $3! = 6$, $4! = 24$, $5! = 120$, $6! = 720, \ldots, 10! = 3628800, \ldots, 20! = 2432902008176640000, \ldots, 50! = 3.0414093201713378043612608166064768844377641568960512 \times 10^{64}$. It turns out that $n!$ grows like $\sqrt{2\pi n}(n/e)^n$; indeed,

$$\lim_{n\to\infty} \frac{n!}{\sqrt{2\pi n}(n/e)^n} = 1.$$

This is called *Stirling's approximation*, and is named after the British mathematician James Stirling (1692–1770).

Let p be a given prime and a be some integer. We denote by $p^m \parallel a$ (p^m *exactly divides* a) as that power p^m of p such that $p^m \mid a$ and $p^{m+1} \nmid a$. We often denote this value of m as $\nu_p(a)$. For example, we see that $3^2 \mid 720$, but $3^3 \nmid 720$; thus, $3^2 \parallel 720$ or $\nu_3(720) = 2$. In 1808, Legendre solved the problem of finding $\nu_p(n!)$. Since only multiples of p can contain the factor p, we can write

$$\nu_p(n!) = \nu_p\left(p \cdot 2p \cdot 3p \cdot \cdots \cdot \left\lfloor \frac{n}{p} \right\rfloor p\right);$$

hence,

$$\nu_p(n!) = \left\lfloor \frac{n}{p} \right\rfloor + \nu_p\left(\left\lfloor \frac{n}{p} \right\rfloor !\right).$$

Since (Chap. 4 Page 47) $\lfloor \lfloor n/p \rfloor / p \rfloor = \lfloor n/p^2 \rfloor$, we get

$$\nu_p(n!) = \left\lfloor \frac{n}{p} \right\rfloor + \left\lfloor \frac{n}{p^2} \right\rfloor + \nu_p\left(\left\lfloor \frac{n}{p^2} \right\rfloor !\right).$$

By continuing the process, we find that

$$\nu_p(n!) = \left\lfloor \frac{n}{p} \right\rfloor + \left\lfloor \frac{n}{p^2} \right\rfloor + \left\lfloor \frac{n}{p^3} \right\rfloor + \cdots,$$

which can only contain a finite number of terms because

$$\left\lfloor \frac{n}{p^k} \right\rfloor = 0 \quad \text{when} \quad p^k > n.$$

This is sometimes referred to as *de Polignac's formula* after the French mathematician Alphonse de Polignac (1826–1863). Thus, if $p = 2$, then

$$\begin{aligned}\nu_2(20!) &= \left\lfloor \frac{20}{2} \right\rfloor + \left\lfloor \frac{20}{2^2} \right\rfloor + \left\lfloor \frac{20}{2^3} \right\rfloor + \left\lfloor \frac{20}{2^4} \right\rfloor \\ &= 10 + 5 + 2 + 1 = 18.\end{aligned}$$

Also, if $p = 5$, then

$$\nu_5(20!) = \left\lfloor \frac{20}{5} \right\rfloor = 4.$$

We can write the formula for $\nu_p(n!)$ in a different way by considering the base-p representation of n. Suppose

$$n = a_0 p^k + a_1 p^{k-1} + a_2 p^{k-2} + \cdots + a_{k-1} p + a_k,$$

where $a_0 \neq 0$, $0 \leq a_i < p$ $(i = 0, 1, 2, \ldots, k)$. We get

$$\begin{aligned}\left\lfloor \frac{n}{p} \right\rfloor &= a_0 p^{k-1} + a_1 p^{k-2} - \cdots + a_{k-1} \\ \left\lfloor \frac{n}{p^2} \right\rfloor &= a_0 p^{k-2} + a_1 p^{k-3} + \cdots + a_{k-2}, \quad \text{etc.}\end{aligned}$$

It follows that since

$$\sum_{j=0}^{h} p^j = \frac{p^{h+1} - 1}{p - 1},$$

we get

$$\begin{aligned}\left\lfloor \frac{n}{p} \right\rfloor + \left\lfloor \frac{n}{p^2} \right\rfloor + \left\lfloor \frac{n}{p^3} \right\rfloor + \cdots &= \sum_{i=0}^{k} a_i \left(\frac{p^{k-i} - 1}{p - 1} \right) \\ &= \frac{n - \sum_{i=0}^{k} a_i}{p - 1}.\end{aligned}$$

Thus, if $\sigma_p(n)$ denotes the sum of the p-ary digits of n, then

$$\nu_p(n!) = \frac{n - \sigma_p(n)}{p - 1}.$$

Since for $p = 2$, $20 = 2^4 + 2^2$, we get $k = 4$ in this case, $\sigma_2(20) = 2$ and $\nu_2(20!) = \frac{20-2}{2-1} = 18$. Also, if $p = 5$, we have $20 = 4 \cdot 5$ and we get $k = 1$, $\sigma_5(20) = 4$ and $\nu_5(20!) = \frac{20-4}{5-1} = 4$.

We can also use Legendre's observation to show that

$$\nu_p((n+m)!) \geq \nu_p(n!) + \nu_p(m!),$$

because

$$\begin{aligned}\nu_p(n!) + \nu_p(m!) &= \left\lfloor \frac{n}{p} \right\rfloor + \left\lfloor \frac{n}{p^2} \right\rfloor + \left\lfloor \frac{n}{p^3} \right\rfloor + \cdots + \left\lfloor \frac{m}{p} \right\rfloor + \left\lfloor \frac{m}{p^2} \right\rfloor + \left\lfloor \frac{m}{p^3} \right\rfloor + \cdots \\ &= \left\lfloor \frac{n}{p} \right\rfloor + \left\lfloor \frac{m}{p} \right\rfloor + \left\lfloor \frac{n}{p^2} \right\rfloor + \left\lfloor \frac{m}{p^2} \right\rfloor + \left\lfloor \frac{n}{p^3} \right\rfloor + \left\lfloor \frac{m}{p^3} \right\rfloor + \cdots \\ &\leq \left\lfloor \frac{n+m}{p} \right\rfloor + \left\lfloor \frac{n+m}{p^2} \right\rfloor + \left\lfloor \frac{n+m}{p^3} \right\rfloor + \cdots \\ &= \nu_p((n+m)!).\end{aligned}$$

From this result we see that if $b_1, b_2, \ldots, b_k$ are positive integers and $b_1 + b_2 + \cdots + b_k = n$, then

$$\nu_p(b_1!) + \nu_p(b_2!) + \cdots + \nu_p(b_k!) \leq \nu_p((b_1 + b_2 + \cdots b_k)!) = \nu_p(n!).$$

It follows that since $n!$ must be a product of distinct prime powers, we must have

$$\frac{n!}{b_1! b_2! \cdots b_k!}$$

an integer whenever $b_1 + b_2 + \cdots + b_k = n$ and b_i $(i = 1, 2, \ldots, k)$ are positive integers.

Suppose we define

$$\binom{n}{k} = \frac{n!}{(n-k)!k!} \quad (n \geq k)$$

and

$$\binom{n}{k} = 0 \quad \text{when} \quad k > n.$$

When spoken, the symbol $\binom{n}{k}$ is pronounced as "n choose k". Since $n - k + k = n$, we see that $\binom{n}{k}$ is always an integer. These numbers, called the *binomial coefficients*, are of use in the following well-known result, usually referred to as the *binomial theorem*.

Theorem 6.1 *It is possible to expand any nonnegative integer power of $x + y$ into a sum of the form*

$$(x+y)^n = \sum_{k=0}^{n} \binom{n}{k} x^{n-k} y^k.$$

For example,

$$(x+y)^4 = x^4 + 4x^3y + 6x^2y^2 + 4xy^3 + y^4.$$

From the definition of $\binom{n}{k}$ it is easy to see that

$$\binom{n}{k} = \binom{n}{n-k}$$

and

$$\binom{n}{k} = \frac{n}{k}\binom{n-1}{k-1} \quad (k \neq 0).$$

It is also easy to verify *Pascal's identity*

$$\binom{n}{k} = \binom{n-1}{k-1} + \binom{n-1}{k}.$$

It is this result that is behind *Pascal's triangle*, named after the French mathematician and philosopher Blaise Pascal (1623–1662). This is an arrangement of the binomial coefficients, where the nth row consists of $\binom{n}{k}$, $k = 0, 1, 2, \ldots, n$. The first 11 rows of this triangle are given below

$$\begin{array}{lccccccccccccccccccccc}
n=0 & &&&&&&&&&& 1 \\
n=1 & &&&&&&&&& 1 && 1 \\
n=2 & &&&&&&&& 1 && 2 && 1 \\
n=3 & &&&&&&& 1 && 3 && 3 && 1 \\
n=4 & &&&&&& 1 && 4 && 6 && 4 && 1 \\
n=5 & &&&&& 1 && 5 && 10 && 10 && 5 && 1 \\
n=6 & &&&& 1 && 6 && 15 && 20 && 15 && 6 && 1 \\
n=7 & &&& 1 && 7 && 21 && 35 && 35 && 21 && 7 && 1 \\
n=8 & && 1 && 8 && 28 && 56 && 70 && 56 && 28 && 8 && 1 \\
n=9 & & 1 && 9 && 36 && 84 && 126 && 126 && 84 && 36 && 9 && 1 \\
n=10 & 1 && 10 && 45 && 120 && 210 && 252 && 210 && 120 && 45 && 10 && 1
\end{array}$$

By Pascal's identity, we see that any entry in the triangle is sum of the two entries appearing above it. For example,

$$\binom{5}{1} + \binom{5}{2} = 5 + 10 = 15 = \binom{6}{2}.$$

In this way, we can see that the integer entries of Pascal's triangle correspond to the triangle below

$$
\begin{array}{lccccccccccccccccccccc}
n=0 & & & & & & & & & & \binom{0}{0} \\
n=1 & & & & & & & & & \binom{1}{0} & & \binom{1}{1} \\
n=2 & & & & & & & & \binom{2}{0} & & \binom{2}{1} & & \binom{2}{2} \\
n=3 & & & & & & & \binom{3}{0} & & \binom{3}{1} & & \binom{3}{2} & & \binom{3}{3} \\
n=4 & & & & & & \binom{4}{0} & & \binom{4}{1} & & \binom{4}{2} & & \binom{4}{3} & & \binom{4}{4} \\
n=5 & & & & & \binom{5}{0} & & \binom{5}{1} & & \binom{5}{2} & & \binom{5}{3} & & \binom{5}{4} & & \binom{5}{5} \\
n=6 & & & & \binom{6}{0} & & \binom{6}{1} & & \binom{6}{2} & & \binom{6}{3} & & \binom{6}{4} & & \binom{6}{5} & & \binom{6}{6} \\
n=7 & & & \binom{7}{0} & & \binom{7}{1} & & \binom{7}{2} & & \binom{7}{3} & & \binom{7}{4} & & \binom{7}{5} & & \binom{7}{6} & & \binom{7}{7} \\
n=8 & & \binom{8}{0} & & \binom{8}{1} & & \binom{8}{2} & & \binom{8}{3} & & \binom{8}{4} & & \binom{8}{5} & & \binom{8}{6} & & \binom{8}{7} & & \binom{8}{8} \\
n=9 & \binom{9}{0} & & \binom{9}{1} & & \binom{9}{2} & & \binom{9}{3} & & \binom{9}{4} & & \binom{9}{5} & & \binom{9}{6} & & \binom{9}{7} & & \binom{9}{8} & & \binom{9}{9} \\
n=10 & \binom{10}{0} & \binom{10}{1} & \binom{10}{2} & \binom{10}{3} & \binom{10}{4} & \binom{10}{5} & \binom{10}{6} & \binom{10}{7} & \binom{10}{8} & \binom{10}{9} & \binom{10}{10}
\end{array}
$$

Notice also that the sum of all the entries in any given row is a power of 2. This follows immediately from Theorem 6.1 with $x = y = 1$. We get

$$2^n = \sum_{k=0}^{n} \binom{n}{k}.$$

By putting $y = 1$ and differentiating $(x + 1)^n$, we get on putting $x = 1$

$$n2^{n-1} = \sum_{k=0}^{n} k\binom{n}{k},$$

and by differentiating again, we can derive

$$n(n + 1)2^{n-2} = \sum_{k=0}^{n} k^2\binom{n}{k}.$$

It is also entertaining to show that

$$\sum_{k=0}^{m} \binom{n+k}{k} = \binom{m+n+1}{m}.$$

This is called the *Hockey-Stick identity* because of the pattern of a hockey stick which emerges when the numbers are plotted on Pascal's triangle. For example, if $m = 4$ and $n = 3$, we have

$$
\begin{aligned}
\sum_{k=0}^{m} \binom{n+k}{k} = \sum_{k=0}^{4} \binom{3+k}{k} &= \binom{3}{0} + \binom{4}{1} + \binom{5}{2} + \binom{6}{3} + \binom{7}{4} \\
&= 1 + 4 + 10 + 20 + 35 \\
&= 70
\end{aligned}
$$

and

$$\binom{m+n+1}{m} = \binom{8}{4} = 70.$$

Upon plotting these on Pascal's triangle, we see the hockey stick emerge:

1
1 1
1 2 1
1 3 3 1
1 4 6 4 1
1 5 10 10 5 1
1 6 15 20 15 6 1
1 7 21 35 35 21 7 1
1 8 28 56 70 56 28 8 1
1 9 36 84 126 126 84 36 9 1
1 10 45 120 210 252 210 120 45 10 1

There is a *multinomial theorem* generalization of Theorem 6.1. This allows us to express

$$(x_1 + x_2 + x_3 + \cdots + x_m)^n = \sum \binom{n}{k_1, k_2, \ldots, k_m} \prod_{i=1}^{m} x_i^{k_i},$$

where the sum is taken over all nonnegative integers k_i $(i = 1, 2, \ldots, m)$ such that $k_1 + k_2 + \cdots k_m = n$. Here the *multinomial coefficient*

$$\binom{n}{k_1, k_2, \ldots, k_m} = \frac{n!}{k_1! k_2! \cdots k_m!},$$

which, as we have already seen earlier, must be an integer. For example, we find for $n = m = 3$

$$(x + y + z)^3 = x^3 + y^3 + z^3 + 3x^2y + 3y^2x + 3y^2z + 3z^2y + 3x^2z + 3z^2x + 6xyz.$$

Note that the coefficient of y^2x is

$$3 = \binom{3}{1, 2, 0} = \frac{3!}{1!2!0!} = \frac{6}{2}$$

and the coefficient of xyz is

$$6 = \binom{3}{1, 1, 1} = \frac{3!}{1!1!1!}.$$

We have already observed some identity relations involving the binomial coefficients, but it must be stressed that there are a bewildering number of such relationships. The reason for this is that the binomial coefficients play a very significant role in *combinatorial enumeration*, an area of mathematics devoted to counting the number of ways certain patterns can be formed. As an example, consider the following problem, originally due to Euler in 1751: in how many different ways can we divide a polygon of $n + 2$ sides into triangles using only non-intersecting diagonals? Consider $n = 3$; we have the 5 patterns below and no others:

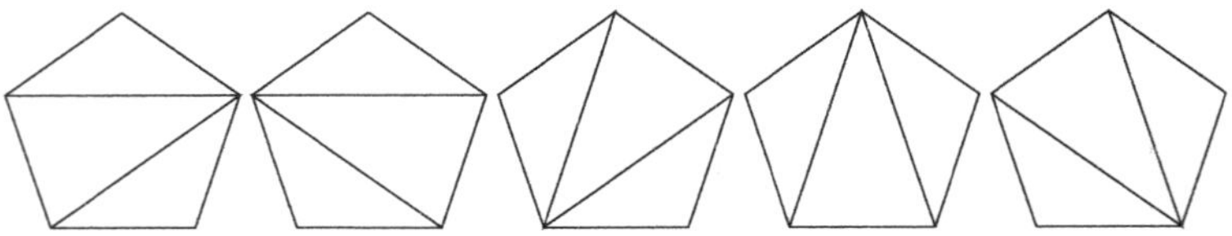

However, when n becomes large the number of such patterns becomes very large indeed. The solution of this problem for a given n is C_n, where

$$C_n = \frac{1}{n+1}\binom{2n}{n} = \binom{2n}{n} - \binom{2n}{n+1}.$$

These numbers are called the *Catalan numbers*; the first few, starting with $C_0 = 1$, are

$$1, 1, 2, 5, 14, 42, 132, 429, 1430, \ldots \quad \text{(A000108 in the OEIS).}$$

They seem first to have been discovered by the Mongolian mathematician Sharabiin Myangat (c. 1692–c. 1793) who worked in the Qing court in China. Today the Catalan numbers, named after the French-Belgian mathematician Eugéne Charles Catalan (1814–1894), have found applications in algebra, analysis, number theory, probability, geometry and many more areas. They are considered by many experts to be the most ubiquitous sequence of numbers in mathematics.

We next turn to the divisibility properties of the binomial coefficients. We first point out that there is a myriad of results concerning this topic; we will only mention a few and direct the reader wanting more information to the relevant works in the references listed at the end of this chapter.

One simple example of such a divisibility property is given below.

Theorem 6.2 *If $d = \gcd(n, k)$, then $\frac{d}{n}\binom{n}{k}$ is an integer.*

Proof Since $\frac{d}{n}\binom{n}{k}$ is a rational number, we must have positive integers r, s such that $\gcd(r, s) = 1$ and $\frac{r}{s} = \frac{d}{n}\binom{n}{k}$. Also, since $\frac{k}{n}\binom{n}{k} = \binom{n-1}{k-1}$ is an integer, we see that both

$$\frac{n}{d}\frac{r}{s} \quad \text{and} \quad \frac{k}{d}\frac{r}{s}$$

are integers. Putting $a = \frac{nr}{ds}$, $b = \frac{kr}{ds}$, we get $ads = nr$, $bds = kr$ or $as = \frac{n}{d}r$, $bs = \frac{k}{d}r$. It follows that s must be a divisor of $\gcd(\frac{n}{d}r, \frac{k}{d}r) = r\gcd(\frac{n}{d}, \frac{k}{d})$. Since $\gcd(\frac{n}{d}, \frac{k}{d}) = 1$, we can only have $s \mid r$, and since $\gcd(r, s) = 1$, we get $s = 1$.

□

Since

$$k\binom{p+1}{k} = (p+1)\binom{p}{k-1},$$

we have the following simple corollary.

Corollary 6.1 *If p is a prime and $p \nmid k$, then $p \mid \binom{p}{k}$ when $k \neq 0, p$. Also, $p \mid \binom{p+1}{k}$ when $k \neq 0, 1, p, p+1$.*

This corollary was noted by Lucas who also observed that since

$$\binom{n}{k} = \frac{n(n-1)(n-2)\cdots(n-k+1)}{k(k-1)(k-2)\cdots 1},$$

we have

$$\binom{p-1}{k} \equiv (-1)^k\frac{k!}{k!} \equiv (-1)^k \pmod{p}$$

when $1 \leq k \leq p-1$.

Lucas also proved the following useful result in 1878.

Theorem 6.3 *Let p be a prime, $n = \left\lfloor \frac{n}{p} \right\rfloor p + r$ and $k = \left\lfloor \frac{k}{p} \right\rfloor p + s$ $(r, s \geq 0)$. We have*

$$\binom{n}{k} \equiv \binom{\lfloor n/p \rfloor}{\lfloor k/p \rfloor}\binom{r}{s} \pmod{p}.$$

Proof (This proof, while not exactly appearing like the proof Lucas provided, is essentially the same.) Put $n_1 = \left\lfloor \frac{n}{p} \right\rfloor$, $k_1 = \left\lfloor \frac{k}{p} \right\rfloor$; we have

$$n = n_1 p + r, \quad k = k_1 p + s$$

and

$$\begin{aligned}(1+x)^n = (1+x)^{pn_1+r} &= (1+x)^{pn_1}(1+x)^r \\ &\equiv (1+x^p)^{n_1}(1+x^r) \pmod{p}\end{aligned}$$

by the above corollary. Thus

$$(1+x)^n \equiv \sum_{i=1}^{n_1}\sum_{j=1}^{r}\binom{n_1}{i}\binom{r}{j}x^{ip+j} \pmod{p}.$$

Notice that in the double sum on the right-hand side each power of x up to x^n occurs just once. In particular, this means that the coefficients of x^{pk_1+s} on both sides of the equality must be the same modulo p. Thus,

$$\binom{n}{k} \equiv \binom{n_1}{k_1}\binom{r}{s} \pmod{p}.$$

□

This result allows us to compute $\binom{n}{k} \pmod{p}$ rather quickly. Consider the following example. Compute

$$\binom{35914}{785} \pmod{13}.$$

We have

$$35914 = 2762 \cdot 13 + 8,$$
$$785 = 60 \cdot 13 + 5;$$

hence,

$$\binom{35914}{785} \equiv \binom{2762}{60}\binom{8}{5} \pmod{13}.$$

We can repeat the same process on $\binom{2762}{60}$, where $2762 = 212 \cdot 13 + 6, 60 = 4 \cdot 13 + 8$. This means that

$$\binom{2762}{60} \equiv \binom{212}{4}\binom{6}{8} \pmod{13}.$$

Since $8 > 6$, we have $\binom{6}{8} = 0$ and find that

$$\binom{35914}{785} \equiv 0 \pmod{13}.$$

It is not difficult to see that if we apply Theorem 6.3 repeatedly, we can establish the following result. Let n and k be positive integers and p be a prime. Suppose $n = (n_0 n_1 n_2 \dots n_i)_p$ and $k = (k_0 k_1 k_2 \dots k_i)_p$ are the base-p representations of n and k, respectively, where one of the representations may have to be padded with zeros on the left. Then

$$\binom{n}{k} \equiv \binom{n_0}{k_0}\binom{n_1}{k_1}\binom{n_2}{k_2}\cdots\binom{n_i}{k_i} \pmod{p}.$$

We conclude this chapter with a beautiful result of the English mathematician Joseph Wolstenholme (1829–1891). In 1862 he proved that for any prime $p \geq 5$, we have

$$\binom{2p-1}{p-1} \equiv 1 \pmod{p^3}.$$

A more general result was proved by the Norwegian mathematician Wilhelm Ljunggren (1905–1973):

$$\binom{ap}{bp} \equiv \binom{a}{b} \pmod{p^3}$$

when p is a prime and $p \geq 5$. A prime which satisfies Wolstenholme's congruence modulo p^4 is called a *Wolstenholme prime*. The only known such primes are $p = 16843$ and $p = 2124679$, and any other must exceed 10^9.

Problems

6.1 (Sum notation) Calculate the sums indicated below:

(a) **[BB]** $\sum_{i=1}^{5} i$

(b) $\sum_{i=1}^{5} i^2$

(c) $\sum_{i=1}^{5} i^3$

6.2 (Product notation) Calculate the products indicated below:

(a) **[BB]** $\prod_{i=1}^{5} i$

(b) $\prod_{i=1}^{5} (i+1)$

(c) $\prod_{i=1}^{5} 4$

6.3 (Stirling's approximation) For the values of n listed below calculate

$$\frac{n!}{\sqrt{2\pi n}(n/e)^n}$$

and round your answers to 5 decimal places:

(a) $n = 3$

(b) **[BB]** $n = 4$

(c) $n = 5$

6.4 (Exactly divides) In the following find m such that $p^m \parallel a$ given p and a; we will use the notation $\nu_p(a)$ to denote this m:
(a) **[BB]** $\nu_3(459)$
(b) $\nu_7(1001)$
(c) $\nu_5(475)$

6.5 (de Polignac's formula) Use de Polignac's formula to solve the following:
(a) $\nu_3(15!)$
(b) **[BB]** $\nu_7(47!)$
(c) $\nu_5(30!)$

6.6 (Binomial theorem) Use Theorem 6.1 to solve:
(a) **[BB]** $(a+b)^5$
(b) the first 3 terms of $(2a+3b)^{16}$
(c) the 5th term of $(a-b)^{11}$

6.7 (Hockey-Stick identity) Verify the hockey-stick identity for given m, n solving for all necessary terms and plotting them on Pascal's triangle:
(a) **[BB]** $m = 3, n = 2$
(b) $m = 4, n = 4$
(c) $m = 5, n = 3$

6.8 (Catalan numbers) Given the nth Catalan number is defined by

$$C_n = \frac{1}{n+1}\binom{2n}{n} = \binom{2n}{n} - \binom{2n}{n+1},$$

find C_n for:
(a) $n = 2$
(b) **[BB]** $n = 4$
(c) $n = 9$

6.9 (Lucas' result) Verify Theorem 6.3 holds in the case of:
(a) **[BB]** $p = 7, n = 25, k = 2$
(b) $p = 13, n = 20, k = 14$
(c) $p = 3, n = 19, k = 7$

Further Reading

Legendre's (de Polignac's) result

Ivan Niven, Herbert S. Zuckerman, and Hugh L. Montgomery, *An Introduction to the Theory of Numbers*, 5th Ed., John Wiley and Sons, Inc., NY, 1991, pp. 182–188.

The binomial theorem and Pascal's triangle

J. L. Coolidge, The story of the binomial theorem, *American Mathematical Monthly*, **56** (1949), 147–157.
John Conway and Richard K. Guy, *The Book of Numbers*, Copernicus, Springer-Verlag, NY, 1996, pp. 65–74.

Identities

John Riordan, *Combinatorial Identities*, John Wiley and Sons, Inc., NY, 1968.

Enumerative combinatorics

Richard P. Stanley, *Enumerative Combinatorics*, Volumes I and II, Cambridge University Press, Cambridge, 1997.

Catalan numbers

Thomas Koshy, *Catalan Numbers with Applications*, Oxford University Press, Oxford, 2009.
Richard P. Stanley, *Catalan Numbers*, Cambridge University Press, NY, 2015.

Euler's problem

Heinrich Dörrie, *100 Great Problems of Elementary Mathematics: Their History and Solution*, Dover Publications, Inc., NY, 1965, pp. 21–27.

Divisibility properties

L. E. Dickson, *History of the Theory of Numbers, Volume I: Divisibility and Primality*, Dover Publications, NY, 2005, pp. 263–278.
Andrew Granville, Arithmetic properties of binomial coefficients. I. Binomial coefficients modulo prime powers, *Organic mathematics (Burnaby, BC, 1995)* CMS Conf. Proc, 20, American Mathematical Society, Providence, RI, 1997.

Wolstenholme primes

R. J. McIntosh and E. L. Roettger, A search for Fibonacci-Wieferich and Wolstenholme primes, *Mathematics of Computation*, **76** (2007), 2087–2094.

Chapter 7
Public-Key Cryptography

Cryptography is the practice of designing and analysing techniques for assuring secure communication in the presence of malicious adversaries. In fact, the word "cryptography" is derived from the Greek *kyrptos* meaning hidden and *graphy* for writing. More informally, it is often referred to as secret writing. It may seem that a chapter on cryptography might be out of place in a work devoted to number theory, but as we shall see, the ideas presented here depend upon number-theoretic principles that we have already developed. No one was more surprised than number theorists, when it was shown that their speciality might actually be beneficial to society at large. Previous to this, many number theory practitioners took a seemingly perverse pride in the likelihood that their subject would be unsullied by application to anything "useful", such as the development of armaments or other artifices that might be employed in warfare or other destructive human activities. During the early seventies, however, it became clear that some kind of rapid and secure means of communicating information had to be developed. This was largely inspired by the requirement of banks to provide cash points, now called ATMs. Since the advent of the Internet in the early eighties, this demand has only increased, and we now have the situation where almost all Internet commerce (or other activity) is conducted by means of public-key cryptography.

As a simple example of cryptography, consider the problem of two schoolchildren who wish to communicate during class without the teacher being able to determine the content of their messages. The children might get together at some point outside of class and exchange a new alphabet. When one sends a message to the other, she (usually, now identified as Alice) writes the message in the new alphabet and sends it to her communicant (usually, now called Bob). Should the teacher (to the students, the malicious adversary) intercept this message, it will appear as gibberish and reveal no information about its content. More generally, in order to establish this kind of communication protocol, we assume the existence of a large number of different transformations, any one of which will render the original message unintelligible to

E. L. F. Roettger and H. C. Williams, *The Enchantment of Numbers*, CMS/CAIMS Books in Mathematics 16, https://doi.org/10.1007/978-3-031-76568-1_7

anyone but the intended recipient. It is usually assumed that these transformations, as a family, are known to any adversary. It is now customary to refer to such an opponent as Eve (Eavesdropper). In order for Bob to be able to read Alice's message, he must know which transformation Alice used. The information that identifies this transformation is called the key. It is this key, not the family of transformations, which is supposed to guarantee the security of this type of cryptographic scheme; thus, it is essential that the key be communicated between Alice and Bob through a secure (inaccessible to Eve) transmission medium. Furthermore, the process by which the keyword is selected should be as random as possible in order to prevent the possibility of easy determination.

To exemplify this idea, we will consider the following communication strategy:

1. A keyword is selected outside of class by Alice and Bob; suppose it is JUPITER.
2. The alphabet is rearranged by writing the keyword underneath the alphabet, followed by all the other letters in alphabetical order

A	B	C	D	E	F	G	H	I	J	K	L	M	N	O	P	Q	R	S	T	U	V	W	X	Y	Z
J	U	P	I	T	E	R	A	B	C	D	F	G	H	K	L	M	N	O	Q	S	V	W	X	Y	Z

In the new transformed alphabet, each letter is replaced by the letter directly below it.

To send the message TEACHER HAS BIG FEET, Alice passes the transformed (encrypted) message QTJPATN AJO UBR ETTQ to Bob. Since Bob knows the keyword, he can easily undo Alice's transformation (i.e., decrypt it) on this communication.

There are many possible variations of this type of cryptographic scheme, called a *one-key cryptosystem*. These depend on the transformation functions chosen and their corresponding key selection process. We give a simple diagram representing a one-key communication scheme below.

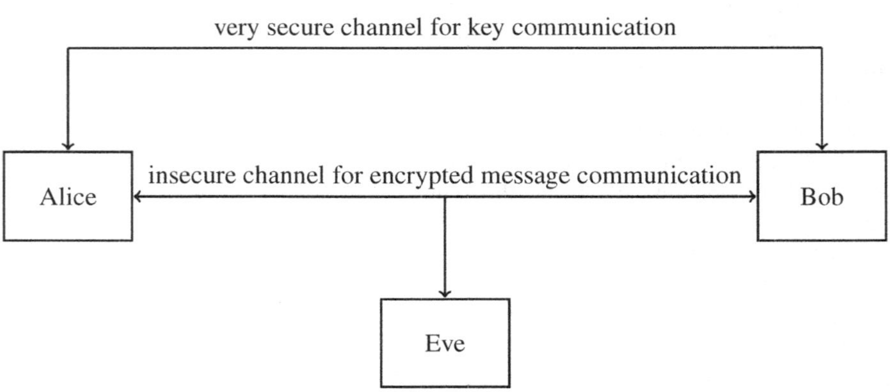

One-key cryptography has a very long history dating back to antiquity, and until the seventies was the only technique available for ensuring communication security.

The main disadvantage of using one-key cryptography is the problem of secure key distribution. In order to communicate, both Alice and Bob have to know the key, and therefore it has to be transmitted from one to the other over a very secure channel, for example, a trusted courier. This can be an expensive and inconvenient process. The great advantage of using one-key cryptography is its fast transmission rate, particularly when modern communication processes are used.

In the early seventies, Whitfield Diffie and Martin Hellman began to collaborate on producing a new paradigm for cryptography, a subject which at that time was not academically fashionable and greatly constrained by the powerful National Security Agency (NSA). They were unhappy with the then current state of cryptography, especially with respect to the problems of secure key distribution and authentication, the process of determining that a received message comes from the stated sender and has not been altered. In a seminal paper published in 1976, they introduced the concept of *public-key cryptography*, produced an elegant solution to the first problem and proposed a scheme for solving the second. They provided an idea for a public-key distribution system that eliminated the need for a separate secure key channel in a one-key cryptosystem. The crux of their proposed method was a reliance on the use of a *one-way function* f, a function that can easily be evaluated, but for which it is difficult to produce an inverse; that is, we can easily compute $y = f(x)$ for a given x, but given only y and f, it is difficult to find x.

We recall that when looking at powers of 2 modulo 19 in Chap. 5, the results followed no particular pattern. This is true in general and because of this, given y, g, m, it seems solving

$$y \equiv g^i \pmod{m}$$

for i is a difficult problem; this is a special case of what is called the *discrete log problem* (DLP); recall this was first introduced at the end of Chap. 4. Due to the apparent difficulty of the DLP and also the ease of computing powers modulo m, provided by Algorithm 2 of Chap. 4, we can see that exponentiation modulo m provides a one-way function, as described by Diffie and Hellman. Hence, we are able to use the above to implement a protocol in which all the communications are performed publicly, and yet at the end the two parties have a shared secret; this can be called *public-key exchange*. This particular type of key exchange was the first ever developed and is called the *Diffie-Hellman key exchange*, although both Diffie and Hellman have recently suggested Ralph Merkle's name be added to the protocol. It has become the norm to describe the key exchange as a conversation between two parties called Alice and Bob, with third party (Eve) listening in. The Diffie-Hellman key exchange is performed in the steps below.

Algorithm 1 Diffie-Hellman key-exchange protocol.

1. Alice and Bob agree on a large prime p and a primitive root g for p. These are both public; that is, Eve knows both p and g.
2. Alice chooses a secret integer a such that $1 < a < p - 1$ and calculates $A \equiv g^a \pmod{p}$. She then shares A publicly with Bob, but keeps a a secret. Thus, Eve knows A, but not a.
3. Bob chooses a secret integer b such that $1 < b < p - 1$ and calculates $B \equiv g^b \pmod{p}$. He then shares B publicly with Alice, but keeps b a secret. Thus, Eve knows B, but not b.
4. Alice computes $s \equiv B^a \pmod{p}$.
5. Bob computes $s \equiv A^b \pmod{p}$.
6. The value s is the shared secret key.

The reason this works is

$$A^b \equiv (g^a)^b = (g^b)^a \equiv B^a \pmod{p}.$$

Its security is based on the believed difficulty of the discrete log problem for significantly large values of p.

For illustrative purposes, let us look at an example of the Diffie-Hellman key exchange with $p = 97$ and $g = 5$.

Step 1: Alice and Bob agree on a large prime $p = 97$ and a primitive root $g = 5$.
Step 2: Alice chooses a secret integer $a = 7$ and calculates $40 \equiv 5^7 \pmod{97}$. She then shares $A = 40$ publicly with Bob.
Step 3: Bob chooses a secret integer $b = 9$ and calculates $30 \equiv 5^9 \pmod{97}$. He then shares $B = 30$ publicly with Alice.
Step 4: Alice computes $51 \equiv 30^7 \pmod{97}$.
Step 5: Bob computes $51 \equiv 40^9 \pmod{97}$.
Step 6: The value $s = 51$ is a shared secret.

Note that security of the above is lost if Eve can solve the discrete log problem for either Alice or Bob's secret exponents; that is, given A, g and p, Eve can solve for unknown a in $A \equiv g^a \pmod{p}$ or given B, g and p, Eve can solve for unknown b in $B \equiv g^b \pmod{p}$. In the following we will look at a few methods for doing exactly this.

We can solve the discrete log problem if the value for p is relatively small. This is possible through the use of an algorithm called *Baby Step, Giant Step* (BSGS) created by Shanks for another purpose. Imagine we want to solve $g^x \equiv c \pmod{p}$, where g is a primitive root of p. We first set $N = \lceil \sqrt{p-1} \rceil$ and then we construct two lists:

1. $g^i \pmod{p}$ for $i = 0, 1, \ldots, N - 1$ (Baby Steps).
2. $c(g^{-1})^{Nj} \pmod{p}$ for $j = 0, 1, \ldots, N - 1$ (Giant Steps).

If we have a match between the two lists, i.e.

$$g^i \equiv c(g^{-1})^{Nj} \pmod{p},$$

then we get

$$g^{i+Nj} \equiv c \pmod{p}.$$

Hence, $x = i + Nj$. In fact, we will always have a match between our two lists since the exponent x satisfies $0 \leq x < p - 1$. Since $p - 1 < N^2$, we can always write x as $x = i + Nj$ with $0 \leq i, j \leq N$. Indeed, our solution is just the base-N representation of x, i.e. $x = i + Nj = (ji)_N$. The complexity of this process is $\tilde{O}(\sqrt{p})$, where the soft O is needed to account for the cost of the list search.

For example, imagine we want to solve the discrete log problem for $5^x \equiv 20 \pmod{503}$. Here, 5 is a primitive root for the prime 503. If we proceed with BSGS, then we can set $N = \lceil\sqrt{503 - 1}\rceil = 23$, we solve for $5^{-1} \equiv 302 \pmod{503}$, and we then make the two following lists (both (mod 503)):

i	5^i (mod 503)	i	5^i (mod 503)	i	5^i (mod 503)	i	5^i (mod 503)
0	1	6	32	12	18	18	73
1	5	7	160	13	90	19	365
2	25	8	297	14	450	20	316
3	125	9	479	15	238	21	71
4	122	10	383	16	184	22	355
5	107	11	406	17	417		

j	$20(302)^{23j}$	j	$20(302)^{23j}$	j	$20(302)^{23j}$	j	$20(302)^{23j}$
0	20	6	319	12	385	18	482
1	435	7	22	13	200	19	172
2	30	8	227	14	326	20	220
3	401	9	33	15	300	21	258
4	45	10	89	16	489	22	330
5	350	11	301	17	450		

Now since we have matching entries for $i = 14$ and $j = 17$, we know then that $x = i + Nj = 14 + 23 \times 17 = 405$.

It is clear then that Eve can recover the secret exponent if the prime used is too small; what is less obvious is she can also recover the secret exponent if $p - 1$ has only small factors. Let the prime power representation of $p - 1$ be

$$p - 1 = q_1^{r_1} q_2^{r_2} \cdots q_i^{r_i},$$

where q_i is the largest prime divisor of $p - 1$ and $q_j < q_{j+1}$ for $j = 1, 2, \ldots, i - 1$. If g is a primitive root for prime p and we are given c and we wish to solve for x in the following:

$$g^x \equiv c \pmod{p},$$

then we can equivalently solve for

$$x \equiv \operatorname{ind}_g c \pmod{q_i^{r_i}},$$

for each i and then recover the value for x using the CRT. We can do this by using what is called the *Pohlig-Hellman algorithm*; the concept for which was first developed by Roland Silver and later refined and published by Stephen Pohlig and Hellman in 1978. The details of the algorithm follow below.

Let q^r be any of the small prime power factors and let

$$(x_{r-1}x_{r-2}\dots x_0)_q$$

bet the base-q representation of x modulo q^r. We solve for each x_i in the order x_0, $x_1, x_2, \dots,$ to yield

$$x \equiv x_0 + x_1 q + x_2 q^2 + x_3 q^3 + \cdots + x_{r-1} q^{r-1} \pmod{q^r},$$

such that this is a solution to

$$x \equiv \operatorname{ind}_g c \pmod{q^r}.$$

First, we note that since g is a primitive root of p, $p - 1$ is the smallest integer such that $g^{p-1} \equiv 1 \pmod{p}$, and consequently $g^n \equiv g^m \pmod{p}$ if and only if $n \equiv m \pmod{p-1}$. This is used to assure unique solutions throughout. Note that since $c \equiv g^x \pmod{p}$, we can raise both sides by $(p-1)/q$ to get

$$c^{(p-1)/q} \equiv g^{x(p-1)/q} \pmod{p}.$$

We will now replace x with $x_0 + x_1 q + x_2 q^2 + x_3 q^3 + \cdots + x_{r-1} q^{r-1}$ in the above and simplify

$$\begin{aligned}
c^{(p-1)/q} &\equiv g^{(x_0+x_1q+x_2q^2+x_3q^3+\cdots+x_{r-1}q^{r-1})(p-1)/q} \\
&\equiv g^{x_0(p-1)/q+(p-1)(x_1+x_2q+x_3q^2+\cdots+x_{r-1}q^{r-2})} \\
&\equiv g^{x_0(p-1)/q} g^{(p-1)(x_1+x_2q+x_3q^2+\cdots+x_{r-1}q^{r-2})} \\
&\equiv g^{x_0(p-1)/q} (g^{(p-1)})^{(x_1+x_2q+x_3q^2+\cdots+x_{r-1}q^{r-2})} \\
&\equiv g^{x_0(p-1)/q} (1)^{(x_1+x_2q+x_3q^2+\cdots+x_{r-1}q^{r-2})} \\
&\equiv g^{x_0(p-1)/q} \pmod{p}.
\end{aligned}$$

Notice that there are only q possible values of x_0 modulo p. Since q is a small prime we can look at all powers of $g^{(p-1)/q}$ modulo p, that is, $g^{j(p-1)/q} \pmod{p}$ for $j = 0, 1, 2, \dots, q-1$ until for some j we have $c^{(p-1)/q} \equiv g^{j(p-1)/q} \pmod{p}$. This value of j will be our value for x_0. We could also use BSGS to find x_0 if q is one of the larger values of the q_i.

Next, if $q^2 \mid p - 1$, we set

$$c_1 \equiv cg^{-x_0} \pmod{p}.$$

Notice then that

$$\begin{aligned} c_1 &\equiv cg^{-x_0} \equiv g^x g^{-x_0} \equiv g^{x_1 q + x_2 q^2 + x_3 q^3 - \cdots + x_{r-1} q^{r-1}} \\ &\equiv g^{q(x_1 + x_2 q + x_3 q^2 + \cdots + x_{r-1} q^{r-2})} \pmod{p}. \end{aligned}$$

Hence, if we raise $c_1 \equiv g^{q(x_1 + x_2 q + x_3 q^2 + \cdots + x_{r-1} q^{r-2})} \pmod{p}$ by $(p-1)/q^2$, we see

$$\begin{aligned} c_1^{(p-1)/q^2} &\equiv (g^{q(x_1 + x_2 q + x_3 q^2 + \cdots + x_{r-1} q^{r-2})})^{(p-1)/q^2} \\ &\equiv g^{((p-1)/q)(x_1 + x_2 q + x_3 q^2 + \cdots + x_{r-1} q^{r-2})} \\ &\equiv g^{x_1 (p-1)/q + (p-1)(x_2 + x_3 q + \cdots + x_{r-1} q^{r-3})} \\ &\equiv g^{x_1 (p-1)/q} (g^{(p-1)})^{(x_2 + x_3 q + \cdots + x_{r-1} q^{r-3})} \\ &\equiv g^{x_1 (p-1)/q} \pmod{p}. \end{aligned}$$

Again, since there are so few numbers to check, we can solve for x_1 by exhaustion.

Similarly, if $q^3 \mid p - 1$,we put

$$c_2 \equiv c_1 g^{-x_1 q} \pmod{p}$$

and we see

$$\begin{aligned} c_2 &\equiv c_1 g^{-x_1 q} \equiv cg^{-x_0} g^{-x_1 q} \equiv g^x g^{-x_0} g^{-x_1 q} \\ &\equiv g^{x_2 q^2 + x_3 q^3 + \cdots + x_{r-1} q^{r-1}} \pmod{p}. \end{aligned}$$

Now, we raise $c_2 \equiv g^{x_2 q^2 + x_3 q^3 + \cdots + x_{r-1} q^{r-1}} \pmod{p}$ by $(p-1)/q^3$ and simplify

$$\begin{aligned} c_2^{(p-1)/q^3} &\equiv (g^{x_2 q^2 + x_3 q^3 + \cdots + x_{r-1} q^{r-1}})^{(p-1)/q^3} \\ &\equiv (g^{q^2 (x_2 + x_3 q + \cdots + x_{r-1} q^{r-3})})^{(p-1)/q^3} \\ &\equiv g^{x_2 (p-1)/q} (g^{p-1})^{x_3 q + \cdots + x_{r-1} q^{r-4}} \\ &\equiv g^{x_2 (p-1)/q} \pmod{p}. \end{aligned}$$

As before, x_2 can be found easily.

We continue letting

$$c_i \equiv c_{i-1} g^{-x_{i-1} q^{i-1}} \pmod{p},$$

and hence using

$$c_i^{(p-1)/q^{i+1}} \equiv g^{x_i (p-1)/q}$$

to find x_i until you have found $x_0, x_1, \ldots, x_{r-1}$. Since $x \equiv x_0 + x_1 q + x_2 q^2 + x_3 q^3 + \cdots + x_{r-1} q^{r-1} \pmod{q^r}$, we then have $x \pmod{q^r}$.

We do the above for all q_i. From this we get a set of equations $x \equiv a_i \pmod{q_i^{r_i}}$ for all i. We now recover $x \pmod{p-1}$ using the CRT. The complexity of this process is $\tilde{O}((\log p)^3)$ if $p-1$ has only small prime factors.

For example, previously we used $p = 97$ in our Diffie-Hellman key exchange. Eve can now use the above Pohlig-Hellman algorithm to find Alice's (or Bob's) secret exponent and hence break the system. Eve notices that $p - 1 = 96 = 2^5 \times 3$, i.e. $p-1$ has many small factors. She sets $g = 5$ and $c = 40$ (both of these are publicly available). She first considers $q^r = 2^5$, so here

$$x = x_0 + 2x_1 + 4x_2 + 8x_3 + 16x_4 \pmod{32}.$$

Eve then calculates

$$c^{(p-1)/q} \equiv 40^{48} \equiv -1 \pmod{97}$$

and

$$g^{(p-1)/q} \equiv 5^{48} \equiv -1 \pmod{97}.$$

Since

$$c^{(p-1)/q} \equiv g^{x_0(p-1)/q} \pmod{97},$$

it must be that $x_0 = 1$.

Next Eve solves

$$c_1 \equiv cg^{-x_0} \equiv 40 \times 5^{-1} \equiv 40 \times 39 \equiv 8 \pmod{97}.$$

Eve then employs c_1 to see

$$c_1^{(p-1)/q^2} \equiv 8^{24} \equiv -1 \equiv g^{x_1(p-1)/q} \equiv (-1)^{x_1} \pmod{97};$$

thus, $x_1 = 1$.

Continuing in this fashion

$$c_2 \equiv c_1 g^{-qx_1} \equiv 8 \times (5^{-1})^2 \equiv 8 \times 39^2 \equiv 43 \pmod{97}.$$

Here we have

$$c_2^{(p-1)/q^3} \equiv 43^{12} \equiv -1 \equiv g^{x_2(p-1)/q} \equiv (-1)^{x_2} \pmod{97}$$

and therefore $x_2 = 1$.

Now

$$c_3 \equiv c_2 g^{-q^2 x_2} \equiv 43 \times (5^{-1})^4 \equiv 43 \times 39^4 \equiv 1 \pmod{97}.$$

Hence,

$$c_3^{(p-1)/q^4} \equiv 1^6 \equiv 1 \equiv g^{x_3(p-1)/q} \equiv (-1)^{x_3} \pmod{97}$$

and $x_3 = 0$.

Finally, for $q = 2$

$$c_4 \equiv c_3 g^{-q^3 x_3} \equiv 1 \times (5^{-8})^0 \equiv 1 \pmod{97}.$$

This yields

$$c_4^{(p-1)/q^5} \equiv 1^3 \equiv 1 \equiv g^{x_4(p-1)/q} \equiv (-1)^{x_4} \pmod{97};$$

and we see $x_4 = 0$. Thus,

$$x \equiv x_0 + 2x_1 + 4x_2 + 8x_3 + 16x_4 \equiv 1 + 2 \times 1 + 4 \times 1 + 8 \times 0 + 16 \times 0 \equiv 7 \pmod{32}.$$

The case of $q^r = 3$ is less work as the power of r is only 1 and thus $x \equiv x_0 \pmod 3$. We see that

$$c^{(p-1)/q} \equiv 40^{32} \equiv 35 \pmod{97} \quad \text{and} \quad g^{(p-1)/q} \equiv 5^{32} \equiv 35 \pmod{97}.$$

Since

$$c^{(p-1)/q} \equiv g^{x_0(p-1)/3} \rightarrow 35 \equiv 35^{x_0} \pmod{97},$$

we must have $x_0 = 1$, or just $x \equiv 1 \pmod 3$.

We can now perform the CRT on the two equations using the same notation as in Chap. 3, noting

$$x \equiv a_1 \pmod{m_1} \rightarrow x \equiv 1 \pmod 3,$$
$$x \equiv a_2 \pmod{m_2} \rightarrow x \equiv 7 \pmod{32}.$$

Here $M = m_1 m_2 = 96$, $M_1 = M/m_1 = m_2 = 32$ and $M_2 = M/m_2 = m_1 = 3$. Using EEA we can find y_1 and y_2 such that

$$y_1 M_1 \equiv 1 \pmod{m_1} \quad \text{and} \quad y_2 M_2 \equiv 1 \pmod{m_2}.$$

It is easy to verify that $y_1 = 2$ and $y_2 = 11$ satisfy

$$y_1 32 \equiv 1 \pmod 3 \quad \text{and} \quad y_2 3 \equiv 1 \pmod{32}.$$

Thus, x (Alice's secret exponent) must be

$$x \equiv y_1 a_1 M_1 + y_2 a_2 M_2 \equiv 2 \times 1 \times 32 + 11 \times 7 \times 3 \equiv 7 \pmod{96}.$$

A seemingly more *ad hoc* method of solving for x in $c \equiv g^x \pmod{p}$, where g is a primitive root of the prime p, can be achieved through the use of a technique called *index calculus*. This procedure seems first to have been discovered by the Russian-born Belgian engineer Maurice Kraitchik (1882–1957) as early as 1922.

We first observe that for integers a and b we have (Theorem 4.7)

$$\mathrm{ind}_g\, a + \mathrm{ind}_g\, b \equiv \mathrm{ind}_g\, ab \pmod{p-1}$$

when p is any prime with primitive root g. Suppose we want to find the base g index (discrete logarithm) of b for a prime p with primitive root g. We first perform a precomputation, the most computationally expensive part of the index calculus algorithm. Let FB (the factor base) be the set $\{r_1, r_2, r_3, \ldots, r_k\}$ of all primes less than a preselected bound B. Select an integer t_i at random from $\{1, 2, 3, \ldots, p-2\}$ and compute $x_i \equiv g^{t_i} \pmod{p}$. We then trial-divide x_i by the primes in FB to determine whether x_i can be represented as

$$x_i = \prod_{j=1}^{k} r_j^{\alpha_{ij}}$$

where $\alpha_{ij} \geq 0$. Thus,

$$t_i = \mathrm{ind}_g\, x_i \equiv \sum \alpha_{ij} \cdot \mathrm{ind}_g\, r_j \pmod{p-1}.$$

If we find a sufficiently large number of relations ($\geq k+1$) like these, we can use linear algebra to solve for $\mathrm{ind}_g\, r_j \pmod{p-1}$ for $j \in \{1, 2, \ldots, k\}$. The next step is to compute $y_i \equiv g_{s_i} b \pmod{p}$ where s_i is selected at random from $\{0, 1, 2, \ldots, p-2\}$ until we find some y_i such that

$$y_i = \prod_{j=1}^{k} r_j^{\beta_j}.$$

We have

$$\mathrm{ind}_g\, b + s_i \equiv \sum_{j=1}^{k} \beta_j \cdot \mathrm{ind}_g\, r_j \pmod{p-1}.$$

Since $\beta_j \cdot \mathrm{ind}_g\, r_j$ ($j \in \{1, 2, \ldots, k\}$) and the s_i are known, we can easily solve for $\mathrm{ind}_g\, b$.

In order to discuss the computational complexity of the index calculus technique, we need to introduce the following initially intimidating expression:

$$L_n[a, b] = e^{b(\log n)^a (\log\log n)^{1-a}}.$$

Here a is a constant satisfying $0 \le a \le 1$ and b is often used as an abbreviation for $b + o(1)$, where b is a fixed positive real number. Notice that if $a = 0$, then $L_n[a, b] = O(\log n)^b$ which is polynomial in $\log n$. Also, if $a = 1$, then $L_n[a, b] = O(n^b) = O(e^{b \log n})$ which is exponential in $\log n$. Thus, if $0 < a < 1$, we see that $L_n[a, b]$ is a function larger asymptotically than any polynomial function of $\log n$, but smaller than any exponential function of $\log n$. In this case we say that $L_n[a, b]$ is *subexponential*. In 1987 Carl Pomerance pointed out that it is possible to make the index calculus method execute in running time $L_p[1/2, \sqrt{5/2}]$.

The index calculus method is easiest to understand by looking at an example. Like BSGS, we do not need $p - 1$ to have many small divisors. For example, we will consider $p = 503, g = 5$ and $c = 43$, that is, we will solve for x in $c \equiv g^x \pmod{p} \rightarrow 43 \equiv 5^x \pmod{503}$. We will select our bound $B = 20$. Calculating $5^k \pmod{503}$ for $1 \le k \le 50$ yields the following useful k:

$$
\begin{aligned}
5^1 &\equiv 5 \pmod{503},\\
5^6 &\equiv 2^5 \pmod{503},\\
5^{15} &\equiv 2 \times 7 \times 17 \pmod{503},\\
5^{23} &\equiv 2 \times 7 \times 19 \pmod{503},\\
5^{26} &\equiv 2^2 \times 13 \pmod{503},\\
5^{28} &\equiv 2 \times 3 \times 7^2 \pmod{503},\\
5^{30} &\equiv 2^2 \times 7 \times 11 \pmod{503},\\
5^{42} &\equiv 11 \pmod{503}.
\end{aligned}
$$

From these we have

$$
\begin{aligned}
1 &\equiv \operatorname{ind}_5 5 \pmod{502},\\
6 &\equiv 5 \operatorname{ind}_5 2 \pmod{502},\\
15 &\equiv \operatorname{ind}_5 2 + \operatorname{ind}_5 7 + \operatorname{ind}_5 17 \pmod{502},\\
23 &\equiv \operatorname{ind}_5 2 + \operatorname{ind}_5 7 + \operatorname{ind}_5 19 \pmod{502},\\
26 &\equiv 2 \operatorname{ind}_5 2 + \operatorname{ind}_5 13 \pmod{502},\\
28 &\equiv \operatorname{ind}_5 2 + \operatorname{ind}_5 3 + 2 \operatorname{ind}_5 7 \pmod{502},\\
30 &\equiv 2 \operatorname{ind}_5 2 + \operatorname{ind}_5 7 + \operatorname{ind}_5 11 \pmod{502},\\
42 &\equiv \operatorname{ind}_5 11 \pmod{502}.
\end{aligned}
$$

We immediately have $\operatorname{ind}_5 5 = 1$ and $\operatorname{ind}_5 11 = 42$, and we can recover the others as follows. First, since $\gcd(5, 502) = 1$, we can use the EEA to see $5^{-1} \equiv 201 \pmod{502}$, and thus

$$
\begin{aligned}
6 \equiv 5 \operatorname{ind}_5 2 \pmod{502} &\rightarrow 6 \times 201 \equiv 201 \times 5 \operatorname{ind}_5 2 \pmod{502}\\
&\rightarrow 202 \equiv \operatorname{ind}_5 2 \pmod{502}.
\end{aligned}
$$

We can now use known $\text{ind}_5 2$ and $\text{ind}_5 11$ to find $\text{ind}_5 7$, as

$$30 \equiv 2\,\text{ind}_5 2 + \text{ind}_5 7 + \text{ind}_5 11 \equiv 2 \times 202 + \text{ind}_5 7 + 42 \pmod{502}$$
$$\rightarrow 86 \equiv \text{ind}_5 7 \pmod{502}.$$

Now that we have $\text{ind}_5 7$ we can use it and $\text{ind}_5 2$ to find $\text{ind}_5 3$, by using the $k = 28$ case

$$28 \equiv \text{ind}_5 2 + \text{ind}_5 3 + 2\,\text{ind}_5 7 \equiv 202 + \text{ind}_5 3 + 2 \times 86 \pmod{502}$$
$$\rightarrow 156 \equiv \text{ind}_5 3 \pmod{502}.$$

We also can easily get $\text{ind}_5 13$ with just the $k = 26$ case and $\text{ind}_5 2$, as

$$26 \equiv 2\,\text{ind}_5 2 + \text{ind}_5 13 \equiv 2 \times 202 + \text{ind}_5 13 \pmod{502}$$
$$\rightarrow 124\,\text{ind}_5 13 \pmod{502}.$$

In a similar fashion we see

$$15 \equiv \text{ind}_5 2 + \text{ind}_5 7 + \text{ind}_5 17 \equiv 202 + 86 + \text{ind}_5 17 \pmod{502}$$
$$\rightarrow 229 \equiv \text{ind}_5 17 \pmod{502}$$

and

$$23 \equiv \text{ind}_5 2 + \text{ind}_5 7 + \text{ind}_5 19 \equiv 202 + 86 + \text{ind}_5 19 \pmod{502}$$
$$\rightarrow 237 \equiv \text{ind}_5 19 \pmod{502}.$$

Upon calculating $cg^r \pmod{p}$ for the first few cases we see for both $r = 4$ and $r = 8$ that $cg^r \equiv \prod p_i^{b_i} \pmod{p}$ where each p_i is less than our bound B. In the case where $r = 4$, we have $cg^r \equiv 43 \times 5^4 \equiv 2^3 \times 3^3 \pmod{503}$. Hence, since $\text{ind}_g c \equiv -r + \sum b_i \,\text{ind}_g p_i \pmod{p-1}$, we get

$$\begin{aligned}\text{ind}_5 43 &\equiv -4 + 3\,\text{ind}_5 2 + 3\,\text{ind}_5 3 \pmod{502}\\ &\equiv -4 + 3 \times 202 + 3 \times 156 \pmod{502}\\ &\equiv 66 \pmod{502}.\end{aligned}$$

Equivalently, when $r = 8$, $cg^r \equiv 43 \times 5^8 \equiv 2^2 \times 7^2 \pmod{503}$. In a similar manner we have

$$\begin{aligned}\text{ind}_5 43 &\equiv -8 + 2\,\text{ind}_5 2 + 2\,\text{ind}_5 7 \pmod{502}\\ &\equiv -8 + 2 \times 202 + 2 \times 86 \pmod{502}\\ &\equiv 66 \pmod{502}.\end{aligned}$$

Either way we have solved for $x = 66$.

To solve the authentication problem, Diffie and Hellman suggested the use of a particular type of public-key cryptography, which is often referred to today as *two-key cryptography*. The idea here is that each member of a group of individuals wishing to exchange information securely among themselves should each have a public key and a private key unique to that individual. If, for example, Alice wishes to communicate with Bob, she selects his public key (call it k_1 here) from a public directory and encrypts a message m by computing $c = E_{k_1}(m)$, where E_k is the encryption function or transformation indexed by the key k. She transmits c to Bob over a public channel, and on receipt of c Bob uses his private key (call it k_2) to compute $m = D_{k_2}(c)$, where D_k is the decryption algorithm indexed by k. Such a cryptosystem should possess several properties:

(1) given k_1 and k_2, both $E_{k_1}(m)$ and $D_{k_2}(c)$ are easy to compute;
(2) $D_{k_2}(E_{k_1}(m)) = m$;
(3) unless k_2 is known, it is *computationally infeasible* (the computation would take more time than is available) to determine m from knowledge of k_1, E_{k_1} and $c = E_{k_1}(m)$;
(4) $E_{k_1}(D_{k_2}(c)) = c$.

The properties (1), (2) and (3) make this scheme a secure communication system, where it is not necessary to distribute a key securely, but the properties (1), (3) and (4) allow participants to authenticate messages. For if Alice wishes to send a message m to Bob that he can authenticate, she applies her private decryption algorithm to m and sends the result (call it $D_A(m)$) to Bob. When Bob receives this, he applies Alice's public encryption algorithm E_A to it to obtain m (property (4)). Because only Alice knows D_A, Bob is convinced that Alice sent m and he can save the signature $D_A(m)$ as proof that the message came from Alice. Indeed, when proper care is taken in implementing a two-key cryptosystem, it can be used to provide the following important services: data integrity, data origin authentication, confidentiality, entity authentication and nonrepudiation. In today's Internet environment these services are essential.

Unfortunately, having devised this very clever scheme, Diffie and Hellman were unable to provide an example of such a system. However, within a year's time, Ron Rivest, the Israeli mathematician Adi Shamir and Leonard Adleman, then all at MIT, produced a very simple example, now commonly referred to as the RSA cryptosystem. This system relies for its security on the presumed difficulty of the integer factoring problem (see Chap. 15). Incidentally, the names Alice and Bob were first used in the 1978 paper which introduced RSA.

In the case of RSA we exchange a message m encoded as an integer, where $1 < m < n - 1$, and n is the product of two large primes p and q. To denote the encryption we will write $E(m) = c$ and decryption $D(c) = m$; hence, $D(E(m)) = c$. The setup of RSA goes as follows.

Algorithm 2 RSA algorithm to calculate the secret key.

1. Alice selects two random primes p, q, such that $p \neq q$ and the size of p is approximately the same as the size of q. These primes are kept secret.
2. Alice computes both $n = pq$ and $\phi(n) = (p-1)(q-1)$. The value of n will be made public and the value of $\phi(n)$ will be kept private.
3. Alice selects a random integer e such that $1 < e < \phi(n)$ and $\gcd(e, \phi(n)) = 1$. The value of e will be made public and will be called the *public exponent*.
4. Alice uses the EEA to calculate d such that $1 < d < \phi(n)$ and $ed \equiv 1 \pmod{\phi(n)}$. The value of d is kept private and will be called the *private exponent*.
5. Alice makes the pair (n, e) public; this is called the *public key*. The private exponent d will be Alice's *private key*.

Now, if Bob wishes to send a message to Alice he does the following.

Algorithm 3 Bob uses Alice's public key to send a message to Alice.

1. Bob acquires Alice's public key (n, e).
2. To encrypt m Bob calculates $E(m) \equiv m^e \equiv c \pmod{n}$. (Bob uses methods described in Chap. 4).
3. Bob sends c to Alice.

Alice can now recover m by calculating $D(c) \equiv c^d \equiv m \pmod{n}$. Notice that since $ed \equiv 1 \pmod{\phi(n)}$, we know $ed = k\phi(n) + 1$ for some nonnegative integer k; thus,

$$D(c) \equiv c^d \equiv m^{ed} \equiv m^{k\phi(n)+1} \equiv (m^{\phi(n)})^k m \equiv 1^k m \equiv m \pmod{n}$$

by Euler's theorem in the case where $\gcd(m, n) = 1$. If it happens to be that $\gcd(m, n) \neq 1$, then it must be the case that either p or q divides m, but not both. Without loss of generality, imagine $p \mid m$ and $q \nmid m$. Since $p \mid m$, we have $m \equiv 0 \pmod{p}$ and thus $m^{ed} \equiv 0 \equiv m \pmod{p}$. Also, since $q \nmid m$, we can see that

$$m^{ed} \equiv m^{k\phi(n)+1} \equiv (m^{(p-1)(q-1)})^k m \equiv (m^{(q-1)})^{(p-1)k} m \equiv 1^{(p-1)k} m \equiv m \pmod{q};$$

hence, $m^{ed} \equiv m \pmod{pq} \rightarrow m^{ed} \equiv m \pmod{n}$.

For example, imagine Alice chooses the primes $p = 79$ and $q = 89$, in which case $n = pq = 79 \cdot 89 = 7031$ and $\phi(n) = (p-1)(q-1) = 78 \cdot 88 = 6864$. Further, assume Alice randomly selects the secret exponent $e = 17$. She can then find d by applying the EEA, from Chap. 2, to $\phi(n)$ and e as follows:

$$\begin{aligned}
6864 = 403 \cdot 17 + 13 \rightarrow 13 &= 6964 - 403 \cdot 17 = \phi(n) - 403e \\
17 = 1 \cdot 13 + 4 \rightarrow 4 &= 17 - 13 = e - (\phi(n) - 403e) = -\phi(n) + 404e \\
13 = 3 \cdot 4 + 1 \rightarrow 1 &= 13 - 3 \cdot 4 = (\phi(n) - 403e) - 3(-\phi(n) + 404e) \\
&= 4\phi(n) - 1615e.
\end{aligned}$$

Thus, since $1 = 4\phi(n) + (-1615)e$, we have

$$1 \equiv (-1615)e \pmod{\phi(n)} \rightarrow 1 \equiv 5249e \pmod{\phi(n)}.$$

Alice has calculated her private exponent $d = 5249$ and she makes $(n, e) = (7031, 17)$ public.

Bob can now use this public information to send Alice the message $m = 7$ in secret. He does so by calculating $m^e \equiv c \pmod{n}$, using Algorithm 2 from Chap. 4, and sending c to Alice. Bob begins by converting $e = 17$ to binary, i.e. $17 = (10001)_2$; hence, in the algorithm $(b_0 b_1 b_2 b_3 b_4)_2 = (10001)_2$. He can then set $r_0 = 7$ and execute the steps below

- $b_1 = 0 \rightarrow r_1 = r_0^2 = 7^2 \equiv 49 \pmod{7031}$
- $b_2 = 0 \rightarrow r_2 = r_1^2 = 49^2 \equiv 2401 \pmod{7031}$
- $b_3 = 0 \rightarrow r_3 = r_2^2 = 2401^2 \equiv 6412 \pmod{7039}$
- $b_4 = 1 \rightarrow r_4 = r_3^2 m = 6412^2 \times 7 \equiv 3316 \pmod{1009}$.

Thus, $c = 3316$ is sent to Alice. To decrypt Alice uses the same algorithm to yield $m \equiv c^d \equiv 3316^{5249} \equiv 7 \pmod{7031}$.

In the case where n is the product of two primes, the problem of factoring n is equivalent to determining $\phi(n)$. Clearly, if you can factor n, then $\phi(n)$ follows immediately. On the other hand, imagine you know $\phi(n)$ and that $n = pq$, but you do not know either p or q. This is enough information to deduce p and q as follows. We begin by observing

$$\phi(n) = (p-1)(q-1) = pq - p - q + 1 = n - p - q + 1.$$

Rearranging this gives us $p + q = n - \phi(n) + 1$; since $n = pq$, we have $q = n/p$ and thus

$$p + n/p = n - \phi(n) + 1.$$

Multiplying both sides of the above by p yields

$$p^2 + n = p(n - \phi(n) + 1) \rightarrow p^2 - p(n - \phi(n) + 1) + n = 0.$$

We can then use the quadratic formula to find p and q as

$$p = \frac{(n - \phi(n) + 1) \pm \sqrt{(n - \phi(n) + 1)^2 - 4n}}{2}$$

and $q = n/p$. Thus, if we know both n and $\phi(n)$, we can easily factor n.

The point of the above is that, if Eve can factor n, then she can calculate $\phi(n)$ and then run the EEA on the public exponent e and $\phi(n)$ to also recover the private exponent d. Furthermore, if by some process she can find the value of $\phi(n)$, she can also find d. The puzzle of breaking RSA provided a great deal of motivation to the study of the problem of integer factorization. In general we will say an integer N is not trivially factorable when it has at least two large prime factors. The problem of integer factorization will be more fully discussed in Chap. 15.

We observe that the basis of the security of the Diffie-Hellman-Merkle or the RSA scheme is the assumption that the DLP or the integer factoring algorithm is hard. It must be stressed here that we do not truly know that either of these problems is hard (there is no proof), but a lot of very clever people have tried to solve these problems, and while there has been enormous progress since 1977, both problems still resist a truly practical solution. Also, it is difficult to find a hard problem upon which to base a public-key cryptosystem, but several have been proposed. In particular, in 1985 it was discovered almost simultaneously by Neal Koblitz and Victor Miller that the theory of elliptic curves could be enlisted to produce a secure key distribution scheme.

Elliptic curves are frequently represented by equations of the form

$$y^2 = ax^3 + bx^2 + cx + d,$$

where a, b, c and d are selected to be in a certain domain, like the reals or the rationals or even the integers. A point (x, y) is said to be on this curve when both x and y are in the chosen domain and satisfy the above equation. What is of special interest to cryptographers is that it is possible to "add" two points on an elliptic curve to get a third point which is also on the elliptic curve. This means that the points on an elliptic curve have an arithmetic structure which can be analysed. These elliptic curves are of great importance in number theory and constitute a major area of current mathematical research, but a study of them here is well beyond the scope of this book. In addition to being applied to cryptography, we will see later that they can be applied to the problems of primality testing and integer factoring.

Since 1977, public-key cryptography has stimulated a great deal of interest in the subjects of computational and algorithmic number theory. For the first time since antiquity, there is a possible application for work done in these formerly esoteric areas of research, precincts of study formerly occupied by a few dedicated academics and enthusiasts. As a result of this surge of interest and, more importantly, the intrinsic significance of the material, research in computational number theory and allied subjects has increased considerably and continues to do so. This upsurge in activity has also been abetted by enormous improvements in computer technology and progress in the computer science subject of complexity theory.

Problems

7.1 (Diffie-Hellman) In the following you play the role of Alice in the Diffie-Hellman key exchange. Use the information provided to calculate the shared secret. State what information is sent to Bob and the final shared secret s.
(a) **[BB]** Alice and Bob agree on the prime $p = 73$ and primitive root $g = 5$. Alice chooses secret integer $a = 10$ and receives $B = 15$ from Bob.
(b) Alice and Bob agree on the prime $p = 101$ and primitive root $g = 2$. Alice chooses secret integer $a = 5$ and receives $B = 68$ from Bob.
(c) Alice and Bob agree on the prime $p = 103$ and primitive root $g = 5$. Alice chooses secret integer $a = 6$ and receives $B = 51$ from Bob.

7.2 (Baby Step, Giant Step) Use the the provided tables and BSGS to solve the following discrete logarithm problems:
(a) **[BB]** $7^x \equiv 13 \pmod{239}$
(b) $7^x \equiv 230 \pmod{239}$
(c) $7^x \equiv 160 \pmod{239}$

i	$7^i \pmod{239}$	i	$7^i \pmod{239}$	i	$7^i \pmod{239}$	i	$7^i \pmod{239}$
0	1	4	11	8	121	12	136
1	7	5	77	9	130	13	235
2	49	6	61	10	193	14	211
3	104	7	188	11	156	15	43

j	$13(205)^{16j}$	j	$13(205)^{16j}$	j	$13(205)^{16j}$	j	$13(205)^{16j}$
0	13	4	199	8	215	12	129
1	112	5	115	9	69	13	137
2	156	6	237	10	190	14	114
3	149	7	185	11	111	15	210

j	$230(205)^{16j}$	j	$230(205)^{16j}$	j	$230(205)^{16j}$	j	$230(205)^{16j}$
0	230	4	138	8	35	12	21
1	235	5	141	9	228	13	89
2	131	6	222	10	181	14	13
3	191	7	19	11	107	15	112

j	$160(205)^{16j}$	j	$160(205)^{16j}$	j	$160(205)^{16j}$	j	$160(205)^{16j}$
0	160	4	96	8	201	12	25
1	18	5	202	9	169	13	197
2	8	6	196	10	22	14	61
3	216	7	34	11	116	15	213

7.3 (Baby Step, Giant Step) Use the the provided tables and BSGS to solve the following discrete logarithm problems:

(a) $5^x \equiv 163 \pmod{383}$
(b) $5^x \equiv 379 \pmod{383}$
(c) $5^x \equiv 87 \pmod{383}$
(d) $5^x \equiv 349 \pmod{383}$

i	5^i (mod 383)	i	5^i (mod 383)	i	5^i (mod 383)	i	5^i (mod 383)
0	1	5	61	10	274	15	245
1	5	6	305	11	221	16	76
2	25	7	376	12	339	17	380
3	125	8	348	13	163	18	368
4	242	9	208	14	49	19	308

j	$163(230)^{20j}$	j	$163(230)^{20j}$	j	$163(230)^{20j}$	j	$163(230)^{20j}$
0	163	5	159	10	141	15	60
1	164	6	355	11	257	16	199
2	212	7	188	12	80	17	360
3	218	8	215	13	10	18	45
4	123	9	362	14	97	19	245

j	$379(230)^{20j}$	j	$379(230)^{20j}$	j	$379(230)^{20j}$	j	$379(230)^{20j}$
0	379	5	365	10	302	15	210
1	191	6	285	11	325	16	122
2	359	7	275	12	280	17	111
3	380	8	178	13	35	18	349
4	239	9	118	14	148	19	283

j	$87(230)^{20j}$	j	$87(230)^{20j}$	j	$87(230)^{20j}$	j	$87(230)^{20j}$
0	87	5	200	10	134	15	220
1	346	6	25	11	304	16	219
2	139	7	51	12	38	17	171
3	161	8	150	13	292	18	165
4	68	9	306	14	228	19	260

j	$349(230)^{20j}$	j	$349(230)^{20j}$	j	$349(230)^{20j}$	j	$349(230)^{20j}$
0	349	5	230	10	269	15	253
1	283	6	316	11	273	16	271
2	179	7	231	12	82	17	369
3	166	8	364	13	106	18	94
4	308	9	237	14	109	19	299

7.4 (Pohlig-Hellman) Use the Pohlig-Hellman algorithm to solve the discrete logarithms below:
(a) **[BB]** $2^x \equiv 29 \pmod{37}$

(b) $3^x \equiv 79 \pmod{89}$
(c) $6^x \equiv 96 \pmod{109}$

7.5 (Index calculus) For the prime $p = 191$ with primitive root $g = 19$, a search of values for g^k modulo p was conducted and values with prime factorizations that have the largest prime less than the bound $B = 12$ were selected and put in the table below

k	$g^k \pmod p$	k	$g^k \pmod p$
11	$3 \cdot 11$	31	$5 \cdot 7$
20	$2 \cdot 3 \cdot 5$	44	2
25	$2 \cdot 7$	50	5

(a) **[BB]** Solve $19^x \equiv 29 \pmod{191}$ given $29 \cdot 19^3 \equiv 2^4 \cdot 5 \pmod{191}$
(b) Solve $19^x \equiv 97 \pmod{191}$ given $97 \cdot 19^3 \equiv 2 \cdot 5 \cdot 7 \pmod{191}$
(c) Solve $19^x \equiv 97 \pmod{191}$ given $97 \cdot 19^{12} \equiv 3^4 \pmod{191}$
(d) Solve $19^x \equiv 110 \pmod{191}$ given $110 \cdot 19 \equiv 2^2 \cdot 3^2 \cdot 5 \pmod{191}$

7.6 (Index calculus) For the prime $p = 673$ with primitive root $g = 5$, a search of values for g^k modulo p was conducted and values with prime factorizations that have the largest prime less than the bound $B = 20$ were selected and put in the table below

k	$g^k \pmod p$	k	$g^k \pmod p$
1	5	39	$2^4 \cdot 11$
7	$3 \cdot 19$	52	$2^5 \cdot 7$
19	$13 \cdot 19$	76	3^5
37	$3 \cdot 7 \cdot 17$	78	$2 \cdot 3^2$

(a) Solve $5^x \equiv 368 \pmod{673}$ given $368 \cdot 5 \equiv 2 \cdot 13 \cdot 19 \pmod{673}$
(b) Solve $5^x \equiv 397 \pmod{673}$ given $397 \cdot 5^5 \equiv 2 \cdot 11 \cdot 13 \pmod{673}$
(c) Solve $5^x \equiv 397 \pmod{673}$ given $397 \cdot 5^6 \equiv 2^2 \cdot 3 \cdot 7 \pmod{673}$
(d) Solve $5^x \equiv 207 \pmod{673}$ given $207 \cdot 5^{12} \equiv 2^5 \cdot 7 \pmod{673}$

7.7 (RSA) In the following you play the role of Alice in the RSA public-key system. Using the primes p, q provided and public value e, calculate the private exponent d and use it to decrypt the given ciphertext c.
(a) **[BB]** Use $p = 17, q = 19, e = 173$, to find d. Use this d to decrypt $c = 268$
(b) Use $p = 23, q = 29, e = 411$, to find d. Use this d to decrypt $c = 296$
(c) Use $p = 53, q = 59, e = 431$, to find d. Use this d to decrypt $c = 2422$

7.8 (Factoring n with $\phi(n)$) Given n is a product of two primes use the provided $\phi(n)$ to factor n:
(a) **[BB]** $n = 202379, \phi(n) = 201480$
(b) $n = 395497, \phi(n) = 394240$
(c) $n = 742613, \phi(n) = 740460$

Further Reading

Examples of one-key cryptography

Laurence Dwight Smith, *Cryptography: The Science of Secret Writing*, Dover Publications, Inc., 1955.

Paul B. Janeczko, *Loads of Codes and Secret Ciphers*, Macmillan Publishing Co., NY, 1984.

History of cryptography

David Kahn, *The Codebreakers: The story of Secret Writing*, Scribner, NY, 1996.

Steven Levy, *Crypto: How the Code Rebels Beat the Government–Saving Privacy in the Digital Age*, Viking Penguin, NY, 2001.

Public-key cryptography

W. Diffie and M. Hellman, New directions in cryptography, *IEEE Transactions in Information Theory* **IT-22**, (1976), 644-654.

Joshua Holden, *The Mathematics of Secrets: Cryptography from Caesar Ciphers to Digital Encryption*, Princeton University Press, Princeton, NJ, 2017, pp. 201–240.

Johannes A. Buchmann, *Introduction to Cryptography*, Springer-Verlag, NY, 2001, pp. 127–259.

Douglas R. Stinson and Maura B. Paterson, *Cryptography: Theory and Practice*, 4th Ed., CRC Press, Boca Raton, FL, 2019. This book has become, over the years that it has been in publication, a standard text containing a variety of topics concerning cryptography.

Discrete logarithm problem

Johannes Buchmann, Michael J. Jacobson, Jr. and Hugh C. Williams, The influence of public-key cryptography on mathematics, *Democratizing Cryptography: The Work of Whitfield Diffie and Martin Hellman*, Chapter 4, Rebecca Slayton, ed., ACM books#42, Morgan & Claypool Publishers, San Rafael CA, 2022, 77–156.

Elliptic curve cryptography

Darrel Hankerson, Alfred Menezes and Scott Vanstone, *Guide to Elliptic Curve Cryptography*, Springer-Verlag, NY, 2004.

Chapter 8
Fibonacci and Lucas Numbers

We begin this chapter with what appears to be an ordinary rectangle:

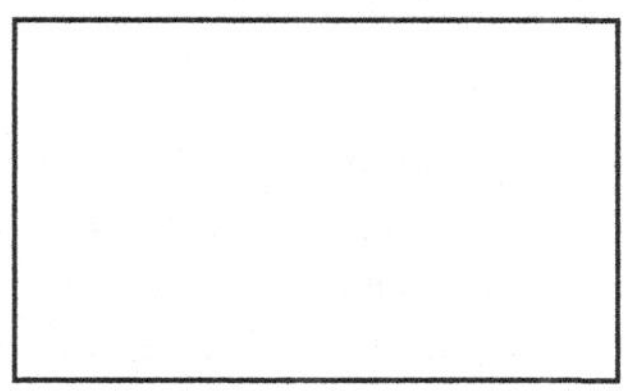

This likely looks like many rectangles you have seen before—think of picture frames—and this is perhaps with good reason, as some consider a rectangle with this ratio of sides to be particularly pleasing. Geometrically what makes this rectangle interesting is that if we were to partition the rectangle as a square and a lone smaller remaining rectangle, the remaining rectangle has the same ratio of sides as the original. Such a rectangle is called a *golden rectangle*. A curious person might wonder about what ratio of the sides creates such a rectangle. This can easily be determined; we begin by labelling the short side by a, and then the long side is $a + b$, where b is the length of the long side less a, as in the diagram:

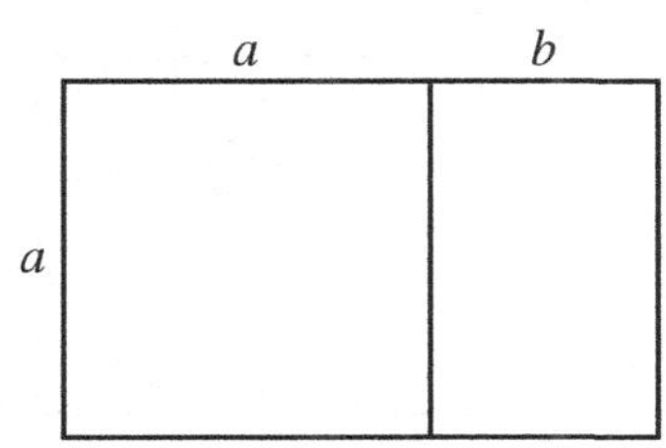

E. L. F. Roettger and H. C. Williams, *The Enchantment of Numbers*, CMS/CAIMS Books in Mathematics 16, https://doi.org/10.1007/978-3-031-76558-1_8

We can then note that

$$\frac{a+b}{a} = \frac{a}{b}.$$

If we let $\phi = \frac{a}{b}$, we can solve for ϕ by noting

$$\frac{a+b}{a} = \frac{a}{b} \to \frac{a}{a} + \frac{b}{a} = \frac{a}{b} \to 1 + \frac{1}{\phi} = \phi \to \phi + 1 = \phi^2 \to \phi^2 - \phi - 1 = 0.$$

Thus from the quadratic formula we see that $\phi = \frac{1 \pm \sqrt{5}}{2}$ and since $\phi > 0$ we get $\phi = \frac{1+\sqrt{5}}{2}$. This ratio is now most often called the *golden ratio* and is, in fact, now widely denoted by ϕ. This notation was suggested in 1915 by the American polymath Mark Barr (1871–1950). The golden ratio was also studied by Euclid who referred to it as the *extreme and mean ratio.* He did this by considering a cut to a line segment such that the ratio of the long line to the short after the cut was the same as the ratio of the entire line segment to the long segment after the cut:

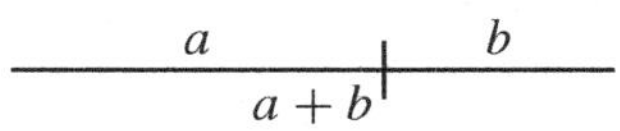

The solutions of both Euclid's problem, and of the one stated initially, have the same steps; however, the golden rectangle is particularly attractive because when we repeat the process of partitioning off squares in a counter clockwise pattern from the remaining rectangles, we are left with the following image:

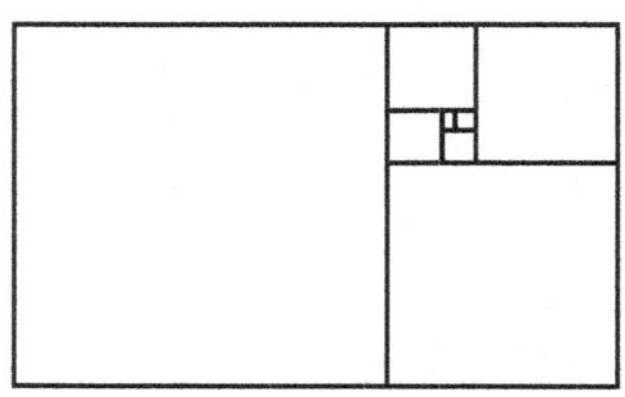

This image can then be modified by drawing a spiral through the infinite number of squares. This is called the *golden spiral* and is most often compared in nature to the shell of the chambered nautilus.

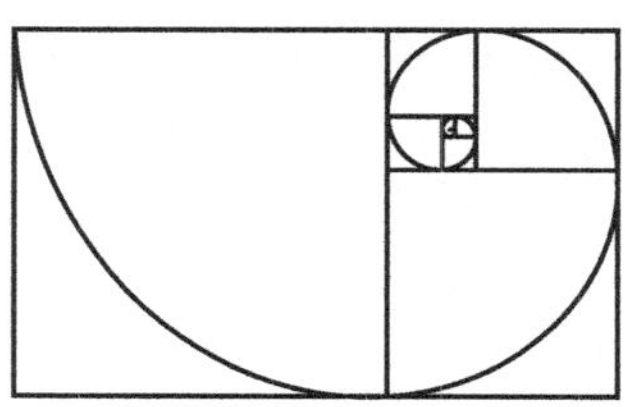

Since $\phi^2 = \phi + 1$, we see that

$$\phi^3 = \phi\phi^2 = \phi(\phi + 1) = \phi^2 + \phi = 2\phi + 1.$$

Indeed, for any integer $n \geq 1$, there must exist integers a_n and b_n such that

$$\phi^n = a_n\phi + b_n,$$

where $a_0 = 0$, $b_0 = 1$, $a_1 = 1$, $b_1 = 0$, $a_2 = 1$, $b_2 = 1$, $a_3 = 2$, $b_3 = 1$. Now

$$a_{n+1}\phi + b_{n+1} = \phi^{n+1} = \phi(a_n\phi + b_n) = (a_n + b_n)\phi + a_n;$$

hence,

$$a_{n+1} = a_n + b_n \quad (n \geq 0),$$
$$b_{n+1} = a_n \quad (n \geq 0).$$

Since $b_n = a_{n-1}$ $(n \geq 1)$, we get

$$a_{n+1} = a_n + a_{n-1} \quad (n \geq 1),$$

where $a_0 = 0$, $a_1 = 1$. Putting $F_n = a_n$, we find that

$$\phi^n = F_n\phi + F_{n-1},$$

where $F_0 = 0$, $F_1 = 1$ and the sequence $(F_k) = F_0, F_1, F_2, \ldots$ is generated by the recursive formula

$$F_k = F_{k-1} + F_{k-2} \quad (k \geq 2).$$

This sequence is now called the *Fibonacci sequence,* and the F_k $(k = 0, 1, 2, \ldots)$ are called the *Fibonacci numbers.* Here, and in what follows, we will use the symbol (X_n) to denote an ordered set of integers or *sequence*

$$X_0, X_1, X_2, \ldots, X_n, \ldots.$$

From the above, it is clear the first few terms of the Fibonacci sequence are

$$0, 1, 1, 2, 3, 5, 8, 13, 21, 34, 55, 89, 144, 233, 377, \ldots.$$

This sequence is named after Leonardo Pisano (1170–1250), which translates to Leonardo of Pisa, so it may seem odd that we say Fibonacci; however, Fibonacci is an abbreviation or nickname of Filio Bonacij, which translates to son of the Bonacij family. Fibonacci's name was attached to these particular numbers because of a problem he posed in his *Liber Abbaci,* originally published in 1202. This problem can be stated as follows: a pair of rabbits is placed in an enclosure under the assumptions that each adult pair of rabbits produces an additional pair each month and that it takes one month for rabbits to reach adulthood. How many rabbits are in the enclosure after one year? The answer to this problem is easily captured in the following table:

Month	Adult pairs	Youth pairs	Total pairs
0	1	0	1
1	1	1	2
2	2	1	3
3	3	2	5
4	5	3	8
5	8	5	13
6	13	8	21
7	21	13	34
8	34	21	55
9	55	34	89
10	89	55	144
11	144	89	233
12	233	144	377

We can see that each column for adult pairs, youth pairs and total pairs are just the Fibonacci numbers with different initial conditions. Thus, the answer to the problem is 377.

An interesting question is whether we can find the nth Fibonacci number without calculating the previous ones, that is, is there a closed-form formula for F_n? This is answered by the following theorem, which gives us *Binet's formula,* named after the French mathematician and physicist Jacques Philippe Marie Binet (1786–1856), but it was known to earlier investigators.

Theorem 8.1 *If we let* $\phi = \frac{1+\sqrt{5}}{2}$ *and* $\bar{\phi} = \frac{1-\sqrt{5}}{2}$, *then*

$$F_n = \frac{\phi^n - \bar{\phi}^n}{\phi - \bar{\phi}} = \frac{\phi^n - \bar{\phi}^n}{\sqrt{5}}.$$

Proof We observe that $\bar{\phi} = -1/\phi$ and as it and ϕ are the two roots of $x^2 - x - 1 = 0$, we have

$$\phi^n = F_n\phi + F_{n-1}$$

and

$$\bar{\phi}^n = F_n\bar{\phi} + F_{n-1}.$$

By subtracting these, we get

$$F_n = \frac{\phi^n - \bar{\phi}^n}{\phi - \bar{\phi}} = \frac{\phi^n - \bar{\phi}^n}{\sqrt{5}}.$$

□

From this result it is a simple exercise to deduce that $F_n = \left\lfloor \frac{\phi^n}{\sqrt{5}} + \frac{1}{2} \right\rfloor$; thus, the Fibonacci sequence (F_n) grows exponentially in n. Since $-1 < \bar{\phi} < 0$, we can also show that $F_n \geq \phi^{n-2}$.

It can also be shown that as we let n get arbitrarily large, the ratio F_n/F_{n-1} gets as close as we like to ϕ. Using limit notation we would denote this as

$$\lim_{n\to\infty} \frac{F_n}{F_{n-1}} = \phi.$$

In order to discuss this in any further detail, we need to make use of an old identity (1660) of the Italian-born French astronomer Giovanni Cassini (1625–1712). This identity is

$$F_{n-1}^2 - F_n F_{n-2} = (-1)^n,$$

and it can be easily verified by using the Binet formulation for the Fibonacci numbers. Notice that we can easily deduce from this identity that consecutive Fibonacci numbers are relatively prime, i.e. $\gcd(F_n, F_{n-1}) = 1$. This identity is also behind what is called the *chessboard paradox,* a puzzle that goes back at least to 1862.

The puzzle goes as follows: imagine we are given the following 4 puzzle pieces:

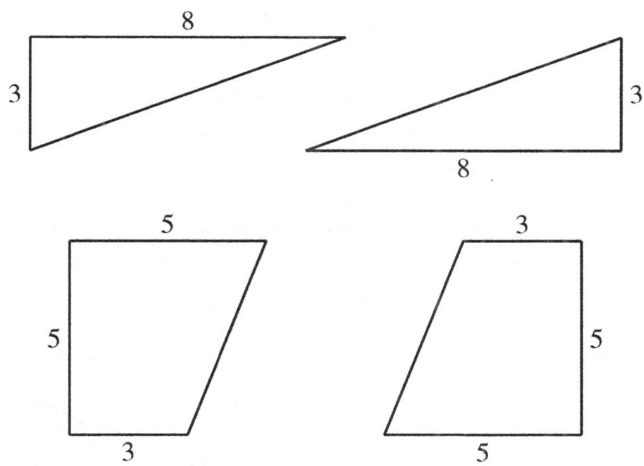

Using basic geometry we can see each right-angled triangle has area 12, and each right-angled trapezoid has area 20. Hence, the combined area of the four puzzle pieces is 64. Further, if we put the pieces together, we see they fit exactly in an 8 by 8 square, hence the word "chessboard" in the name of the puzzle:

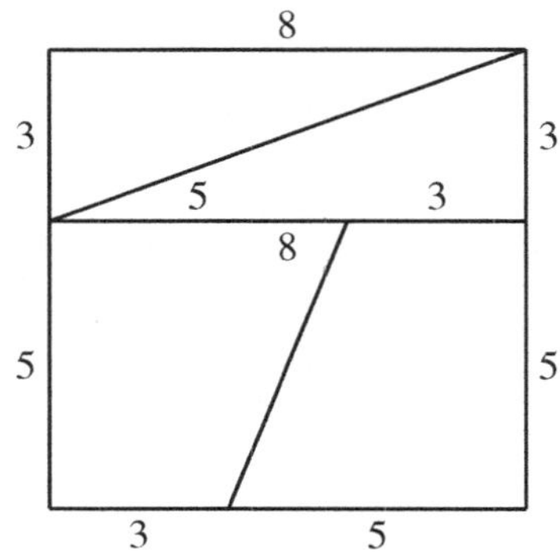

Now let us rearrange the same pieces as below

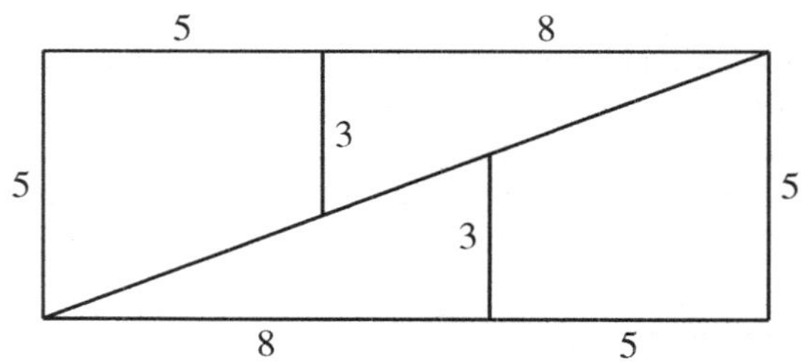

This figure is now 5 by 13 and hence has area 65. But since it is constructed from the same 4 puzzle pieces, it seems that we have shown that $64 = 65$; however, we know this cannot be the case. The confusion is introduced here by a somewhat dishonest diagram. The diagram above was drawn such that the diagonal is a straight line, as it might appear to be if we were using physical puzzle pieces. But if we were to plot these points honestly, and zoom in we would see that these pieces produce a gap in the middle in the shape of a parallelogram:

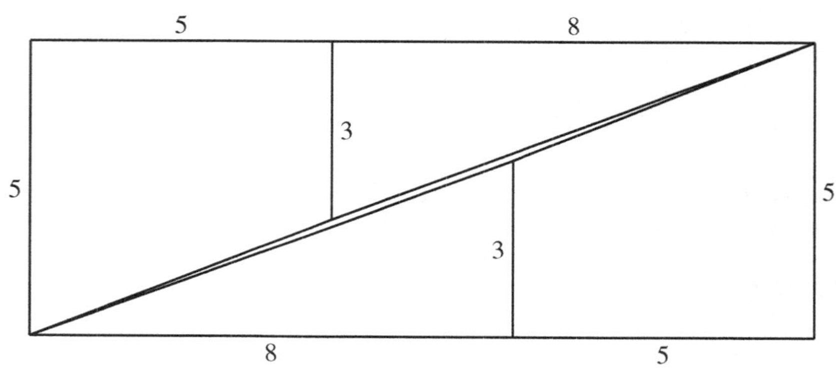

There is no doubt that the area of this parallelogram is 1 as the puzzle pieces have area 64, and the area of the total rectangle is 65. From the fact that the integers 5 ($= F_5$), 8 ($= F_6$), 13 ($= F_7$) are consecutive Fibonacci numbers, we see from Cassini's identity with $n = 7$ that the small parallelogram is of area exactly one square unit. We can use any three consecutive Fibonacci numbers F_{n-3}, F_{n-2}, F_{n-1} $(n \geq 7)$, to make similar puzzles. We see that if we have two puzzle pieces as two right triangles with sides F_{n-1}, F_{n-3}, and two trapezoids labelled similar to before with sides F_{n-3}, F_{n-2}, F_{n-2}, we can arrange these as a square, of area F_{n-1}^2, and again as a rectangle, of area $F_{n-2}F_n$, with the enclosed parallelogram. By Cassini's identity we have for odd n

$$F_{n-1}^2 - F_n F_{n-2} = -1,$$

and the analogous four pieces come together form a gap of area exactly one, as is the case with the chessboard paradox when $n = 7$. However, when n is even

$$F_{n-1}^2 - F_n F_{n-2} = 1,$$

in this case, rather than having a gap between the puzzle pieces, there is an overlap and at that overlap has area exactly 1. The chessboard paradox was a favourite puzzle of Charles Dodgson (1832–1898), better known as Lewis Carroll, the author of *Alice's Adventures in Wonderland*.

We next observe that

$$\phi = \phi^n / \phi^{n-1} = \frac{F_n\phi + F_{n-1}}{F_{n-1}\phi + F_{n-2}} \quad (n \geq 2).$$

Thus,

$$\phi - F_n/F_{n-1} = \frac{F_{n-1}^2 - F_n F_{n-2}}{F_{n-1}(F_{n-1}\phi + F_{n-2})} = \frac{(-1)^n}{F_{n-1}(F_{n-1}\phi + F_{n-2})}$$

by Cassini's identity. Since

$$F_{n-1}\phi + F_{n-2} > F_{n-1}\phi \quad (n \geq 3),$$

we find that

$$|\phi - F_n/F_{n-1}| < 1/(\phi F_{n-1}^2) \quad (n \geq 3).$$

It follows from this that not only does $\lim_{n\to\infty} F_n/F_{n-1} = \phi$, but that F_n/F_{n-1} is an excellent rational approximation to ϕ.

There are many more interesting properties of the Fibonacci numbers; however, to explore them we need to introduce another sequence: the *Lucas numbers* named after the French mathematician François Édouard Anatole Lucas (1842–1891), mentioned in Chaps. 1 and 6. Lucas defined the sequence (L_n) by

$$L_n = F_{2n}/F_n = \phi^n + \bar{\phi}^n.$$

Like the Fibonacci numbers these numbers satisfy a second order linear recurrence. With the initial values $L_0 = 2$ and $L_1 = 1$, L_n can be determined by

$$L_n = L_{n-1} + L_{n-2} \quad (n \geq 2).$$

Much like the trigonometric functions *sine* and *cosine* work as a pair for identities such as

$$\sin^2\theta + \cos^2\theta = 1,$$

so do the sequences F_n and L_n. For example, we can easily derive

$$L_n^2 - 5F_n^2 = 4(-1)^n,$$

and from this we see that

$$\gcd(F_n, L_n) = 1 \text{ or } 2.$$

Also, by definition of L_n, we have

$$F_{2n} = F_n L_n$$

and

$$\begin{aligned} L_{2n} &= \phi^{2n} + \bar{\phi}^{2n} = \phi^{2n} + 2(\phi\bar{\phi})^n + \bar{\phi}^{2n} - 2(-1)^n \\ &= (\phi^n + \bar{\phi}^n)^n - 2(-1)^n = L_n^2 - 2(-1)^n. \end{aligned}$$

Furthermore, we can derive *addition formulas*. For $m \geq 1$ we have

$$F_{m+n} = F_m F_{n+1} + F_n F_{m-1} \quad \text{and} \quad L_{m+n} = L_m F_{n+1} + F_n L_{m-1}.$$

It is even possible to write any F_n entirely in terms of Lucas numbers and L_n in terms of Fibonacci numbers, as follows:

$$5F_n = 2L_{n+1} - L_n = L_{n+1} + L_{n-1} \quad \text{and} \quad L_n = 2F_{n+1} - F_n = F_{n+1} + F_{n-1}.$$

We have

$$\sum_{i=0}^{n} F_i = F_{n+2} - 1, \quad \sum_{i=0}^{n} F_i^2 = F_n F_{n+1} \quad \text{and} \quad \sum_{i=1}^{2n-1} F_i F_{i+1} = F_{2n}^2.$$

We also have the rather pretty relationship between the binomial coefficients and the Fibonacci numbers

$$\sum_{k=0}^{\lfloor n/2 \rfloor} \binom{n-k}{k} = F_{n+1},$$

due to Lucas in 1878. On looking at Pascal's triangle, this identity can be illustrated by looking at shallow diagonals and summing:

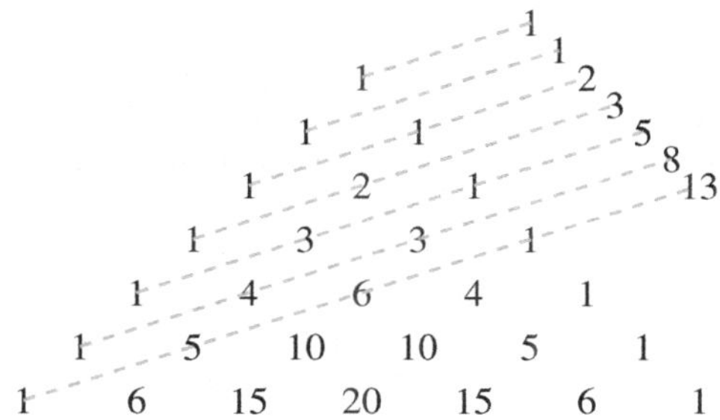

We also have

$$L_n = \sum_{k=0}^{\lfloor n/2 \rfloor} \frac{n}{n-k}\binom{n-k}{k},$$

$$\sum_{i=0}^{n}\binom{n}{i}F_i = F_{2n}, \quad \sum_{i=0}^{n}\binom{n}{i}L_i = L_{2n},$$

$$\sum_{i=0}^{n}(-1)^{i+1}\binom{n}{i}F_i = F_n, \quad \sum_{i=0}^{n}(-1)^{i}\binom{n}{i}L_i = L_n.$$

There are many, many more identities connecting the Fibonacci numbers, Lucas numbers and binomial coefficients, but these few will give some idea of how rich this theory is.

The Fibonacci numbers and the golden ratio do not only appear in mathematics, but also in nature. Botanists have defined *phyllotaxis* as the pattern of petals, seeds or leaves in a plant. Perhaps the most classic example of Fibonacci phyllotaxis is the arrangement of seeds in the spiral of a sunflower, called the *golden spiral*, introduced earlier. The same pattern appears in pinecones, spiral aloe, pineapples and others. Although this pattern is a relatively optimal packing for seeds or for petals to capture light, it is an open problem in Biology to explain why plants evolved in such a way. Since this golden ratio is considered pleasing to the eye, it has been used in such human endeavours as architecture and art.

To complete this chapter, we will look at two applications of the Fibonacci numbers. The first of these is a lesser known theorem that has a habit of frequently being rediscovered, called Lamé's theorem. It is stated below.

Theorem 8.2 *The number of steps in the Euclidean algorithm for positive integers a, b such that $a > b$ is at most 5 times the number of digits of b.*

French mathematician Gabriel Lamé (1795–1870) proved this result in 1844, but it had been discovered earlier. The proof is omitted here, but we point out that it relies on some properties of the Fibonacci numbers and, in fact, shows that the longest possible run time for this algorithm occurs when a and b are consecutive Fibonacci numbers. Another interesting thing about this result is that it is one of the earliest known attempts to analyse the complexity of an algorithm. Because of this result, it was common in France between 1844 and 1877 to call the sequence of Fibonacci numbers the *series of Lamé.* It was Lucas who rediscovered Fibonacci's earlier work and then changed the name of this sequence away from Lamé's sequence, although it should be emphasized that this sequence appears in Indian literature as early as 200 BCE.

The second application we will look at here makes use of the Fibonacci numbers for primality testing, and was discovered by Lucas in 1876. He was aware that the sequence of Fibonacci numbers is an example of a *divisibility sequence,* that is, if $n \mid m$, then $F_n \mid F_m$ (this is not the case for the Lucas numbers). We can demonstrate this important feature of the Fibonacci numbers by making use of a proof process called *mathematical induction.* This is a method of proving that some statement $P(n)$ is true for all integers n greater than or equal to some integer m. The basic idea is first to establish that $P(m)$ is true, and next prove that if $P(k)$ is true for any given $k \geq m$ (the induction hypothesis), then $P(k+1)$ must be true.

Suppose we put $m = hn$. We know that $F_n \mid F_{hn}$ for $h = 1$ and $h = 2$. Suppose next that $F_n \mid F_{kn}$ and $F_n \mid F_{(k-1)n}$ for some given $k \geq 2$ (the induction hypothesis). This is certainly the case for $k = 2$. We know by the first addition formula that

$$F_{(k+1)n} = F_{kn+n} = F_{kn}F_{n+1} + F_nF_{kn-1}.$$

Since, by hypothesis, we have $F_n \mid F_{kn}$, we must have both $F_n \mid F_{(k+1)n}$ and, trivially, $F_n \mid F_{kn}$. We now see that our initial induction hypothesis holds for h increased by 1. As there is no limit to the number of times this argument can be repeated, we find that $F_n \mid F_{hn}$ (and $F_n \mid F_{(h-1)n}$) for all $h \geq 2$.

By using properties of the binomial coefficients $\binom{p-1}{k}$ and $\binom{p+1}{k}$ modulo an odd prime p, Lucas was also able to show from the Binet formula and the binomial theorem that if we define $\epsilon(n)$ to be the Legendre symbol $(n|5)$, then

$$2F_{p+1} \equiv 1 + \epsilon(p) \pmod{p} \quad \text{and} \quad 2F_{p-1} \equiv 1 - \epsilon(p) \pmod{p};$$

thus, $p \mid F_{p-\epsilon(p)}$. Also, we can use our earlier identity properties ($F_{n+1} = F_n + F_{n-1}$, $L_n = F_{n+1} + F_{n-1}$) to deduce from these results that

$$L_{p-\epsilon(p)} \equiv 2\epsilon(p) \pmod{p}.$$

It follows that for any odd prime p there must exist some factor q of $p - \epsilon(p)$ such that $p \mid F_q$ and q (> 1) is minimal; furthermore, if $p \mid F_m$, then it must be that $q \mid m$. This is because we can put $m = qs + t$, where $0 \leq t < q$, and we then have from the first addition formula

$$F_m = F_{qs+t} = F_{qs}F_{t+1} + F_tF_{qs-1}.$$

Thus, if $p \mid F_m$ and $p \mid F_q$, then because $\gcd(F_{qs}, F_{qs-1}) = 1$ we can only have $p \mid F_t$, and therefore we must have $t = 0$ by definition of q. We can now consider any divisor, say d, of F_n; we will say d is a *primitive* divisor of F_n if $d \nmid F_r$ for all

r such that $1 < r < n$. Further, if p is a prime divisor of F_n, then p is a primitive prime divisor of F_n when $p \nmid F_r$ for all $r \mid n$ and $1 < r < n$. Lucas was able to show that if p is any primitive prime divisor of F_n, then $p \equiv \pm 1 \pmod{n}$.

From these observations Lucas deduced that $F_{29} = 514229$ is a prime. He did this by first noting that $L_n^2 - 5F_n^2 = 4(-1)^n$ and $L_n = F_{2n}/F_n$; thus, when n is odd, if p is any prime divisor of F_n, then $(-1|p) = 1$, i.e. $p \equiv 1 \pmod 4$. Now since 29 is prime, if $p \mid F_{29}$, we know $p \nmid F_r$ for all $1 \le r < 29$. Hence, we must have $p \equiv \pm 1 \pmod{29}$. It follows from the CRT that we have either $p \equiv 1 \pmod{4 \cdot 29}$ or $p \equiv 57 \pmod{116}$. Since $514229 < 718^2$ we need only check primes below 718 which satisfy either of these congruences; these primes are 173, 233, 349 and 521. Since $F_{13} = 233$ and $L_{13} = 521$, we can reject these, as we know they cannot divide F_{29}. Finally, 173 and 349 can be rejected by trial division. Thus, it must be the case that F_{29} is prime. These techniques have been refined considerably since Lucas' work. According to the *PrimePages*, the largest known n such that F_n is prime is currently $n = 148091$ (see also A005478 of the OEIS).

Problems

8.1 The Fibonacci sequence is determined by $F_0 = 0$, $F_1 = 1$ and $F_k = F_{k-1} + F_{k-2}$ $(k \ge 2)$. Calculate the first 25 terms of this sequence.

8.2 (Walking up stairs) If you can walk up a flight of stairs either going up one or two steps with each stride, then how many ways can you walk up a flight of:
(a) **[BB]** 5 steps
(b) 10 steps
(c) 20 steps

8.3 (Approximating the golden ratio) It is well known that $\lim_{n\to\infty} \frac{F_n}{F_{n=1}} = \phi$, where ϕ is the golden ration. Find some rational approximations of ϕ by solving $\frac{F_n}{F_{n-1}}$ for:
(a) **[BB]** $n = 6$
(b) $n = 10$
(c) $n = 20$

8.4 (Miles to kilometres) It happens to be the case that the ratio of 1 mile over 1 kilometre is very close to the golden ratio. Use this and the fact $\frac{F_n}{F_{n-1}} \approx \phi$ to approximate the following distances in miles to kilometres:
(a) 610 miles
(b) **[BB]** 89 miles
(c) 1597 miles

8.5 (Five by five puzzle)
(a) Create a generalization of the chessboard paradox using a grid that is only 5×5.
(b) Justify, by using Cassini's identity, that in this case it seems you lose a square

when rearranging the 5 by 5 square into a 3 by 8 rectangle.
(c) What is the geometric explanation for this “missing” area?

8.6 The Lucas numbers are those in the sequence determined by $L_0 = 2$, $L_1 = 1$ and $L_k = L_{k-1} + L_{k-2}$ ($k \geq 2$). Calculate the first 25 terms of this sequence.

8.7 Verify the identity $L_n^2 - 5F_n^2 = 4(-1)^n$ for the particular cases of:
(a) $n = 2$
(b) **[BB]** $n = 5$
(c) $n = 10$

8.8 Verify the identity $F_{2n} = F_n L_n$ for the particular cases of:
(a) $n = 2$
(b) **[BB]** $n = 5$
(c) $n = 10$

8.9 Verify the identity $L_{2n} = L_n^2 - 2(-1)^n$ for the particular cases of:
(a) **[BB]** $n = 2$
(b) $n = 5$
(c) $n = 10$

8.10 Verify the identities

$$F_{m+n} = F_m F_{n+1} + F_n F_{m-1} \quad \text{and} \quad L_{m+n} = L_m F_{n+1} + F_n L_{m-1}$$

for the particular cases of:
(a) **[BB]** $m = 2$ and $n = 3$
(b) $m = 1$ and $n = 5$
(c) $m = 4$ and $n = 6$

8.11 Verify the identity $\sum_{i=0}^{n} F_i = F_{n+2} - 1$ for the particular cases of:
(a) **[BB]** $n = 2$
(b) $n = 5$
(c) $n = 7$

8.12 Verify the identity $\sum_{i=0}^{n} F_i^2 = F_n F_{n+1}$ for the particular cases of:
(a) **[BB]** $n = 2$
(b) $n = 5$
(c) $n = 7$

8.13 Verify the identity $\sum_{i=1}^{2n-1} F_i F_{i+1} = F_{2n}^2$ for the particular cases of:
(a) **[BB]** $n = 2$
(b) $n = 5$
(c) $n = 7$

8.14 Verify and plot on Pascal's triangle the identity $\sum_{k=0}^{\lfloor n/2 \rfloor} \binom{n-k}{k} = F_{n+1}$, for the particular cases of:
(a) $n = 2$
(b) **[BB]** $n = 5$
(c) $n = 7$

8.15 Verify the identity $\sum_{k=0}^{\lfloor n/2 \rfloor} \frac{n}{n-k}\binom{n-k}{k} = L_n$ for the particular cases of:
(a) $n = 2$
(b) **[BB]** $n = 5$
(c) $n = 7$

8.16 Verify the identity $\sum_{i=0}^{n} \binom{n}{i} F_i = F_{2n}$ for the particular cases of:
(a) **[BB]** $n = 2$
(b) $n = 5$
(c) $n = 7$

8.17 Verify the identity $\sum_{i=0}^{n} \binom{n}{i} L_i = L_{2n}$ for the particular cases of:
(a) **[BB]** $n = 2$
(b) $n = 5$
(c) $n = 7$

8.18 Verify the identity $\sum_{i=0}^{n} (-1)^{i+1} \binom{n}{i} F_i = F_n$ for the particular cases of:
(a) **[BB]** $n = 2$
(b) $n = 5$
(c) $n = 7$

8.19 Verify the identity $\sum_{i=0}^{n} (-1)^{i} \binom{n}{i} L_i = L_n$ for the particular cases of:
(a) **[BB]** $n = 2$
(b) $n = 5$
(c) $n = 7$

8.20 (Lamé's theorem)
(a) Calculate the gcd(21, 13) using the Euclidean algorithm.
(b) How many steps did the Euclidean algorithm take in (a)? How does this compare the maximum number of steps predicted by Theorem 8.2?

8.21 (Divisibility sequence) The Fibonacci numbers are an example of a divisibility sequence, that is, if $n|m$, then $F_n \mid F_m$. Verify this is true for the particular cases of:
(a) **[BB]** $n = 2, m = 4$
(b) $n = 3, m = 6$
(c) $n = 5, m = 10$

8.22 Verify that $p \mid F_{p-\epsilon(p)}$, where $\epsilon(p) = (n|5)$, for primes:
(a) **[BB]** $p = 3$
(b) $p = 5$
(c) $p = 7$

8.23 Verify that $L_{p-\epsilon(p)} \equiv 2\epsilon(p) \pmod{p}$, where $\epsilon(p) = (n|5)$, for primes:
(a) $p = 3$
(b) $p = 5$
(c) **[BB]** $p = 7$

Further Reading

The golden ratio

Mario Livio, *The Golden Ratio*, Broadway Books, NY, 2002.

Fibonacci

Keith Devlin, *Finding Fibonacci*, Princeton University Press, Princeton NJ, 2017.

Fibonacci and Lucas number identities

Thomas Koshy, *Fibonacci and Lucas numbers with Applications*, Volume 1, Second edition, John Wiley and Sons, NY, 2018 pp. 108–113. This is an encyclopedic treatment of the Fibonacci and Lucas numbers and covers all the topics in this chapter.

Phyllotaxis

A. S. Posamentier and I. Lehman, *The (Fabulous) Fibonacci Numbers*, Prometheus Books, NY, 2007, pp. 63–76.

The chessboard paradox

W. W. Rouse Ball and H. S. M. Coxeter, *Mathematical Recreations and Essays*, 13th Edition, Dover Publications, NY, 1987, pp. 84-86.

David Singmaster, Vanishing area puzzles, *Recreational Mathematics Magazine*, **1** (2014), 10–21.

Lamé's theorem

J. O. Shallit, Origins of the analysis of the Euclidean Algorithm, *Historia Mathematica*, **21** (1994), 401–419.

Fibonacci numbers in India

Parmanand Singh, The so-called Fibonacci numbers in ancient and medieval India, *Historia Mathematica*, **12** (1985), 229–244.

Divisibility properties of the Fibonacci numbers

S. Vajda, *Fibonacci & Lucas numbers and the Golden Section: Theory and Applications*, Dover Publications, NY, 2007.

Fibonacci primes

H. Dubner and W. Keller, New Fibonacci and Lucas primes, *Mathematics of Computation*, **68** (1999), 417–427.

Chapter 9
Sociable Numbers

In number theory a function of a positive integer that is related to the divisors of that integer is called a *divisor function*. One particular example of such a function is the *sum of divisors function*, denoted by σ, where $\sigma(n)$ is equal to the sum of all the positive divisors of n. For example,

$$\sigma(12) = \sum_{d|12} d = 1 + 2 + 3 + 4 + 6 + 12 = 28,$$

noting in this case that the restriction on the sum is over all the values of d such that d divides 12. If we have positive integers m, n such that $\gcd(m, n) = 1$, then it is not hard to see that $\sigma(mn) = \sigma(m)\sigma(n)$, that is, σ is a multiplicative function. Notice that since m, n are relatively prime, each d that is a divisor of mn can be uniquely written such that $d = d_m d_n$ where $d_m \mid m$ and $d_n \mid n$ and $\gcd(d_m, d_n) = 1$, thus

$$\sigma(mn) = \sum_{d|mn} d = \sum_{d_m|m, d_n|n} d_m d_n = \sum_{d_m|m} d_m \sum_{d_n|n} d_n = \sigma(m)\sigma(n).$$

We can also see that if p is a prime and n is any positive integer, then the only divisors of p^n are 1 and powers of p up to p^n. From this we have

$$\begin{aligned}\sigma(p^n) &= 1 + p + p^2 + \cdots + p^{n-1} + p^n \\ &= (1 + p + p^2 + \cdots + p^{n-1} + p^n)\frac{(p-1)}{(p-1)} \\ &= \frac{p^{n+1} - 1}{p - 1}.\end{aligned}$$

Thus, if m is a positive integer with prime decomposition

E. L. F. Roettger and H. C. Williams, *The Enchantment of Numbers*, CMS/CAIMS Books in Mathematics 16, https://doi.org/10.1007/978-3-031-76568-1_9

$$m = p_1^{n_1} p_2^{n_2} \cdots p_k^{n_k},$$

then since σ is a multiplicative function and $\sigma(p^n) = \frac{p^{n+1}-1}{p-1}$ we have

$$\sigma(m) = \sigma(p_1^{n_1} p_2^{n_2} \cdots p_k^{n_k}) = \sigma(p_1^{n_1})\sigma(p_2^{n_2}) \cdots \sigma(p_k^{n_k})$$
$$= \frac{p_1^{n_1+1} - 1}{p_1 - 1} \cdot \frac{p_2^{n_2+1} - 1}{p_2 - 1} \cdot \ldots \cdot \frac{p_k^{n_k+1} - 1}{p_k - 1}.$$

Thus, if we have the prime factorization of m we can compute $\sigma(m)$ easily. For example, let $m = 539784$. Since $539784 = 2^3 \cdot 3^4 \cdot 7^2 \cdot 17$, we see that

$$\sigma(539784) = \frac{2^4 - 1}{2 - 1} \cdot \frac{3^5 - 1}{3 - 1} \cdot \frac{7^3 - 1}{7 - 1} \cdot \frac{17^2 - 1}{17 - 1}$$
$$= \frac{15}{1} \cdot \frac{242}{2} \cdot \frac{342}{6} \cdot \frac{288}{16}$$
$$= 15 \cdot 121 \cdot 57 \cdot 18 = 1862190.$$

This might all seem very familiar, as the above divisor function is very similar to the function $s(n)$ from Chap. 2. Recall that $s(n)$ was the sum of aliquot divisors, and hence,

$$s(n) = \sigma(n) - n.$$

An interesting question is whether there exists, given a positive integer m, an n such that $m = s(n)$. We saw in Chap. 2 that $s(60) = 108$, thus $m = 108$ has such a number: $n = 60$. However, we can easily see that if $m = 2$, it is impossible to find an n such that $s(n) = 2$, as 1 is a divisor of all $n > 0$ and the next smallest distinct divisor is 2, but $1 + 2 = 3$, and hence is too large. Another example that is easy to understand occurs for $m = 5$; again we must include 1 and since divisors must be distinct we cannot have $1 + 2 + 2$ nor $1 + 1 + 3$, the only way to write 5 as a sum of distinct positive integers is $1 + 4$; however, if 4 is a divisor, then 2 would also be. We call any positive integer that cannot be expressed as a sum of all aliquot divisors of a positive integer an *untouchable number*. Thanks to the work of the prolific Hungarian mathematician Paul Erdös (1913–1996), we know that there are infinitely many such numbers. Untouchable numbers are also known as *nonaliquot numbers*, and the first few of these interesting numbers are listed as a sequence on the OEIS A001065.

A very old problem in number theory was left to us by the German mathematician Christian Goldbach (1690–1764) in a letter to Euler in 1742. The problem is known today as *Goldbach's conjecture*. It states that every even positive integer greater than 2 can be written as the sum of two prime numbers. For instance, $4 = 2 + 2$, $6 = 3 + 3$ and $8 = 3 + 5$. Although there have been many impressive advances in analytic number theory, this very simple statement remains unproved. If we had a slightly stronger version of Goldbach's conjecture, i.e. one claiming that every even positive integer greater than 6 can be written as the sum of two distinct prime numbers, then under this assumption we could show that 5 is in fact the only odd untouchable number. For, given m odd and greater or equal to 9, then $m - 1$ is even and under the stronger Goldbach's conjecture $m - 1 = p + q$, where $p \neq q$. Thus, $m = p + q + 1$, and then for $n = pq$ we have $s(n) = s(pq) = 1 + p + q = m$. Since we already know 7 is not untouchable, as $s(8) = 7$, under our assumption 5 would be the only odd untouchable number.

Recall from Chap. 2 that we defined $s_0(n) = n$, $s_1(n) = s(n)$, and for any positive integer k

$$s_k(n) = s(s_{k-1}(n)),$$

and thus the example in Chap. 2, began

$$s_0(60) = 60, \quad s_1(60) = 108, \quad s_2(60) = s(s_1(60)) = s(108) = 172,$$

and continuing in this fashion, we saw that this yields the aliquot sequence belonging to 60:

$$60, 108, 172, 136, 134, 70, 74, 40, 50, 43, 1, \ldots.$$

Since a number p is a prime if and only if it has an aliquot sum of 1, we see that any aliquot sequence that arrives at a prime will terminate. However, this is not the only outcome of an aliquot sequence. There is the possibility that the sequence enters what is called an *aliquot cycle*, a set of distinct positive integers, say $\{m_1, m_2, \ldots, m_k\}$, such that $s(m_i) = m_{i+1}$ for $1 \leq k - 1$ and $s(m_k) = m_1$. In this case we call this set of positive integers *sociable*. In the case that the cycle length is 1, that is, $s(m) = m$, we then say m is a *perfect number*. If the cycle length is just 2, and the set $\{m_1, m_2\}$ satisfies $s(m_1) = s(m_2)$ and $s(m_2) = s(m_1)$, we call the numbers m_1, m_2 an *amicable pair*.

For example, consider $n = 12496$, the aliquot sequence belonging to 12496 is

$$12496, 14288, 15472, 14536, 14264, 12496, 14288, 15472, 14536, 14264, \ldots.$$

Thus, the set $\{12496, 14288, 15472, 14536, 14264\}$ is an aliquot cycle of order 5 (this is the only known set of 5 sociable numbers). We can verify this set by noting: $12496 = 2^4 \cdot 11 \cdot 71$, $14288 = 2^4 \cdot 19 \cdot 47$, $15472 = 2^4 \cdot 967$, $14536 = 2^3 \cdot 23 \cdot 79$ and $14264 = 2^3 \cdot 1783$, thus

$$s(12496) = \sigma(12496) - 12496 = \frac{2^5 - 1}{2 - 1} \cdot \frac{11^2 - 1}{11 - 1} \cdot \frac{71^2 - 1}{71 - 1} - 12496 = 14288,$$
$$s(14288) = \sigma(14288) - 14288 = \frac{2^5 - 1}{2 - 1} \cdot \frac{19^2 - 1}{19 - 1} \cdot \frac{47^2 - 1}{47 - 1} - 14288 = 15472,$$
$$s(15472) = \sigma(15472) - 15472 = \frac{2^5 - 1}{2 - 1} \cdot \frac{967^2 - 1}{967 - 1} - 15472 = 14536,$$

$$s(14536) = \sigma(14536) - 14536 = \frac{2^4 - 1}{2 - 1} \cdot \frac{23^2 - 1}{23 - 1} \cdot \frac{79^2 - 1}{79 - 1} - 14536 = 14264,$$

$$s(14264) = \sigma(14264) - 14264 = \frac{2^4 - 1}{2 - 1} \cdot \frac{1783^2 - 1}{1783 - 1} - 14264 = 12496.$$

Currently there are 5410 known sets of sociable numbers with cycles greater than 2; we see in the table below that most of these have cycle length 4:

Cycle length	Number of known sets	Smallest integer in any known set
4	5398	1264460
5	1	12496
6	5	21548919483
8	4	1095447416
9	1	805984760
28	1	14316

Note that there are no known sets of sociable numbers that have a cycle length of 3, 7 or 11; indeed, it is widely believed that there is no set of sociable numbers with a cycle length congruent to 3 modulo 4.

The German mathematician Peter Gustav Lejeune Dirichlet (1805–1859) is credited with a basic counting argument, called the *pigeonhole principle,* which states that if n items are put into m containers where $n > m$, then at least one container must contain more than 1 item. Imagine $s_k(n)$ is bounded as k goes to infinity and the sequence does not terminate with $s_{k'}(n) = 1$ for some k', then by the pigeonhole principle we must for some minimal positive h and some minimal positive $j > h$ have $s_j(n) = s_h(n)$, but this means that $s_{j-h}(m) = m$, for some m $(= s_h(n) > 1)$. Hence, the set $\{m, s(m), \ldots, s_{j-h-1}(m)\}$ is a set of sociable numbers. If $j = h + 1$, we have m a perfect number and if $j = h + 2$ we have an amicable pair.

To summarize, for any aliquot sequence there are 3 possibilities:

(i) The sequence arrives at a prime and then terminates at 1 in the next iteration of s.
(ii) The sequence arrives at a perfect number, amicable pair, or some other sociable number set and then cycles.
(iii) The sequence growth is unbounded.

The difficulty here is that mathematicians have so far been unable to prove or disprove that there exist aliquot sequences that have unbounded growth. Recall in Chap. 2 the Catalan-Dickson conjecture stated that (iii) could not happen. There is, however, a competing conjecture put forward by Richard K. Guy (1916–2020) and John L. Selfridge (1927–2010) in 1975, that not only does there exist an unbounded aliquot sequence, but that there are infinitely many such sequences.

Euclid in his *Elements* proved that if $2^n - 1$ is a prime then $2^{n-1}(2^n - 1)$ is a perfect number. In antiquity only the first four perfect numbers were known; these are

n	$2^n - 1$	$2^{n-1}(2^n - 1)$
2	3	6
3	7	28
5	31	496
7	127	8128

The biconditional that states, an even positive integer is perfect if and only if it has the form $2^{n-1}(2^n - 1)$ where $2^n - 1$ is a prime, is now known as the Euclid-Euler theorem, as it was Euler who was able to prove the converse of Euclid's result.

It is not clear how this ancient concept of numerical perfection originated, but if

$$s(n) = n,$$

then

$$\sum_{d|n,d<n} d = n,$$

and therefore

$$\sum_{d|n,d<n} \frac{d}{n} = 1 \to \sum_{d|n,d<n} \frac{1}{(n/d)} = 1.$$

Thus,

$$\sum_{d|n,d>1} \frac{1}{d} = 1;$$

that is, the sum of all the reciprocals of all the non-unit divisors of n is 1. This could serve as an alternate definition of a perfect number. For example when $n = 6$,

$$\frac{1}{2} + \frac{1}{3} + \frac{1}{6} = 1.$$

As a great deal of Greek mathematical activity at the time of Euclid was being done in the Egyptian city of Alexandria, it is possible that this idea of numerical perfection derives from the ancient Egyptian practice of performing arithmetic using unit fractions (reciprocals of positive integers), because perfect numbers provide representations of 1 as a sum of distinct unit fractions.

As a result of conjectures made about perfect numbers by the French mathematical enthusiast and monk, Marin Mersenne (1588–1648), we now call a number of the form $2^n - 1$ a *Mersenne number,* denoted by M_n. The Italian mathematician Pietro Cataldi (1548–1626) was the first to observe that a Mersenne number can be a prime only in the case that n is a prime; however, this does not guarantee that it is a prime. For instance, $M_{11} = 2^{11} - 1 = 2047 = 23 \cdot 89$. In the case that M_p is a prime for a given prime p, we say that M_p is a *Mersenne prime.* Cataldi was able to show M_{13}, M_{17} and M_{19} to be prime by using trial division up the numbers' respective square roots (the method described in Chap. 2); however, the fact that both M_{13} and M_{17} yield perfect numbers was known to previous investigators. In 1644 Mersenne compiled a list of exponents n for which $2^n - 1$ is prime up to the bound $n = 257$. This list is as follows:

$$2, 3, 5, 7, 13, 17, 19, 31, 67, 127, 257.$$

At this time the list was known to be correct up to $n = 19$. Missing from this list are the values 61, 89, 107, as M_{61}, M_{89}, M_{107} are now known to be prime. Furthermore, 67 and 257 should not be included, as neither M_{67} nor M_{257} is prime. An inquisitive person might be curious as to how Mersenne was selecting numbers for his list. Fortunately we have some indication of Mersenne's thinking from his *Novarum observationum physico-mathematicarum* in a passage that has been reconstructed as follows:

> A useful rule for knowing prime numbers is this: any number is prime which is of the form $2^p - 1$, where p is a prime and differs by three or less from some even power of 2. Thus, $2^7 - 1$ is a prime since $7 = 2^2 + 3$. Further, if three is added to 64, the prime 67 arises, and thus the 67th power, plus [should be minus] 1, will be the prime [later shown to be composite] number that follows, 147573952589676412927, of which this is a property, that multiplied into its half, perfect numbers [clearly a single number] are generated; which is assumed only of prime numbers that are less by one than a number of the binary powers, whence this property does not fit the number 5, but the numbers 3, 7, 31, 127, 8191, 131571, 524287, 2147483647 and all others of this genus.

In short, Mersenne appears to have believed that values of n that produce prime values of $2^n - 1$ must be primes which do not differ by more than 1 from a power of 2 or differ by more than 3 from an even power of 2. This rule is now known to be false. Interestingly, under the assumption of this rule, 61 should have been included in Mersenne's list.

Although his list is incorrect, Mersenne correctly asserted in the *Praefatio generalis* of his *Cogita physico-mathematica* that

> One of the greatest difficulties in mathematics is to exhibit a prescribed number of perfect numbers, and to tell if a given number of 15 or 20 digits is prime or not; all time would not suffice for the test, whatever use is made of what is already known.

Here Mersenne is rightly pointing out the futility of primality testing based on trial divisions when the number is large. We will discuss this in more detail in the next chapter. It was not until Lucas started his study of Mersenne numbers that Mersenne's list could be verified, as Lucas was able to develop novel tests that did not rely on mere trial division. Although we have often avoided proof details, we include them here, as they are very clever and rely on topics we have already covered in Chap. 8.

Suppose we put $r_i = L_{2^i}$, $s_i = F_{2^i}$. From results in Chap. 8, we have $r_1 = L_2 = 3$ and

$$s_{i+1} = r_i s_i = r_i r_{i-1} s_{i-1} = \cdots = r_i r_{i-1} r_{i-2} \cdots r_1.$$

Also,

$$r_{i+1} = r_i^2 - 2(-1)^{2^i} = r_i^2 - 2 \quad (i > 0)$$

and

$$r_{i+1} \equiv r_i \equiv r_{i-1} \equiv \cdots \equiv r_1 \equiv 1 \pmod 2.$$

Thus, all the r_i values must be odd and since $\gcd(F_{2^i}, L_{2^i}) \mid 2$, we find that $\gcd(r_i, s_i) = 1$ for $i = 1, 2, \ldots$. From this, it is easy to deduce that $\gcd(r_i, r_j) = 1$ when $1 \le j < i$. We also observe that since (F_n) is a divisibility sequence, we must have $s_{i-1} \mid s_i$ $(i = 1, 2, \ldots)$.

We are now able to establish Lucas' first primality test for Mersenne numbers. Since the ideas behind this criterion are at the heart of many modern techniques for ascertaining primality, we provide a simple proof of this result.

Theorem 9.1 *Let $N = M_n = 2^n - 1$ with $n \equiv 3 \pmod 4$. If none of the r_i $(i = 1, 2, \ldots, n-1)$ above is divisible by N, then N is not a prime. If j is the least value of i such that $1 \le j \le n-1$ and $N \mid r_j$, then any prime divisor of N must be of the form $2^{j+1}k \pm 1$.*

Notice that this test does not establish the prime character of N. It does, however, determine either that N is composite or that the least prime divisor of N is of the form $2^{j+1}k \pm 1$. If, however, $2j > n$, then the least prime divisor of N must exceed $2^{j+1} - 1 > \sqrt{N}$, which means that in this case N can only be a prime.

Proof If N is a prime, then since

$$\epsilon(N) = (N|5) = (2^n - 1|5) = (2^{4h-3} - 1|5) = (7|5) = (2|5) = -1,$$

we have

$$N \mid F_{N+1}.$$

Since $F_{N+1} = s_n = r_{n-1} r_{n-2} \cdots r_1$, we must have $N \mid r_j$ for some $1 \le j \le n-1$. Thus, if $N \nmid r_j$ for $j = 1, 2, \ldots, n-1$, then N is composite.

Next, suppose that N is composite and $N \mid r_j$. If p is any prime divisor of N, then $p \mid r_j$, which since $\gcd(r_j, s_j) = 1$ means that $p \nmid s_j$. Also, $p \nmid r_{j-1}$. It follows that $p \mid F_{2^{j+1}}$ $(= r_j s_j)$ and $p \nmid F_{2^j}$ $(= r_{j-1} s_{j-1})$. By using the reasoning near the end of Chap. 8, we must have $2^{j+1} \mid p - \epsilon(p)$, which means that $p = 2^{j+1}k \pm 1$. □

Lucas' next primality test for M_n with $n \equiv 3 \pmod 4$ is given as a corollary of this theorem.

Corollary 9.1 *If $N = M_n = 2^n - 1$ with $n \equiv 3 \pmod 4$, then M_n is a prime if and only if $M_n \mid r_{n-1}$, where $r_1 = 3$ and $r_{i+1} \equiv r_i^2 - 2 \pmod{M_n}$.*

Proof By the theorem we see that if $N \mid r_{n-1}$, then since $2(n-1) > n$, N must be a prime. Suppose N is a prime, then $\epsilon(N) = (N|5) = -1$ and $L_{N+1} \equiv 2\epsilon(N) \equiv -2 \pmod N$. Since $-2 \equiv L_{N+1} = (L_{\frac{N+1}{2}})^2 - 2$, we see that $N \mid L_{\frac{N+1}{2}}$ or $N \mid r_{n-1}$. □

As an example of this test we try $M_7 = 2^7 - 1 = 127$. We have $r_1 = 3$ and

$$\begin{aligned} r_2 &\equiv 3^2 - 2 \equiv 7 \pmod{M_7}, \\ r_3 &\equiv 7^2 - 2 \equiv 47 \pmod{M_7}, \\ r_4 &\equiv 47^2 - 2 \equiv 48 \pmod{M_7}, \\ r_5 &\equiv 48^2 - 2 \equiv 16 \pmod{M_7}, \\ r_6 &\equiv 16^2 - 2 \equiv 0 \pmod{M_7}. \end{aligned}$$

Thus, M_7 is a prime.

In the above we see that Lucas did a most extraordinary thing: he developed a primality proof that did not rely on trial division. This provided him with the ability to test the primality of Mersenne numbers and, by consequence, find perfect numbers of magnitudes that could not have been envisioned before. To this day the world's largest known primes are Mersenne primes and are still found using Lucas' methods, albeit now with the aid of computers. It should be stated that initially Lucas did not provide complete proofs for his results, and when he did so later, the proofs contained some errors. These errors were later corrected by Robert Carmichael (1879–1967) in 1913. Furthermore, as we will show in the next chapter, tests like Lucas' were developed further by Lehmer in the 1930s.

Lucas employed his technique to show that M_{127} is a prime. By his test it is enough to show that $r_{126} \equiv 0 \pmod{M_{127}}$, but without the aid of a computer this is still a considerable task. Lucas did this in a rather fun way. He performed the calculations on a 127×127 chessboard and he worked in binary, where a pawn would represent a 1 and an empty space would be zero. He also made use of the fact that

$$2^n \equiv 1 \pmod{2^n - 1} \rightarrow 2^{n+m} \equiv 2^m \pmod{M_n}.$$

The chessboard was populated with pawns based on the squaring of the binary representation of r_i, each row of the chessboard being the result of one positional multiplication, and using the above fact to allow wrap-around. On doing this, we can render the multiplication tableau modulo M_n into an $n \times n$ square, consisting of zeros and ones. The following rules can then be employed to add up these powers of 2 modulo M_n, and also perform the subtraction of 2. He made use of 3 rules:

(i) Take (when possible but only once) a pawn away from the second column of the chessboard. This corresponds to the subtraction of 2 in the theorem. If a pawn never appears in this column, then the subtraction of 2 is performed from the final answer.

(ii) For each pair of pawns in any column remove one from the board and move the other to the column to the left. If the pair is in the leftmost column, move the pawn to the first column, again because $2^{n+m} \equiv 2^m \pmod{M_n}$.

(iii) Move any pawn up to an empty spot in the same column.

The above manoeuvres are applied until only the top row of the chessboard contains pawns.

Rather than working on the M_{127} case, let us perform one iteration of this using a 7×7 board working with M_7 for demonstrative purposes. We will let $r_4 = 48 = 0110000_2$ and then recover r_5 with above method.

```
           0110000
          ×0110000
          --------
           0000000
          0000000
         0000000
        0000000
       0110000
      0110000
     +0000000
```

since $2^{7+i} \equiv 2^i \pmod{M_7}$ → exploit wrap-around

7	6	5	4	3	2	1
0	0	0	0	0	0	0
0	0	0	0	0	0	0
0	0	0	0	0	0	0
0	0	0	0	0	0	0
0	0	0	0	1	1	0
0	0	0	1	1	0	0
0	0	0	0	0	0	0

We can now replace the zeros with empty spaces and the ones with pawns. To indicate the pawns having rules (i) or (ii) applied to them, we will draw x's over them for clarity. In the case that we are applying rule (i), there will only be one pawn crossed out, and in the following step it will be removed from the board. In the case that we are applying rule (ii), two pawns will be crossed out and in the next step a new pawn will appear in the column to the left.

7	6	5	4	3	2	1
0	0	0	0	0	0	0
0	0	0	0	0	0	0
0	0	0	0	0	0	0
0	0	0	0	0	0	0
0	0	0	0	1	1	0
0	0	0	1	1	0	0
0	0	0	0	0	0	0

working with pawns →

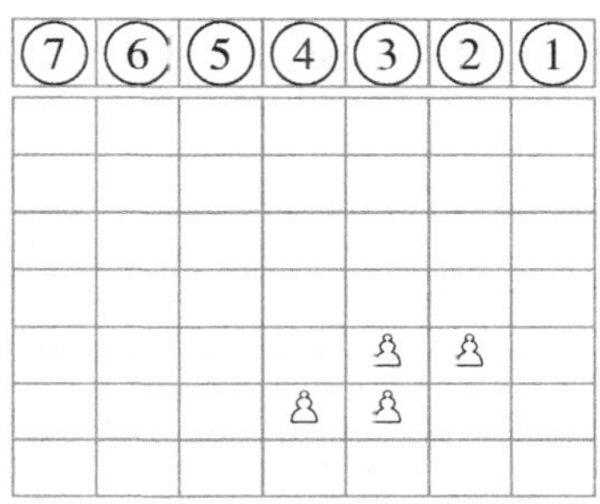

We can apply rule (iii) four times and move all the pawns up.

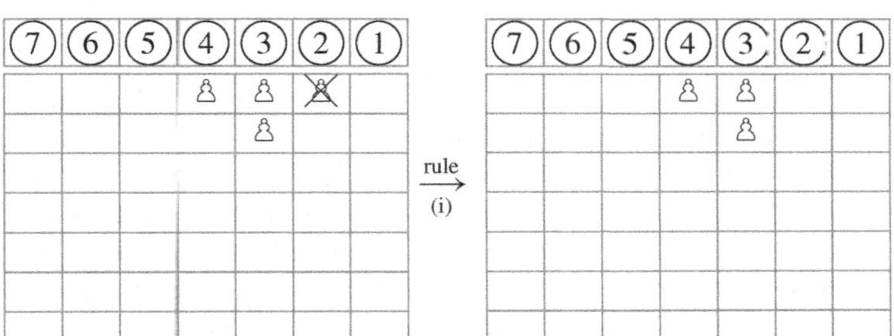

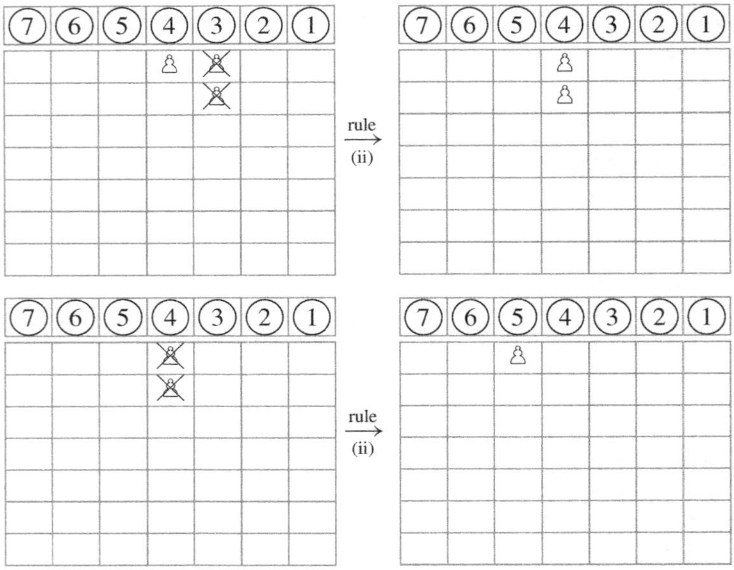

Thus, $r_5 \equiv 2^{5-1} \equiv 16 \pmod{M_7}$. While solving for r_{i+1} in cases that take more iterations it is almost entirely by applying rule (ii) repeatedly. Given $r_3 = 47$, try these rules to show $r_4 = 48$. You will see you become quickly skilled at applying the rule (ii) manoeuvre after just a few iterations.

Lucas used the above method on a 127×127 chessboard to check the primality of M_{127}. After his calculations were complete, he wrote:

> I think I have proved that the number $A = 2^{127} - 1$ is prime...

It appears that Lucas was unsure of the result. Perhaps this is because his method, once employed, leaves no written record of the calculations. However, in either the case that M_n is prime or composite any miscalculation would most likely cause $M_n \nmid r_{n-1}$. Thus, in the event that $M_n \mid r_{n-1}$ one might have expected Lucas to be more confident in his assertion. It is estimated that it took him between 170 and 300 hours to perform the calculations for M_{127} using the method described above. Interestingly, Lucas was not motivated by Mersenne's list; he was only informed of it by the Italian mathematician Angelo Genocchi (1817–1889) after he had already proved M_{127} prime.

Nicomachus of Gerasa (c. 60–c. 120), an ancient Greek mathematician, made the claim that all perfect numbers are even. Because of Euler's work this would mean that all perfect numbers are of the form $M_p(2^{p-1})$ where M_p is a Mersenne prime. To date, no one has been able to prove or disprove this claim. Hence, the existence of an odd perfect number is an open problem. There has been some progress made on this problem by mathematicians proving an ever growing set of restrictions that an odd perfect number would have to satisfy. This started with the English mathematician James Joseph Sylvester (1814–1897) in 1888. He noted such things as no odd perfect number is divisible by 105 and an odd perfect number must have at least 5 distinct prime divisors. His goal was to create a complex web of conditions, now called *Sylvester's web.* Over the years the minimum number of distinct prime divisors necessary for an odd perfect number has continued to grow. Another layer to this web has been finding lower bounds on the largest prime factor of an odd perfect number and now such bounds are over 10^8; hence, any odd perfect number would be very large. Indeed, recent work has shown that an odd perfect number must be greater than 10^{1500} and must be the product of over 100 prime factors.

We now return to the concept of amicable numbers. As mentioned earlier, a pair of numbers m, n are said to be amicable if $s(m) = n$ and $s(n) = m$. That is, as mentioned earlier, they are an aliquot sequence that has a cycle length of 2. The earliest known, and smallest, amicable numbers are the pair 220 and 284. Iraqi mathematician Thābit ibn Qurra (826–901) was able to develop a method by which amicable numbers could be derived. This idea was later rediscovered by both Pierre de Fermat (1607–1665) and René Descartes (1596–1650). Thābit's rule states that if

$$p = 3 \cdot 2^{n-1} - 1, \quad q = 3 \cdot 2^n - 1, \quad \text{and} \quad r = 9 \cdot 2^{2n-1} - 1,$$

where n is an integer $n > 1$ and p, q, r are all prime numbers, then $2^n pq$ and $2^n r$ are a pair of amicable numbers. We can verify that if $n = 2$, then

$$p = 3 \cdot 2^{2-1} - 1 = 5, \quad q = 3 \cdot 2^2 - 1 = 11, \quad \text{and} \quad r = 9 \cdot 2^{2 \cdot 2-1} - 1 = 71,$$

and since these are all prime $2^n pq = 2^2 \cdot 5 \cdot 11 = 220$ and $2^n r = 2^2 \cdot 71 = 284$ for an amicable pair, although one of which we were already aware. Amicable pairs are also formed in the above when $n = 4$ (yields pair 17296, 18416) and $n = 7$ (9363584, 9437056); however, these are the only known pairs of amicable numbers that follow from Thābit's rule.

Euler's rule provides a generalization of Thābit's rule. It states that if

$$p = (2^{n-m} + 1)2^m - 1, \quad q = (2^{n-m} + 1)2^n - 1, \quad \text{and} \quad r = (2^{n-m} + 1)^2 2^{m+n} - 1,$$

where $n > m > 0$ are integers and p, q, r are prime numbers, then $2^n pq$ and $2^n r$ form pair of amicable numbers. All pairs formed by Thābit's rule are also generated with Euler's (here $m = n - 1$). However, with Euler's rule you get additional amicable numbers when $(m, n) = (1, 8)$ and $(29, 40)$; these are respectively

$$\begin{gathered} 2172649216, 2181168896 \quad \text{and} \\ 2724918040393706557785752240819405848576, \\ 2724918040396184856306258038787235905536. \end{gathered}$$

However, many other amicable pairs exist that do not satisfy Thābit's nor Euler's rules. For example, $1184 = 2^5 \cdot 37$ and $1210 = 2 \cdot 5 \cdot 11^2$ is an amicable pair that cannot be generated by either of the above rules.

Problems

9.1 (Sum of divisors function) Recall that the function $\sigma(n)$ is equal to the sum of all the positive divisors on n. Find $\sigma(n)$ for the n values below:
(a) **[BB]** $n = 68962 = 2 \cdot 29^2 \cdot 41$
(b) $n = 222264 = 2^3 \cdot 3^4 \cdot 7^3$
(c) $n = 378125 = 5^5 \cdot 11^2$

9.2 How many untouchable numbers (also known as nonaliquot numbers) are there? Give an example of an untouchable number.

9.3 Is $m = 450$ an untouchable number? Hint calculate $s(270)$.

9.4 Show $m = 9$ is not an untouchable number by finding n such that $s(n) = 9$.

9.5 Verify Goldbach's conjecture up to $n = 30$.

9.6 (Aliquot sequence) Recall the aliquot sequence belonging to n is $s_0(n) = n$, $s_1(n) = s(n)$ and for any positive integer k

$$s_k(n) = s(s_{k-1}(n)).$$

Find the aliquot sequence belonging to (note the below values of n were chosen such that they all terminate or enter a cycle such that you will not need to calculate more than 7 terms of the aliquot sequence):
(a) **[BB]** $n = 95$
(b) $n = 94$,
(c) $n = 2714$,
(d) $n = 1230524$

9.7 (Perfect number) Recall a perfect number is one such that $n = s(n)$. Verify the numbers below are perfect numbers:
(a) **[BB]** $n = 6$
(b) $n = 28$
(c) $n = 496$

9.8 (Perfect number) Recall a perfect number is one such that $n = s(n)$. Given that the values stated below are Mersenne primes find the corresponding perfect number:
(a) **[BB]** $n = 2^{13} - 1$
(b) $n = 2^{17} - 1$
(c) $n = 2^{19} - 1$

9.9 [BB] Use Lucas' Corollary 9.1 to show that $2^{19} - 1$ is a prime.

9.10 Use Lucas' Corollary 9.1 to show that $2^{11} - 1$ is not a prime.

9.11 [BB] Given $r_2 \equiv 7 \pmod{M_7}$ show $r_3 \equiv 47 \pmod{M_7}$ using Lucas' chessboard method.

9.12 Given $r_3 \equiv 47 \pmod{M_7}$ show $r_4 \equiv 48 \pmod{M_7}$ using Lucas' chessboard method.

Further Reading

Prime numbers

David Wells, *Prime Numbers: the Most Mysterious Figures in Math*, John Wiley and Sons, Inc., NJ, 2005.

Sociable numbers

Song Y. Yan, *Perfect, Amicable and Sociable Numbers: a computational approach*, World Scientific, Singapore, 1996.

Aliquot sequences

Richard K. Guy and J. L. Selfridge, What drives an aliquot sequence, *Mathematics of Computation*, **29** (1975), 101–107.

Egyptian fractions

David Reimer, *Count Like an Egyptian*, Princeton University Press, 2014.

Marin Mersenne

Justin Grosslight, Small skills, big networks: Marin Mersenne mathematical intelligencer, *History of Science*, **51-3** (2013), 337–374.

Mersenne numbers

S. Drake, The rule behind Mersenne's numbers, *Physics-Riv. Internanz. Storia Sci.*, **13** (1971), 421–424.

Lucas and M_{127}

H. C. Williams, *Édouard Lucas and Primality Testing*, Wiley-Interscience, John Wiley and Sons, NY, 1998, pp. 58–65.

Odd perfect numbers

Brian D. Beasly, Euler and the ongoing search for odd perfect numbers, *ACMS Conference Proceedings*, **211** (2013), 1–11.

Amicable numbers and early computing

J. Alanen, O. Ore and J. Stemple, Systematic computations on amicable numbers, *Mathematics of Computation*, **21** (1967), 242–245.

Chapter 10
Lucas and Lehmer Sequences

Of particular interest to recreational number theorists is the identification of large primes. At the time of writing, the largest known prime is $M_{136279841}$, a number of 41024320 decimal digits. We now know of 52 Mersenne primes, and it is conjectured that there exists an infinitude of such numbers, but this is very far from being proved. The best we can say is that a *heuristic argument*—mental shortcuts which are not necessarily mathematically rigorous, but are nevertheless intellectually compelling—of Sam Wagstaff suggests that if $N(x)$ is the number of Mersenne primes $\leq x$, then $N(x)$ is approximately $c \log\log x$ for an explicitly given constant c. In this chapter, we will discuss in some detail the simple procedure that is commonly used to find Mersenne primes.

In Chap. 8, we saw that the sequences (F_n) and (L_n) can be defined by

$$F_n = \frac{\phi^n - \bar{\phi}^n}{\phi - \bar{\phi}} \quad \text{and} \quad L_n = \phi^n + \bar{\phi}^n,$$

where ϕ and $\bar{\phi}$ are the roots of $x^2 - x - 1 = 0$. In a series of papers published between 1876 and 1878 Lucas investigated the properties of a pair of functions that are more general than F_n and L_n; we denote these more general functions by U_n and V_n and with Lucas define them by

$$U_n = U_n(P, Q) = \frac{\alpha^n - \beta^n}{\alpha - \beta} \quad \text{and} \quad V_n = V_n(P, Q) = \alpha^n + \beta^n,$$

where α, β are the roots of $x^2 - Px + Q = 0$ for given coprime integers P and Q. In the interest of brevity we will often simply use the symbols U_n and V_n without including the parameters P, Q when it is understood from context what P and Q are or when we are referring to U_n and V_n generally. We now call these functions the

E. L. F. Roettger and H. C. Williams, *The Enchantment of Numbers*, CMS/CAIMS Books in Mathematics 16, https://doi.org/10.1007/978-3-031-76568-1_10

Lucas functions because, although they were known to earlier workers, Lucas was the person who first fully developed their theory and applied it to primality testing.

The notation (U_n), (V_n) is used when we are referring to the *Lucas sequences*. It is easily seen that the sequences (F_n) are (L_n) are special cases of (U_n) and (V_n), respectively, for $P = 1$ and $Q = -1$. In the same way that F_n and L_n work as a pair in many interesting and useful identities, the same is true for U_n and V_n. A handful of such results will be discussed herein; however, these remarkable functions have a rich and extensive history, so what we present below will only scratch the surface.

To begin, since α and β are the zeros of $x^2 - Px + Q$, we have $P = \alpha + \beta$ and $Q = \alpha\beta$. Thus, it is easy to see that $U_0 = 0$, $U_1 = 1$ and $V_0 = 2$, $V_1 = P$ and then by immediate consequence of the definition of the Lucas functions the following second-order recurrences can be verified by working in terms of α and β:

$$U_{n+2} = PU_{n+1} - QU_n \quad \text{and} \quad V_{n+2} = PV_{n+1} - QV_n \quad (n \geq 0).$$

Hence, both (U_n) and (V_n) are integer sequences. The initial values for the Lucas functions can be seen in the following table.

n	U_n	V_n
0	0	2
1	1	P
2	P	$P^2 - 2Q$
3	$P^2 - Q$	$P(P^2 - 3Q)$

We also point out that since

$$U_n = PU_{n-1} - QU_{n-2} \quad (n \geq 2),$$

we find that

$$U_n \equiv PU_{n-1} \equiv P^2U_{n-2} \equiv \cdots \equiv P^{n-1}U_1 \equiv P^{n-1} \pmod{Q}.$$

Thus,

$$\gcd(U_n, Q) = \gcd(P^{n-1}, Q) = 1 \quad (n > 0).$$

Similarly,

$$\gcd(V_n, Q) = 1 \quad (n > 0).$$

Lucas was able to show a surprising relationship between U_n and V_n and the sine and cosine functions; he proved that

$$U_n = \frac{2Q^{n/2}}{\sqrt{-D}} \sin\left(\frac{n\sqrt{-1}}{2}\log\left(\frac{\alpha}{\beta}\right)\right) \quad \text{and} \quad V_n = 2Q^{n/2}\cos\left(\frac{n\sqrt{-1}}{2}\log\left(\frac{\alpha}{\beta}\right)\right),$$

where $D = (\alpha - \beta)^2 = P^2 - 4Q$ (we also set $\delta = \alpha - \beta$). Notice that

$$U_{2n} = \frac{\alpha^{2n} - \beta^{2n}}{\alpha - \beta} = \frac{(\alpha^n - \beta^n)}{(\alpha - \beta)}(\alpha^n + \beta^n) = U_n V_n;$$

thus, $U_{2n} = U_n V_n$ corresponds to the trigonometric identity $\sin(2x) = \sin(x)\cos(x)$. Lucas seems to have used standard trigonometric identities to guide him in producing identities involving the Lucas sequences.

The identities involving the Fibonacci and Lucas numbers mentioned in Chap. 8 all have their analogues holding for the the Lucas sequences. For example, Cassini's identity becomes

$$U_{n-1}^2 - U_{n-2}U_n = Q^n \quad (n \geq 1).$$

Thus, $\gcd(U_n, U_{n-1}) \mid Q^n$ and since $\gcd(U_n, Q) = 1$, we find that

$$\gcd(U_n, U_{n-1}) = 1 \quad (n \geq 1).$$

We also have

$$V_{2n} = V_n^2 - 2Q^n$$

and

$$V_n^2 - DU_n^2 = 4Q^n \quad (n \geq 0).$$

The addition formulas are now:

$$U_{n+m} = U_m U_{n+1} - QU_n U_{m-1}, \quad V_{n+m} = V_m U_{n+1} - QU_n V_{m-1}.$$

Furthermore,

$$DU_n = V_{n+1} - QV_{n-1} = PV_n - 2QV_{n-1} = 2V_{n+1} - PV_n.$$
$$V_n = U_{n+1} - QU_{n-1} = PU_n - 2QU_{n-1} = 2U_{n+1} - PU_n.$$

In correspondence with the trigonometric identities

$$\sin(x + y) = \sin x \cos y + \sin y \cos x,$$
$$\cos(x + y) = \cos x \cos y - \sin x \sin y,$$

we have the *addition formulas* of Lucas:

$$2U_{m+n} = U_m V_n + V_m U_n,$$
$$2V_{m+n} = V_m V_n + DU_m U_n.$$

Lucas also derived the *multiplication formulas* for (U_n) and (V_n):

$$2^{m-1}U_{mn} = \sum_{k=0}^{\lfloor (m-1)/2 \rfloor} \binom{m}{2k+1} D^k U_n^{2k+1} V_n^{m-2k-1},$$

$$2^{m-1}V_{mn} = \sum_{k=0}^{\lfloor m/2 \rfloor} \binom{m}{2k} D^k U_n^{2k} V_n^{m-2k}.$$

Broadly speaking, Lucas' addition and multiplication formulas are identities respectively of the form $2X_{m+n} = \ldots$ and $2^{m-1}X_{mn} = \ldots$, where the right-hand sides are polynomial expressions involving the mth and nth terms of the sequences (U_n) and (V_n). There is a veritable pageant of identities involving the Lucas sequences, but these few will suffice for the purposes of this chapter. However, we should point out that the simple identity

$$U_n^2 - QU_{n-1}^2 = U_{2n-1}$$

generalizes the identity mentioned in the first example in Chap. 1.

We are now able to use the same reasoning as that employed in Chap. 8 to prove that (U_n) is a divisibility sequence, i.e. $U_n \mid U_{mn}$ whenever $n, m \geq 1$. Next suppose that p is any odd prime and for any odd n put $\epsilon(n)$ equal to the Jacobi symbol $(D|n)$. If $p \nmid Q$ and $\epsilon = \epsilon(p)$, we find from the multiplication formulas and the divisibility properties (mod p) of the binomial coefficients that

$$U_p \equiv D^{\frac{p-1}{2}} \equiv \epsilon(p) \pmod{p} \quad \text{and} \quad V_p \equiv P \pmod{p}.$$

Thus, on appealing to the identities relating V_n, U_n to V_{n-1} and U_{n-1} or V_{n+1} and U_{n+1}, we get

$$p \mid U_{p-\epsilon} \quad \text{and} \quad V_{p-\epsilon} \equiv 2Q^{\frac{1-\epsilon}{2}} \pmod{p}.$$

Thus, there must exist a least positive integer r $(= r(p))$ such that $p \mid U_r$. Also, if $p \mid U_m$, then by the methods in Chap. 8 we must have $r \mid m$. Lucas called r the "rang du apparition" (*rank of appearance*) of p in (U_n). He showed that if m is any positive integer such that $\gcd(m, Q) = 1$, then there must exist some positive integer n such that $m \mid U_n$. Thus, there must exist a least positive integer r $(r = r(m))$ such that $m \mid U_r$. He called this r the rank of appearance of m in (U_r), and showed that if $m = p^k$, then $r(m) \mid p^{k-1}(p - \epsilon(p))$. Furthermore, if $m \mid U_n$, then $r(m) \mid n$. From this it follows that if the prime power representation of m is given by

$$m = \prod_{i=1}^{j} p_i^{k_i}$$

where p_i $(i = 1, 2, \ldots, j)$ are distinct primes, then $r(m) \mid \Phi(m)$, where

$$\Phi(m) = \prod_{i=1}^{j} p_i^{k_i-1}(p_i - \epsilon(p_i)).$$

Thus, $\Phi(m)$ is the Lucas sequence analogue of Euler's $\phi(m)$. Indeed, if D is a perfect square and $\gcd(m, DQ) = 1$, then $\phi(m) = \Phi(m)$.

Now suppose we return to the problem of testing M_n for primality. Since

$$M_{2k} = (2^k + 1)(2^k - 1)$$

is composite when $k > 1$, we may assume that n is odd and $n \geq 3$. Lucas suggested putting

$$P = 2^{\frac{n+1}{2}}, \quad Q = -1, \quad \text{then} \quad D = P^2 - 4Q = 2^{n+1} + 4 = 4(2^{n-1} + 1).$$

If we define

$$r_1 = V_2 = P^2 - 2Q = 2^{n+1} + 2,$$

then $r_1 \equiv 4 \pmod{M_n}$, and if $r_i = V_{2^i}$, we have

$$r_{i+1} = r_i^2 - 2.$$

If we use $s_i = U_{2^i}$, we can now prove the following result.

Theorem 10.1 *Let $N = M_n = 2^n - 1$, where n is odd and $n \geq 3$. If none of the r_i $(i = 1, 2, \ldots, n-1)$ is divisible by N, then N is not a prime. If j is the least value of i such that $1 \leq j \leq n-1$ and $N \mid r_j$, then any prime divisor of N must be of the form $2^{j+1}k \pm 1$.*

Proof If N is a prime, then

$$\begin{aligned}\epsilon = \epsilon(N) = (D|N) &= (2^{n-1} + 1|2^n - 1) = (2^n - 1|2^{n-1} + 1) \\ &= (2(2^{n-1} + 1) - 3|2^{n-1} + 1) = (-3|2^{n-1} + 1) = (3|2^{n-1} + 1) \\ &= (2^{n-1} + 1|3) = (2|3) = -1.\end{aligned}$$

Thus, $N \mid U_{N+1}$ or $N \mid s_n$. The remainder of the proof follows along the lines of the proof of Theorem 9.1. □

Corollary 10.1 *If $N = M_n = 2^n - 1$, where n is odd and $n \geq 3$, then N is a prime if and only if $N \mid r_{n-1}$, where $r_1 = 4$ and $r_{i+1} \equiv r_i^2 - 2 \pmod{N}$.*

Proof If N is a prime, then $V_{N+1} \equiv 2Q^{\frac{1-\epsilon}{2}} \equiv -2 \pmod{N}$. The remainder of the proof follows just as the proof of Corollary 9.1. □

Corollary 10.1 is the famous Lucas-Lehmer test for the primality of Mersenne numbers. Lucas had actually assembled all the ideas needed to prove it, but it is not at all clear that he was aware of Corollary 10.1 as stated. It was not until several years later that Lehmer provided a different approach and proof of this result.

Lucas was aware that his methods for primality testing could be made more general. In fact, he provided some insight to how tests could be developed for numbers of the form

$$N = Ar^n + \gamma,$$

where r is a prime, $r \nmid A$, $\gamma \in \{1, -1\}$ and N is odd. For example, if $A = 1$, $r = 2$ and $\gamma = -1$, then N is the Mersenne number $2^n - 1$.

More precisely Lucas presented three basic tests for N, one when $r = 2$, another for $r = 3$ and a third test when $r = 5$ (In this latter test he only considered $\gamma = 1$.). We can express these tests as follows:

1. Select P, Q such that the Jacobi symbol $(D|N) = \gamma$. In Lucas' theorems, he tended to use $P = 1$, $Q = -1$, so that (U_n) is the Fibonacci sequence. This, incidentally, explains why he could only have $\gamma = 1$ when $r = 5$ because in this case we can only have $(D|N) = (5|N) = (N|5) = 1$. Nevertheless, any P, Q such that $(D|N) = \gamma$ would suffice.
2. He next defines a sequence $\{s_i\}$, where

$$s_0 = U_A(P, Q), \quad s_1 = U_{rA}(P, Q) \quad \text{for } (r = 5) \text{ or}$$
$$s_1 = U_{rA}(P, Q)/u_A(P, Q) \quad \text{for } (r = 2, 3)$$

and

$$s_{i+1} \equiv f_r(s_i) \pmod{N} \quad (i \geq 1).$$

Here $f_r(x)$ is a polynomial with integer coefficients whose degree and coefficients depend only on the value of r. For example, he used $f_2(x) = x^2 - 2$, $f_3(x) = x^3 + 3x - 3$ and $f_5(x) = x^5 + 5x^3 + 5x$.
3. The statements of the tests differed slightly for $r = 2$, $r = 3$ and $r = 5$ but in the last case he asserted that N is a prime if the first term of $\{s_i\}$ which is divisible by N is s_n, that N is composite if none of the terms $s_0, s_1, s_2, \ldots, s_n$ is divisible by N and finally if s_α ($\alpha \leq n$) is the first term of the sequence that N divides, then the prime divisors of N must have the form $k5^\alpha \pm 1$. In fact, Lucas wrote $Ak5^\alpha \pm 1$, but this cannot be justified. The test in the case that $r = 2$ is correct once the same error is corrected.

Unfortunately, all of the above tests as given by Lucas contain some errors. Further, since Lucas did not provide any proofs for his tests it is difficult to see exactly what his thinking was; however, these tests do provide valuable ideas for primality testing and have since been made rigorous with some modifications.

In his Ph.D. thesis, published in 1930, Lehmer investigated a set of functions that provided an extension to the Lucas functions. These functions are now called *Lehmer functions*, and we will denote the corresponding *Lehmer sequences* by $(\bar{U}_n)$ and $(\bar{V}_n)$. To define his functions he considered the equation $x^2 - \sqrt{R}x + Q = 0$ where R and Q are coprime integers; essentially the P from the Lucas functions is replaced by $\sqrt{R}$. In this formulation we now assign α, β to be the roots of $x^2 - \sqrt{R}x + Q = 0$ and we set

$$U_n = U_n(\sqrt{R}, Q) = \frac{\alpha^n - \beta^n}{\alpha - \beta} \quad \text{and} \quad V_n = V_n(\sqrt{R}, Q) = \alpha^n + \beta^n.$$

Under these new conditions it is no longer the case that U_n and V_n are integers for $n \geq 0$; however, when we set

$$\bar{U}_n = \bar{U}_n(R, Q) = \begin{cases} U_n/\sqrt{R}, & \text{if } 2 \mid n; \\ U_n, & \text{if } 2 \nmid n \end{cases} \quad \text{and} \quad \bar{V}_n = \bar{V}_n(R, Q) = \begin{cases} V_n, & \text{if } 2 \mid n; \\ V_n/\sqrt{R}, & \text{if } 2 \nmid n, \end{cases}$$

we then have $\bar{U}_n$ and $\bar{V}_n$ as integers for $n \geq 0$. It is also the case that now $D = R - 4Q$, $\alpha + \beta = \sqrt{R}$ and $\alpha\beta = Q$. The first few terms of $(\bar{U}_n)$ and $(\bar{V}_n)$ are provided in the following table.

n	$\bar{U}_n$	$\bar{V}_n$
0	0	2
1	1	1
2	1	$R - 2Q$
3	$R - Q$	$R - 3Q$

Both $(\bar{U}_n)$ and $(\bar{V}_n)$ satisfy the fourth order linear recurrence

$$X_{n+4} = (R - 2Q)X_{n+2} - Q^2 X_n.$$

In view of the initial values given above, we can see by induction that $\bar{U}_n$ and $\bar{V}_n$ will always be integers whenever $n \geq 0$.

We call the sequences $\bar{U} = (\bar{U}_n(R, Q))$ and $\bar{V} = (\bar{V}_n(R, Q))$ the *Lehmer sequences*. In 1976 these sequences were, in turn, extended to what are called *generalized Lehmer sequences*. Since the identity relations given earlier hold for non-integral values of P and Q, it is possible to derive several properties of the Lehmer sequences which are analogous to corresponding results concerning the Lucas sequences. For example we easily find that

$$\begin{aligned}
\bar{U}_{2n}(R, Q) &= \bar{V}_n(R, Q)\bar{U}_n(R, Q); \\
\bar{V}_{2n}(R, Q) &= \bar{V}_n(R, Q)^2 - 2Q^n, \quad \text{if } 2 \mid n; \\
\bar{V}_{2n}(R, Q) &= R\bar{V}_n(R, Q)^2 - 2Q^n, \quad \text{if } 2 \nmid n.
\end{aligned}$$

It is also easy to derive

$$R\bar{V}_n^2 - D\bar{U}_n^2 = 4Q^n,$$

whenever n is odd, and

$$\bar{V}_n^2 - D\bar{U}_n^2 = 4Q^n,$$

whenever n is even.

Lehmer also pointed out that $(\bar{U}_n)$ is a divisibility sequence and if $r = r(m)$ is the rank of appearance of the integer m in $(\bar{U}_n)$, then $r \mid n$ whenever $m \mid \bar{U}_n(R, Q)$. He also showed that if p is an odd prime and $p \nmid Q$, then

$$\bar{U}_p \equiv (D|p) \pmod{p}, \quad \bar{V}_p \equiv (R|p) \pmod{p}.$$

From these congruences, he deduced that

$$p \mid \bar{U}_{p-\epsilon\sigma}(R, Q) \quad \text{and} \quad \bar{V}_{p-\epsilon\sigma}(R, Q) \equiv 2\sigma Q^{\frac{1-\epsilon\sigma}{2}} \pmod{p},$$

where $\epsilon = (D|p)$ and $\sigma = (R|p)$. If $p \nmid RQ$, we see that $r(p) \mid p - \epsilon\sigma$.

Furthermore, Lehmer proved the following theorem.

Theorem 10.2 *Define $\tau = (Q|p)$. If p is an odd prime and $p \nmid QRD$, then $p \mid V_{\frac{p-\epsilon\sigma}{2}}$ if and only if $\sigma = -\tau$.*

If we put $R = 2$ and $Q = -1$, then $D = R - 4Q = 6$. If n is odd, $n \geq 3$ and M_n is a prime, then

$$\epsilon = \epsilon(M_n) = (6|M_n) = (3|M_n) = -(M_n|3) = -1,$$
$$\sigma = (2|M_n) = 1 \quad \text{and} \quad \tau = (-1|M_n) = -1.$$

Thus, by applying the same ideas as in the proof of Theorem 10.1 to the Lehmer sequences, Lehmer produced another proof of Corollary 10.1. This test, because of its speed and easy applicability, continues being used to find the largest known primes. Indeed, except for a brief period between 1989 and 1992, the largest known prime, since 1876, has always been a Mersenne prime. All these numbers larger than M_{127} were proved prime by the Lucas-Lehmer test. A table of these primes can be found at the GIMPS website mentioned below.

If we compare the tests given in Corollaries 9.1 and 10.1, we see that they are the same except for the choice of r_1 (the seed). Lehmer proved that there exists an infinitude of choices other than 4 for r_1 such that Corollary 10.1 holds. For example, $r_1 = 10, 52, 724, 970, \ldots$ will also all work. Notice that by Corollary 9.1 the seed $r_1 = 3$ can be used for 1/2 of all the odd values of n ($n \equiv 3 \pmod{4}$). It is an entertaining exercise to find other values of r_1 such that this is the case.

In 1996 George Woltman established the Great Internet Mersenne Prime Search (GIMPS) website; this is a distributed computing project in which anyone with access to computing power can participate. Since the website's inception 18 Mersenne primes have been identified and all prime Mersenne exponents less than 124399361 have been tested at least once.

Problems

10.1 (Lucas sequences) The Lucas functions $U_n = U_n(P, Q)$ and $V_n = V_n(P, Q)$ can be defined by the second order recurrence

$$U_{n+2} = PU_{n+1} - QU_n \quad \text{and} \quad V_{n+2} = PV_{n+1} - QV_n \quad (n \geq 0),$$

where $U_0 = 0$, $U_1 = 1$, $V_0 = 2$, and $V_1 = P$. Find the first 15 terms of the following Lucas sequences (U_n) and (V_n) for:

(a) **[BB]** $P = 3, Q = 2$
(b) $P = -1, Q = 4$
(c) $P = 5, Q = 7$

10.2 (Addition formulas) Verify the the addition formulas

$$U_{n+m} = U_m U_{n+1} - QU_n U_{m-1}, \quad V_{n+m} = V_m U_{n+1} - QU_n V_{m-1}$$

for the particular cases of:

(a) **[BB]** $P = 3, Q = 2, n = 2, m = 3$
(b) $P = -1, Q = 4, n = 1, m = 4$
(c) $P = 5, Q = 7, n = 3, m = 5$

10.3 (Addition formulas of Lucas) Verify the addition formulas

$$2U_{m+n} = U_m V_n + V_m U_n, \quad 2V_{m+n} = V_m V_n + DU_m U_n$$

for the particular cases of:
(a) **[BB]** $P = 3, Q = 2, n = 2, m = 3$
(b) $P = -1, Q = 4, n = 1, m = 4$
(c) $P = 5, Q = 7, n = 3, m = 5$

10.4 Verify that $p \mid U_{p-\epsilon(p)}$, where $\epsilon(p) = (D|p)$, for the particular cases of:

(a) **[BB]** $P = 3, Q = 2, p = 5$
(b) $P = -1, Q = 4, p = 7$
(c) $P = 5, Q = 7, p = 11$

10.5 Verify that $V_{p-\epsilon(p)} \equiv 2Q^{\frac{1-\epsilon(p)}{2}} \pmod{p}$, where $\epsilon(p) = (D|p)$, for the particular cases of:
(a) **[BB]** $P = 3, Q = 2, p = 5$
(b) $P = -1, Q = 4, p = 7$
(c) $P = 5, Q = 7, p = 11$

10.6 [BB] Use Corollary 10.1 (the Lucas-Lehmer test) to show that $2^{13} - 1$ is a prime.

10.7 Use Corollary 10.1 (the Lucas-Lehmer test) to show that $2^{17} - 1$ is a prime.

10.8 Use Corollary 10.1 (the Lucas-Lehmer test) to show that $2^{33} - 1$ is not a prime.

Further Reading

Mersenne primes

Samuel S. Wagstaff, Jr., Divisors of Mersenne numbers, *Mathematics of Computation*, **40** (1983), 385–397.

The Lucas sequences

Christian J.-C. Ballot and Hugh C. Williams, *The Lucas Sequences: Theory and Applications*, CMS/CAIMS Books in Mathematics, Vol. 8, Springer Nature Switzerland AG, 2023, pp. 11–65.
Paulo Ribenboim, *The Book of Prime Number Records*, 2nd Edition, Springer-Verlag, NY, 1989, pp. 41–70.
H. C. Williams, *Édouard Lucas and Primality Testing*, Wiley-Interscience, John Wiley and Sons, NY, 1998, pp. 69–93.

Lucas' primality tests

E. Lucas, Théorie des fonctions numériques simplement périodiques, *American Journal of Mathematics*, **1** (1878), 189-240, 289-321.
E. L. Roettger and H. C. Williams and R. K. Guy, Some Primality Tests that Eluded Lucas, *Designs, Codes and Cryptography*, **77** (2015), 515–539.

Lehmer functions

D. H. Lehmer, An extended theory of the Lucas' functions, *Annals of Mathematics*, **31** (1930), 419–448.

Generalized Lehmer sequences

H. C. Williams, A generalization of Lehmer's functions, *Acta Arithmetica*, **29** (1976), 315–341.

GIMPS

https://www.mersenne.org

History

Constance Reid, *From Zero to Infinity*, 5th ed., A. K. Peters, Wellesley, MA 2006, pp. 85–98. This is a wonderful book for a novice and is highly recommended.

H. C. Williams, *Édouard Lucas and Primality Testing*, Wiley-Interscience, John Wiley and Sons, NY, 1998, pp. 107–117, 145–156, 196–201, 336–338.

Chapter 11
Primality

This chapter is devoted to the problem of determining when a given odd positive integer N is a prime. This question goes back to antiquity and is today a favourite activity of recreational number theorists. In the previous chapter we derived a simple necessary and sufficient test for the primality of Mersenne numbers, and it would be reasonable to wonder whether a necessary and sufficient test for the primality of N exists for any N. In fact, such a test exists and is due to the English mathematician and judge John Wilson (1741–1793) in 1770.

Theorem 11.1 *The integer N (> 1) is a prime if and only if N divides* $(N-1)!+1$.

Proof The theorem is certainly true for $N \le 5$, so we may assume that $N > 5$. Suppose N is a prime. If a is any element of the RSR (mod N) such that $1 < a < N-1$, then by the EEA we know there must exist a^{-1} (mod N), where a^{-1} is in the RSR (mod N) and

$$aa^{-1} \equiv 1 \pmod{N}.$$

Also $a \ne a^{-1}$, for if this were the case $a^2 \equiv 1 \pmod{N}$, which is impossible by selection of a. It follows that the $N-3$ integers $\not\equiv \pm 1 \pmod{N}$ in the RSR (mod N) can be arranged into $(N-3)/2$ unique pairs (a, a^{-1}); hence

$$(N-2)! \equiv \prod_{i=2}^{N-2} i \equiv \prod_{i=2}^{(N-3)/2} aa^{-1} \equiv 1 \pmod{N}$$

and therefore

$$(N-1)! \equiv N-1 \equiv -1 \pmod{N}.$$

Next, suppose N is not a prime and $N \mid (N-1)!+1$. In this case we can write $N = ab$, where $1 < a, b < N$. Since $1 < a < N$, we must have $a \mid (N-1)!$ as a

E. L. F. Roettger and H. C. Williams, *The Enchantment of Numbers*, CMS/CAIMS Books in Mathematics 16, https://doi.org/10.1007/978-3-031-76568-1_11

must appear among the integers whose product is $(N-1)!$. Since $a \mid (N-1)!+1$ and $a \mid (N-1)!$ we must have $a \mid 1$, which is impossible. □

At first glance, Wilson's theorem appears to give us a very convenient test for primality. Unfortunately, it is not useful in primality testing because of the computational difficulty in finding $(N-1)! \pmod N$.

In Chap. 2, we mentioned that a simple method for testing an integer N for primality is to trial divide it by all the primes less than or equal to $\sqrt{N}$. To analyse the efficiency of this process, it is necessary to have some estimate of how many primes are $\leq \sqrt{N}$. This is supplied by the Prime Number Theorem (PNT), which provides the asymptotic distribution of the prime numbers. We say that two functions $f(x)$ and $g(x)$ are asymptotically equivalent if

$$\lim_{x\to\infty} \frac{f(x)}{g(x)} = 1,$$

and we denote this by

$$f(x) \sim g(x).$$

The exact statement of the PNT is somewhat technical, but for our purposes it states that if $\pi(x)$ denotes the number of primes less than x, then

$$\pi(x) \sim \frac{x}{\log x}.$$

Furthermore, thanks to the work of J. Barkley Rosser (1907–1989) and Lowell Schoenfeld (1920–2002), we know that

$$\frac{x}{\log x} < \pi(x) < 1.25506\frac{x}{\log x}.$$

The constant 1.25506 is a five decimal approximation of $\frac{30\log 113}{113}$, as $\frac{\pi(x)\log x}{x}$ has its maximum value at $x = 113$ and $\pi(113) = 30$.

A more precise version of the PNT asserts that

$$\pi(x) \sim \mathrm{li}(x),$$

where $\mathrm{li}(x)$ is the *logarithmic integral function* defined for all real $x \neq 1$ by

$$\mathrm{li}(x) = \int_0^x \frac{dt}{\log t}.$$

As the function $\frac{dt}{\log t}$ has a singularity at $t = 1$, we often write

$$\operatorname{li}(x) = \lim_{\epsilon \to 0^+} \left(\int_0^{1-\epsilon} \frac{dt}{\log t} + \int_{1+\epsilon}^{x} \frac{dt}{\log t} \right)$$

for $x > 1$. We can relate $\frac{x}{\log x}$ to $\operatorname{li}(x)$ through the asymptotic expansion

$$\operatorname{li}(x) = \frac{x}{\log x} \left(\sum_{k=0}^{\infty} \frac{k!}{(\log x)^k} \right).$$

If we write

$$\operatorname{Li}(x) = \int_2^x \frac{dt}{logt},$$

we can avoid the singularity at $t = 1$. This *Eulerian logarithmic integral* is related to $\operatorname{li}(x)$ by

$$\operatorname{li}(x) = \operatorname{Li}(x) + \operatorname{li}(2),$$

where $\operatorname{li}(2) = 1.045163780\ldots$.

The values of $\pi(10^n)$ have been tabulated for all positive integers $n \leq 29$. While this may seem a rather modest achievement, a great deal of work and ingenuity has gone into finding these results. They can be found in A006880 of the OEIS. To get some idea of how close $\operatorname{li}(x)$ is to $\pi(x)$, consult the table of values of

$$\lfloor \operatorname{li}(10^n) \rfloor - \pi(10^n)$$

in A057752 of the OEIS. Notice that it always seems that

$$\operatorname{li}(x) > \pi(x),$$

but this is not always the case. Indeed, it was shown in 1914 that $\operatorname{li}(x) < \pi(x)$ must occur infinitely often. In 1933 the South African mathematician Stanley Skewes (1899–1988) used the Riemann hypothesis (RH) (see Chap. 21) to show that

$$\pi(x) > \operatorname{li}(x)$$

for some x such that

$$x < 10^{10^{10^{34}}}.$$

As late as 1973, this number, now called *Skewes' number*, was thought to be the largest integer that has ever appeared in a mathematical proof, even though in 1955 Skewes showed without recourse to the RH that the sign of $\pi(x) - \operatorname{li}(x)$ must change for some

$$x < 10^{10^{10^{963}}}.$$

Now, however, the large numbers mentioned in Chap. 3 are inconceivably larger than either of Skewes' numbers. Today we also know

$$\pi(x) > \mathrm{li}(x)$$

for some $x < 1.4 \times 10^{316}$, a much lower limit than Skewes' number.

As calculations have revealed, the approximation of $\mathrm{li}(x)$ to $\pi(x)$ is actually very good. Indeed, if the RH holds, it has been shown that

$$\pi(x) = \mathrm{li}(x) + O(\sqrt{x}\log x).$$

An effective version of this result, proved by Lowell Schoenfeld under the RH in 1976, is

$$|\pi(x) - \mathrm{li}(x)| < \frac{\sqrt{x}\log x}{8}$$

for $x \geq 2657$.

Suppose a and d are positive integers such that $\gcd(a, d) = 1$. If we let $\pi_{d,a}(x)$ denote the number of primes of the form $a + dn$ less than x, then a more general version of the PNT, due to Dirichlet, asserts that

$$\pi_{d,a}(x) \sim \frac{1}{\phi(d)}\frac{x}{\log x}.$$

This result, while proved many years earlier in 1896, is a consequence of an even more general result of Nikolaĭ Chebotarëv (1894–1947) in 1923.

From the PNT it follows that the number of primes $\leq \sqrt{N}$ is approximately

$$\frac{\sqrt{N}}{\log\sqrt{N}} = \frac{2\sqrt{N}}{\log N}.$$

When N is even a moderately large number, say 20 digits, this is approximately

$$\frac{2 \cdot 10^{10}}{20\log 10} = \frac{10^9}{\log 10} \approx 4.34 \times 10^8,$$

which represents a lot of trial divisions. If N is a much larger number, the number of trial divisions becomes prohibitively large, rendering this test for primality essentially useless (even on modern computers) for numbers of 60 digits or more. Nevertheless, it is frequently employed to eliminate candidates for primality for the simple reason that if N is not a prime, it is very likely to be divisible by a small prime. However, it is still necessary to store all the small primes in a computer, and this is often impractical.

We now illustrate a method of circumventing this difficulty. We let

$$d_1, d_2, \ldots, d_m$$

be the $m = 5760 = \phi(30030)$ positive integers which are less than $30030 = 2 \cdot 3 \cdot 5 \cdot 7 \cdot 11 \cdot 13$ and relatively prime to it. Note that $d_1 = 1$ and $d_m = 30030 - 1$. If N has a prime divisor, it is either 2, 3, 5, 7, 11, or 13 or it must have the form

$$p(k, i) = 30030k + d_i.$$

Putting $\delta_i = d_{i+1} - d_i$, we see that

$$\begin{aligned} p(0, 1) &= d_1 = 1, \\ p(k, i + 1) &= p(k, i) + \delta_i, \\ p(k + 1, 1) &= p(k, m) + 2. \end{aligned}$$

Thus, to find all the prime divisors of N less than a given bound $\mathcal{B}$, we need only trial divide N by all $p(k, i) < \mathcal{B}$. These methods are called *wheel* techniques. Each time k is increased by 1, we say (in this case) the 5760 spoke wheel has been turned. Naturally, these ideas can be extended to larger moduli than 30030.

Suppose now we consider the integer 349 and observe that

$$349 = 910 - 561 = 2 \cdot 5 \cdot 7 \cdot 13 - 3 \cdot 11 \cdot 17.$$

Since $349 < 361 = 19^2$ and each prime less than 19 occurs in exactly one of the two terms above, we see immediately that 349 must be a prime. We say that 349 is a *prime at a glance*. More generally, suppose for a given N we have integers B and L subject to the following four conditions:

(i) $N = B - L$,
(ii) $B > |L|$,
(iii) $\gcd(B, L) = 1$,
(iv) $p \mid BL$ for all primes $p \leq \sqrt{N}$.

Theorem 11.2 *If $N > 1$, then N is a prime if and only if integers B and L exist such that conditions (i)–(iv) hold.*

Proof Certainly, if $N > 1$ and satisfies the above conditions, then N must be a prime as it has no prime divisors $p \leq \sqrt{N}$. Also, if N is a prime, then split the product of all primes less than $\sqrt{N}$ into two factors u and v with $\gcd(u, v) = 1$. Note that $\gcd(N, uv) = 1$. We can use the EEA to find x and y such that $x > 0$ and

$$xu - yv = 1.$$

Put $B = u(Nx + v)$, $L = v(Ny + u)$. We have $N = B - L$ and since $\gcd(N, B) = \gcd(N, uv) = 1$, we have $\gcd(B, L) = 1$. Thus conditions (i)–(iv) must all hold.

□

With the above ideas in mind, we are now able to prove a simple primality test for N. We let p_k be any kth prime such that

$$p_k < N < p_{k+1}^2$$

and put $M = p_1 p_2 \cdots p_k$ with $M = qN + r$, where $0 \le r < N$. If $r = 0$, then N must be composite and have no prime divisor exceeding p_k. If $r \ne 0$, let $g = \gcd(N, r)$. Notice that $g = 1$ if and only if N is a prime.

This rather elegant test for primality is, unfortunately, not very practical for large N because M becomes prohibitively large as N increases due to the PNT. However, for small values of N, it can often be useful.

For larger values of N, we can make use of the *Fermat congruence*

$$a^N \equiv a \pmod{N},$$

which by Corollary 4.1 must hold for prime values of N and any integer a. Thus, if the Fermat congruence does not hold for a given N and some a, then N cannot be a prime. Unfortunately for primality proving, it is possible for a composite value of N to satisfy the Fermat congruence for some a. For example, consider $a = 2$ and $N = 2701 = 37 \cdot 73$, an example known to Lucas. Thus, a direct converse of Fermat's theorem does not exist. Indeed, there exists an infinitude of composite values of N, called *Carmichael numbers*, for which the Fermat congruence holds for all values of a such that $\gcd(a, N) = 1$. If, for a given N, we concern ourselves with values of a such that $\gcd(a, N) = 1$, the Fermat congruence can be written as

$$a^{N-1} \equiv 1 \pmod{N}.$$

Any integer N which satisfies such a congruence is called a *base-a probable prime* (PRP).

It was Lucas who found a converse of Fermat's theorem, and subject to an observation of Lehmer, we can write it as follows.

Theorem 11.3 *Suppose N is odd. If for some a we have $a^{N-1} \equiv 1 \pmod{N}$ and*

$$a^{(N-1)/p} \not\equiv 1 \pmod{N}$$

for each distinct prime divisor p of $N - 1$, then N is a prime.

Application of this result requires us to know the complete factorization of $N - 1$, which is often difficult to acquire for a large value of N. In 1975 John Brillhart (1930–2022), Lehmer and Selfridge developed algorithms which required only the partial factorization of $N - 1$ and $N + 1$, but even these techniques quickly reached their limit of applicability. Since 1975 a variety of primality testing algorithms have been proposed, many of them suggesting that there might be a polynomial time technique for verifying primality. This was finally provided in 2002 when three Indian computer scientists, Manindra Agrawal, Neeraj Kayal and Nitin Saxena announced their deterministic polynomial time algorithm for primality proving. Subsequent to this Hendrik Lenstra, Jr. and Carl Pomerance proved that a modification of this method could establish primality in $\tilde{O}((\log N)^6)$ elementary operations.

Although this work is of great theoretical importance, it is still not as useful in practical primality testing as might be hoped. For large N, $(\log N)^6$ still represents a lot of operations; as a result of this, most primality proving of large N today is done by using a technique referred to as fastECPP or fast elliptic curve primality proving. Although it is difficult to establish the computational complexity of this procedure, it is usually very fast, can be applied to any N and has been used to prove the primality of numbers up to 10000 digits.

It is still unknown what the true complexity of primality testing is, but we can get some sense of what it might be by making use of an idea of Selfridge and Peter Weinberger. We define the *pseudosquare* L_p (p a prime) as follows:

(a) $L_p \equiv 1 \pmod 8$;
(b) The Legendre symbol $(L_p|q) = 1$ for all primes $q \leq p$;
(c) L_p is the least positive non-square integer satisfying (a) and (b).

We will discuss pseudosquares further in Chap. 16. Selfridge and Weinberger showed that if N satisfies

(i) $N < L_p$;
(ii) $p_i^{(N-1)/2} \equiv \pm 1 \pmod N$ for all primes p_i such that $2 \leq p_i \leq p$;
(iii) $2^{(N-1)/2} \equiv -1 \pmod N$ when $N \equiv 5 \pmod 8$ or
$2^{(N-1)/2} \equiv 1 \pmod N$ when $N \equiv 1 \pmod 8$;

then N must be a prime or be a power of a prime. Notice that if N is a prime, then both (ii) and (iii) must hold. Also, we know by a result of Dan Bernstein that it is possible to determine that N is not a perfect integral power in time $\tilde{O}(\log N)$.

It can be shown under the GRH that

$$L_p > \sqrt{p}/2,$$

but by making use of heuristic reasoning it has been shown that we would expect

$$L_p \sim c 2^{p/\log p} \log p$$

for a constant $c = 3.56214\ldots$. If we let n denote the value of $\pi(p)$, it can be deduced from this and the PNT that

$$n = \tilde{O}(\log N).$$

To perform step (ii) we need to compute n modular exponentiations (modulo N) and only one modular exponentiation to perform step (iii). Thus, since the cost of a modular exponentiation is $\tilde{O}((\log N)^2)$, the time complexity of executing the Selfridge-Weinberger test is $\tilde{O}((\log N)^3)$, assuming that we are given a value of L_p such that (i) holds. Since the complexity of performing a Lucas-Lehmer primality test is $\tilde{O}((\log N)^2)$, the complexity measure of $\tilde{O}((\log N)^3)$ may be best possible for primality proving of a general N. Indeed, some empirical (numerical) work suggests that

$$\log L_p > \frac{p \log 2}{\log p}.$$

If this can be proved, we can dispense with the costly process of computing the table of L_p in order to perform the Selfridge-Weinberger test. For if we select n such that

$$n = \pi(p) > \frac{1.25506 \log N}{\log 2},$$

then since

$$n < \frac{1.25506p}{\log p},$$

we have

$$\log N < \frac{n \log 2}{1.25506} < \frac{p \log 2}{\log p} < \log L_p$$

and we then know that $N < L_p$.

Modern primality proving techniques, while powerful, are mathematically sophisticated and well beyond the scope of this work. Thus, we will confine our attention in what follows to numbers of special forms, like the Mersenne numbers, for which we can find relatively simple primality tests. We have already mentioned that Lucas dealt with numbers of the form

$$N = Ar^n + \gamma,$$

where r is a prime and $\gamma \in \{1, -1\}$, but his tests were only sufficient for N to be prime. For the special case of

$$N = A2^n - 1$$

Lehmer proved the following necessary and sufficient test.

Theorem 11.4 *Suppose $N = A2^n - 1$, where $n > 1$, A is odd and $A < 2^n$. If $(D|N) = (Q|N) = -1$, then N is a prime if and only if $N \mid V_{\frac{N+1}{2}}$.*

This result, however, is not effective in that it does not provide an explicit recipe for finding the needed values of P and Q. However, in the simple case where $3 \nmid AN$ and $n \geq 3$, we can always use $P = -2$, $Q = N - 2$. When $3 \mid A$, the problem of finding values for P and Q becomes more difficult, but it can be handled as long as $A \neq 4^m - 1$.

When $r = 3$, we have a result similar to Theorem 11.4.

Theorem 11.5 *Let $N = A3^n + \gamma$, where $3 \nmid A$, $n \geq 1$, $A < 3^n$ and $\gamma \in \{1, -1\}$. Suppose q is a prime such that $q \equiv 1 \pmod 3$ and $N^{\frac{q-1}{3}} \not\equiv 1 \pmod q$. Let $4q = t^2 + 27u^2$ where $t \equiv 1 \pmod 3$. If $\gcd(N, qt) = 1$, then N is a prime if and only if*

$$V_{2(N-\gamma)/3}(t, q) \equiv -q^{(N-\gamma)/3} \pmod N.$$

When q is a prime such that $q \equiv 1 \pmod 3$, it can be shown that values for t and u always exist. For example, if $q = 67$, then

$$4 \cdot 67 = (-5)^2 + 27 \cdot 3^2.$$

By making use of generalized Lehmer sequences, further results, similar to Theorem 11.4, can also be produced for a general odd prime r. Like Theorem 11.5 these results are not effective, as they require the determination of a prime q such that $q \equiv 1 \pmod r$ and

$$N^{(q-1)/r} \not\equiv 1 \pmod q.$$

While in practice, it is not difficult to find such a q, it still remains an unsolved problem to produce an efficient and effective algorithm to do it.

As an example of a special case of Theorem 11.5 consider $N_n = 2 \cdot 3^n - 1$ ($A = 2$, $\gamma = -1$) and put $Q = q = 7$, $P = t = 1$. We first observe that if $n \equiv 4 \pmod 6$, then $7 \mid N_n$ and N_n cannot be prime. If we put

$$W_n \equiv V_{2n}(Q^{-1})^n \equiv V_{2n}Q^{-n} \pmod{N_n},$$

it is easy to prove from the Lucas sequence identity

$$V_{3n} = V_n^3 - 3Q^n V_n$$

that

$$W_{3n} \equiv W_n^3 - 3W_n \pmod N.$$

By the theorem, if

$$N_n^2 = N_n^{(q-1)/3} \not\equiv 1 \pmod{N_n},$$

then $N = N_n$ is a prime if and only if

$$W_{3^{n-1}} \equiv -1 \pmod N.$$

Putting

$$r_0 \equiv W_1 \equiv V_2/Q \equiv Q^{-1} - 2 \equiv 7^{-1} - 2 \pmod N$$

and

$$r_{j+1} \equiv r_j(r_j^2 - 3) \pmod N,$$

we see that

$$r_i \equiv W_{3^i} \pmod N.$$

We now have the following result:

Corollary 11.1 *If* $n \equiv a \pmod 6$*, where* $a = 1, 2, 3, 5$ *and* r_0 *is selected from the table below, then* $N = 2 \cdot 3^n - 1$ *is a prime if and only if*

$$r_{n-1} \equiv -1 \pmod N,$$

where

$$r_{i+1} \equiv r_i(r_i^2 - 3) \pmod N.$$

a	1	2	3	5
r_0	$4(N-5)/7+1$	$2(N-3)/7-1$	$5(N-4)/7+1$	$3(N-2)/7-1$

Thus, if $6 \nmid n$, we get a result concerning the primality of $2 \cdot 3^n - 1$ that is similar to the Lucas-Lehmer test (Corollary 10.1) for the primality of $2^n - 1$.

In spite of the theoretically difficult problem of finding a suitable q, it is often easy to solve the problem for certain $N = Ar^n + \gamma$. For example, consider

$$N = 4 \cdot 3^n - 1.$$

It is an entertaining exercise to show that if $C = \{7, 13\}$, then

$$N^{(q-1)/3} \not\equiv 1 \pmod q$$

for some $q \in C$. For a fixed r and γ a set like C always seems to exist for most values of A. Indeed, it is often possible to enlist the Lucas sequences to produce effective necessary and sufficient test for the primality of N of certain forms. For example, this is always possible for a generalization $N = (r-1)r^n - 1$ of the Mersenne numbers when $2 \le r < 10^7$, but not for $r = 3$. This also applies to certain primes of the form $2r^n - 1$, but as we have seen, such a test exists when $r = 3$ as long as $6 \nmid n$. However, there are primes of the form $2 \cdot 3^n - 1$ when $6 \mid n$; for example, $2 \cdot 3^{12} - 1 = 1062881$ is such a prime. It seems to be very difficult to find a small set S of primes, each congruent to 1 modulo 3, like C for $N = 4 \cdot 3^n - 1$, such that for any n

$$(2 \cdot 3^n - 1)^{(q-1)/3} \not\equiv 1 \pmod q$$

for some q in S. However, we can show that there exists such a set containing 71 primes, each ≤ 11953, as long as $k \nmid n$, where

$$k = 2^5 \cdot 3^5 \cdot 5^2 \cdot 7 \cdot (\#97) > 3 \times 10^{42}$$

(recall from Chap. 2 page 9 that #97 is the primorial of 97).

We have examined the primality of numbers of the form $2^n - 1$, but what about numbers of the form $2^n + 1$? For such numbers we see that $3 \mid 2^n + 1$ when n is odd. More generally, if $n = 2^k m$, where m is odd, then

$$2^{2^k} + 1 \mid 2^n + 1.$$

Thus, if $2^n + 1$ is to be a prime, we must have $n = 2^k$. In an unfortunate coincidence of notation, number theorists use the same symbol F_k to denote the *Fermat number*

$$F_k = 2^{2^k} + 1$$

and the kth Fibonacci number. To avoid confusion, we will restrict our use of F_k for the kth Fermat number in this chapter. In subsequent chapters, it should be clear from context what F_k denotes. It was the French polymath Pierre de Fermat (1607–1665) who discovered the Fermat numbers; he also conjectured that all such numbers were prime. It was not until 1733 when Euler discovered that $641 \mid F_5$ that this conjecture was disproved. Today it is widely believed (but not yet proved) that the only prime values of F_k occur for $k = 0, 1, 2, 3, 4$. There is a simple test for the primality of F_k due to the French mathematician Father Théophile Pepin S. J. (1826–1904):

$$F_k \text{ is a prime if and only if } 3^{(F_k-1)/2} \equiv -1 \pmod{F_k}.$$

Unfortunately, it is difficult to apply this very simple test for even modestly sized values of k because of the very rapid growth of F_k. It was Lucas who proved that if $k \geq 2$ and p is a prime divisor of F_k, then p must have the form $m2^{k+2} + 1$. The largest Fermat number known to be composite is $F_{18233954}$, which is divisible by the prime $7 \cdot 2^{18233956} + 1$.

You might well wonder how this result was obtained, given that $F_{18233954}$ is a number of over $10^{5488967}$ decimal digits, and therefore impossible to write down, let alone try to divide by another. However, if for a given m we define $r_1 = 2^2 = 4$ and define the sequence (r_i) by

$$r_{i+1} \equiv r_i^2 \pmod{m},$$

then by induction it is easy to show that

$$r_j \equiv 2^{2^j} \pmod{m};$$

hence,

$$r_n \equiv 2^{2^n} \equiv F_n - 1 \pmod{m}.$$

Thus, we can show that m divides F_n simply by computing r_n modulo m in n squaring operations and then determining that

$$r_n \equiv -1 \pmod{m}.$$

Numbers of the form $m2^n + 1$ can be proved prime by the following test:
Suppose $N = A2^n + 1$ where A is odd, $A < 2^n$. If a is some integer such that the

Jacobi symbol $(a|N) = -1$, *then* N *is a prime if and only if*

$$a^{\frac{N-1}{2}} \equiv -1 \pmod{N}.$$

This result is due to the French number enthusiast François Proth (1852–1879).

Suppose we now change our point of view and try to find values of A such that $A2^n + 1$ can never be a prime for any n. Such values of A are called Sierpiński numbers for the Polish mathematician Wacław Sierpiński (1882–1969), who in 1960 showed that there exists an infinitude of such numbers. In 1962 Selfridge showed that 78557 is a Sierpiński number and it has been conjectured that it is the least such number. This is likely a problem that can only be solved by using computers. At the time of writing only five candidates for A remain as possible counterexamples to this conjecture; these are $A = 21181$, 22699, 24737, 55459 and 67607. Before embarking on a quest to eliminate any of these numbers, the reader should be aware that a recently eliminated candidate is 10223 which was only possible when it was discovered that

$$10223 \cdot 2^{31172165} + 1,$$

a number of over 9 million digits, is a prime. All numbers of the form $A2^n + 1$ for each of the 5 remaining possible values of A have been tested for all $n \leq 37408811$ and no prime value of $A2^n + 1$ has been found.

In the next chapter we will discuss other special prime numbers, some of which have efficient primality proofs.

Problems

11.1 (Wilson's theorem) Use Wilson's theorem (Theorem 11.1) to test the primality of the following integers:

(a) **[BB]** $N = 7$
(b) $N = 9$
(c) $N = 11$
(d) $N = 13$

11.2 What potential problems would make using Wilson's theorem not practical for testing the primality of $N = 7919$.

11.3 (Prime Number Theorem) Verify that

$$\frac{x}{\log x} < \pi(x) < 1.25506\frac{x}{\log x},$$

where log is the natural logarithm, for:

(a) **[BB]** $x = 20$
(b) $x = 30$
(c) $x = 50$
(d) $x = 100$

11.4 (Wheel techniques) We let

d_1	d_2	d_3	d_4	d_5	d_6	d_7	d_8
1	7	11	13	17	19	23	29

be the $m = 8 = \phi(30)$ positive integers less than $30 = 2 \cdot 3 \cdot 5$ and relatively prime to it. If N has a prime divisor, it is either 2, 3, 5 or it must have the form

$$p(k, i) = 30k + d_i.$$

Find all the prime divisors of the given N below by trial dividing by 2, 3, 5 and all $p(k, i) < \sqrt{N}$ for:

(a) **[BB]** $N = 1891$
(b) $N = 101$
(c) $N = 137$
(d) $N = 2773$

11.5 (Prime at a glance) Use Theorem 11.2 to generate a prime at a glance using the given maximal prime p (solutions not unique):

(a) **[BB]** $p = 11$
(b) $p = 13$
(c) $p = 19$
(d) $p = 23$

11.6 (Converse of Fermat's theorem) Use Theorem 11.3 to test the following N for primality:

(a) **[BB]** $p = 31$, use $a = 3$
(b) $p = 41$, use $a = 6$
(c) $p = 59$, use $a = 2$
(d) $p = 101$ use $a = 2$

11.7 (Fermat primes) For the Fermat number $F_k = 2^{2^k} + 1$ it is known that F_k is a prime if and only if $3^{(F_k-1)/2} \equiv -1 \pmod{F_k}$. Use this to verify the primality of:

(a) **[BB]** F_2
(b) F_3

Further Reading

Primality testing (methods and ideas)

Richard Crandall and Carl Pomerance, *Prime Numbers: A Computational Perspective*, Springer-Verlag, NY, 2001.

Carl Pomerance, Primality testing: variations on a theme of Lucas, *Congressus Numerantium*, **201** (2010), 301–312.

Primality testing (history)

H. C. Williams, *Édouard Lucas and Primality Testing*, Wiley-Interscience, John Wiley and Sons, NY, 1998.

The prime number theorem

Marcus du Sautoy, *The Music of the Primes*, HarperCollins, NY, 2003, pp. 46–101.

Tom M. Apostol, *Introduction to Analytic Number Theory*, Springer-Verlag, NY, 1976, pp. 146–156.

Peter Stevenhagen and H. W. Lenstra, Jr., Chebotarëv and his density theorem, *Mathematical Intelligencer*, **18** (1996), 26–37.

Estimates for $\pi(x)$

Christian Axler, Estimates for $\pi(x)$ for large values of x and Ramanujan's prime counting inequality, *Integers*, **18** (2018), #A61.

The wheel

M. C. Wunderlich and J. L. Selfridge, A design for a number theory package with an optimized trial division routine, *Communications of the ACM*, **17** (1974), 272–276.

Primes at a glance

R. K. Guy, C. B. Lacampagne and J. L. Selfridge, Primes at a glance, *Mathematics of Computation*, **48** (1987), 183–202.

Carmichael numbers

Richard Crandall and Carl Pomerance, *Prime Numbers: A Computational Perspective*, Springer-Verlag, NY, 2001, pp. 121-123.

The AKS primality test

Martin Dietzfelbinger, *Primality Testing in Polynomial Time*, Springer-Verlag, Berlin, 2004.
Andrew Granville, It is easy to determine whether a given integer is prime, *Bulletin of the AMS*, **42** (2004), 3–38.

The Selfridge-Weinberger test

Johannes Buchmann, Michael J. Jacobson, Jr. and Hugh C. Williams, The influence of public-key cryptography on mathematics, *Democratizing Cryptography: the Work of Whitfield Diffie and Martin Hellman*,Rebecca Slayton Editor, ACM, 2022, pp. 85–87.

Primality proving for large integers

Preda Mihăilescu, Recent developments in primality proving, *Mathematics and Computers in Simulation*, **49** (1999), 193–204.
François Morain, Implementing the asymptotically fast version of the elliptic curve primality proving algorithm, *Mathematics of Computation*, **76** (2007), 493–505.

Primes of special forms

H. C. Williams, *Édouard Lucas and Primality Testing*, Wiley-Interscience, John Wiley and Sons, NY, 1998, pp. 423–428.

Generalized Mersenne numbers

Andreas Stein and H. C. Williams, Explicit primality criteria for $(p-1)p^n-1$, *Mathematics of Computation*, **69** (2000), 1721–1734.
E. L. Roettger and H. C. Williams, Some primality conditions for $N = 2p^n - 1$, *Integers*, **22** (2022), #A76.

Fermat numbers

Michal Křížek, Florian Luca and Lawrence Somer, *17 Lectures on Fermat Numbers: From number theory to Geometry*, Springer-Verlag, NY, 2001.
Wilfrid Keller, Prime Factors of Fermat Numbers, http://www.prothsearch.com/fermat.html

Sierpiński numbers

Richard K. Guy, *Unsolved Problems in Number Theory*, 3rd Edition, Springer, NY, 2004, pp. 119–121.
Prime pages, https://t5k.org/glossary/page.php?sort=SierpinskiNumber

Chapter 12
Prime Curios

In this chapter, we will discuss primes that are in some way amusing or peculiar. For instance, if a prime has some unusual digit pattern in its base-10 or decimal representation, it may be considered entertaining. We include such special primes in a class of primes we call *prime curios*. Since this term will cover an enormous variety of primes, it is impossible to give a cohesive definition of what constitutes such a number beyond its being prime and having some additional characteristic(s) that make it diverting. We have already seen that the Mersenne primes are very interesting, as they possess a number of unusual characteristics: they are one less than a prime power of two, for each Mersenne prime there corresponds an even perfect number, and there exist efficient tests for proving their primality. Hence, it is fair to say that Mersenne primes belong to the prime curio designation.

Only a few such curiosities will be covered herein, but a more extensive discussion of prime curios can be found in Caldwell and Honaker's book *Prime Curios! The Dictionary of Prime Number Trivia*. Caldwell also maintained a companion website which contained the latest findings of such remarkable primes.

In this chapter, we will attempt to group together various types of prime curios because of shared features. We note that sometimes a particular prime can satisfy more than one characteristic that would make it a curio. For example, in the 1980s the computer engineer and mathematician Samuel Yates (1919–1991) defined a *titanic prime* to be a prime with 1000 digits or more in its base-10 representation; hence, all Mersenne primes from the 19th onward are certainly curios because they are both Mersenne primes and also titanic primes. In what follows, when considering interesting patterns or values of a prime curio, we will restrict ourselves only to considering the base-10 representation, even though we recognize that such a restriction is arbitrary.

We will begin with a class of prime curio in which the primes are considered special because of the shapes of base 10 digits of which they are composed. A prime whose only digits are 1, 4, or 7 has been named a *straight-digit prime*. Along the

E. L. F. Roettger and H. C. Williams, *The Enchantment of Numbers*, CMS/CAIMS Books in Mathematics 16, https://doi.org/10.1007/978-3-031-76568-1_12

same lines a prime that only contains the digits 0, 3, 6, 8, or 9 is called a *curved-digit prime* or the digits 0, 6, 8, or 9 is called a *circular-digit prime*. If we restrict the digits to be the composite single digit numbers, i.e., 4, 6, 8, or 9, then a prime composed of these is a *composite-digit prime*. We also have the concept of a *prime-digit prime*; a prime that only contains the digits 2, 3, 5, or 7. Some examples: 1447 is the smallest straight-digit prime that uses each possible digit at least once, 89 is a composite-digit prime, a curved-digit prime and a circular-digit prime and

$$72323252323272325252 \cdot (10^{3120} - 1)/(10^{20} - 1) + 1$$

is a 3120 digit prime-digit prime found by the computer engineer Harvey Dubner (1928–2019) in 1992. A *Yarborough prime* is a prime that does not contain the digits 0 or 1 and an *anti-Yarborough prime* is a prime that only contains the digits 0 or 1. Rather than restrict the digits, suppose we instead insist that each digit be included at least once and the number be a prime, then we have a *pandigital prime*. The smallest pandigital prime is 10123457689. There are other primes of the same ilk that are considered interesting without being given a standard definition; for example, 1973 is unique as the last digit of any prime must be 1, 3, 7, or 9, and this is the smallest prime that can be formed by using each of these digits. All of these types of primes may be considered somewhat frivolous by serious number theorists, but they are fun.

We also have a group of prime curios that rely on the prime having a special form. We include the well discussed Mersenne primes in this class, but there are many others. A *Woodall prime* is a prime of the form $n2^n - 1$. At the time of writing this book the known Woodall primes have $n = 2, 3, 6, 30, 75, 81, 115, 123, 249, 362, 384,$ 462, 512, 751, 822, 5312, 7755, 9531, 12379, 15822, 18885, 22971, 23005, 98726, 143018, and 151023. It is an open problem whether there are infinitely many such primes. Similar to the Woodall primes, we have *Cullen primes*, those of the form $n2^n + 1$. The set of Cullen primes is conjectured to be infinite; however, the only known examples are those that result from $n = 1$, 141, 4713, 5795, 6611, 18496, 32292, 32469, 59656, 90825, 262419, 361275, 481899, 1354828, 6328548, and 6679881.

A *cuban prime* is a prime that is the difference of consecutive cubic numbers; i.e., it is of the form $(n + 1)^3 - n^3$ and is a prime. It is easy to see that the smallest such prime is $7 = 2^3 - 1^3$, but there is likely an infinitude of such primes. Given a prime p, if the number of the form $\frac{2^p+1}{3}$ is also prime, then we call it a *Wagstaff prime*. It is easy to see that exponents 3, 5, and 7 yield the Wagstaff primes 3, 11 and 43, respectively. Currently, there is no efficient test known to prove a Wagstaff number prime; hence, the OEIS separates the list of known Wagstaff primes from Wagstaff probable primes (OEIS A000979 and A000978). As we mentioned in Chap. 11, a *Fermat prime* is a prime of the form $2^{2^n} + 1$; we have seen that there are only 5 such primes known and it is conjectured that there are only these 5. This is a difficult problem as no efficient test, like that for the Mersenne numbers, exists to test these for primality. A Fibonacci number which happens to also be prime is a *Fibonacci prime* and a Lucas number which is prime is a *Lucas prime*.

Next, we consider a class of prime curios that come in pairs or sets. Perhaps most famously we have the *twin primes*; these are a pair of primes that differ by 2. For example, 17 and 19 constitute a pair of twin primes. Similarly we have *cousin primes* and *sexy primes*, these are prime pairs that differ by 4 and 6, respectively. Another way to look at this is as twin primes have a prime gap of 2, cousin primes have a prime gap of 4 and sexy primes have a prime gap of 6. It is an open problem whether there exists an infinitude of twin, cousin or sexy primes. Until 2013 it was not even known if there was an infinite set of prime pairs with a fixed prime gap. Remarkably, in 2013 the Chinese-American mathematician Yitang Zhang proved that there is an infinite set of pairs of primes that differ by less than 70 million. Since Zhang published his work there has been a burst of activity in this area, and the prime gap now has been lowered from 70 million to now just 246. However, a prime gap of 6 is the best case scenario using current methods. Hence, a proof of the *twin prime conjecture*, that there exists an infinitude of twin primes, is likely a long way in the future. The largest twin primes currently known are the 388342 digit pair

$$2996863034895 \cdot 2^{1290000} \pm 1.$$

Historically there has been interest in primes p where p^2 divides some particular expression. For instance, we know from Theorem 11.1 that p must divide $(p-1)!+1$; a *Wilson prime* is a prime p such that $p^2 \mid (p-1)!+1$. Similarly, a *Wieferich prime* is a p such that $p^2 \mid 2^{p-1}-1$. Also a *Wall-Sun-Sun prime* or *Fibonacci-Wieferich prime* is a prime p, such that $p>5$ and $p^2 \mid F_{p-(p|5)}$. It is believed that there exists an infinitude of Wilson primes and of Wieferich primes, but we only know 3 of the former (5, 13 and 563) and 2 of the latter (1093 and 3511). Extensive computer searches have yet to reveal a Wall-Sun-Sun prime; nevertheless, it is still thought by many that there must exist an infinitude of them. This is called the *Wall-Sun-Sun conjecture*.

The study of Wieferich primes is in fact motivated by the famous Fermat's last theorem, which states that no three positive integers x, y, and z satisfy the equation $x^n + y^n = z^n$ for any integer $n \geq 2$. The German mathematician and teacher Arthur Wieferich (1884–1954) showed if solutions were to exist for $x^p + y^p = z^p$ for integers x, y, and z such that p is an odd prime and $p \nmid xyz$, then it must be the case that $p^2 \mid 2^{p-1}-1$. Despite Fermat's last theorem having been proved and the original motivation now no longer existing, because of this history there is still much interest in such primes.

We next define some primes based on characteristics of the pattern of the digits of such numbers. These may result from reversing, reflecting, removing or flipping the numbers. A prime, such as 13331, that is the same when the digits are written is reverse order is called a *palindromic prime*. Similarly, a *reflectable prime* is a prime that remains unchanged when reflected upon the line it sits on; hence, it can only contain digits 0, 1, 3, or 8. If a prime is both palindromic and reflectable, then we call it *palindromic reflectable prime* or *triadic prime*. For example, 101 is triadic as

$$\frac{101}{\text{IOI}}.$$

Another interesting type of prime is one whose digits are all the same; hence, it must be all 1's, in which case we call it a *repunit prime*, a term coined by Albert Beiler in 1964; as we have seen in Chap. 1, these numbers can be written in the form $R_n = \frac{10^n-1}{9}$. The first 4 primes of this type are 11 ($n = 2$), 11111111111111111 ($n = 17$), 11111111111111111111111 ($n = 23$), and the next is a decimal number consisting of 317 digits ($n = 317$). Notice that Mersenne primes are repunits in base-2. In 1985, it was shown that the next possible repunit prime candidate ($n = 1031$) is indeed a prime. It wasn't until 2002 that the electrical engineer Harvey Dubner (1928–2019) proved that the next possible candidate for a prime R_n occurs for $n = 49081$. He showed that R_{49081} is likely a prime by finding that it is a base a probable prime (base unspecified). It was not until 2022 that Paul Underwood was able to use a version of elliptic curve primality proving to prove that R_{49081} is a prime. We now know that the next possible repunit primes have n: 86453, 109297, 270343, 5794777, and 8177207. These all have passed a probable prime test, but it will likely be many years before any is proved a prime.

Thus, it seems that repunit primes are quite rare, but we can define a *near-repdigit prime* to be a prime whose digits are all the same except for one. For instance, both 337 and 1181 are near-repdigit primes. We also have *circular primes* numbers that remain prime when we rotate the digits for all rotations. For example, 197, 971, and 719 are all primes and circular rotations of the digits 1, 9, and 7 as seen in the diagram:

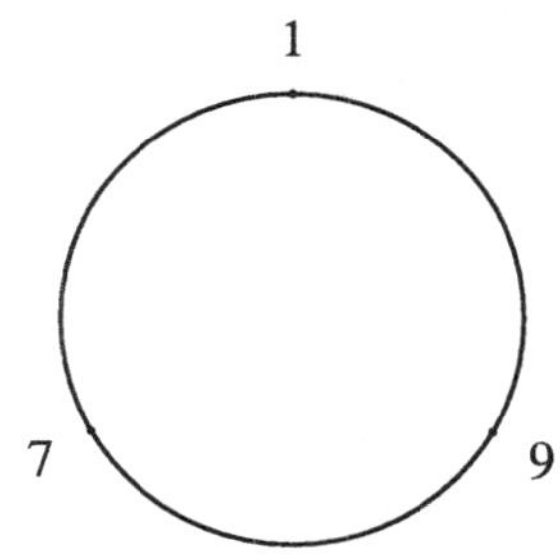

Also a prime curio that relies on moving the digits around is the much more difficult to obtain *absolute prime*, a prime that remains prime for all possible permutations of its digits. Hence, since any of the digits need to make a prime when in the last position, the only allowable digits here are 1, 3, 7, and 9. We can see that 337 is an absolute prime as both 373 and 733 are also primes. Rather than permuting digits we can think of changing digits; in this case, a prime that is *weakly prime* is a prime that upon changing any of its digits becomes composite. The smallest weakly prime number is 294001, to verify this we need to show that all 472392 numbers that result from changing a single digit are composite.

A *strobogrammatic prime* is a prime that reads the same when rotated through 180° about an axis which is perpendicular to the plane on which the number is written; hence, it can only contain the digits 0, 1, 6, 8, or 9. Note that 19861 is a prime and rotated through 180° we have 19861, which reads the same. The playfully named *emirp* is a prime that is also a prime when its digits are reversed, but is not a palindromic prime. For example, the prime 113 is an emirp as 311 is also a prime. Similarly, an *iccanobiF prime* is a prime that becomes a Fibonacci number when its digits are reversed. Note that if this number is also prime, then we have a *iccanobiF emirp* . There are also primes that can be arranged in shapes consisting of identical digits nested about the centre, when recorded in the form of a symmetrical figure or polygon. Such primes are called *congruent primes*. As an example, consider the square array:

1	1	1
1	8	1
1	1	1

a rendering of the stobogrammatic square-congruent prime 111181111. Next consider the 5×5 array:

1	1	1	1	1
1	8	8	8	1
1	8	8	8	1
1	8	8	8	1
1	1	1	1	1

This is a rendering of the strobogrammatic and square-congruent prime

1111118881188811888111111.

A more extreme example is the square-congruent prime:

9	9	9	9	9	9	9	9	9	9	9	9	9	9	9
9	9	9	9	9	9	9	9	9	9	9	9	9	9	9
9	9	9	9	9	9	9	9	9	9	9	9	9	9	9
9	9	9	9	9	9	9	9	9	9	9	9	9	9	9
9	9	9	9	9	9	9	9	9	9	9	9	9	9	9
9	9	9	9	9	9	9	9	9	9	9	9	9	9	9
9	9	9	9	9	9	9	9	9	9	9	9	9	9	9
9	9	9	9	9	9	9	5	9	9	9	9	9	9	9
9	9	9	9	9	9	9	9	9	9	9	9	9	9	9
9	9	9	9	9	9	9	9	9	9	9	9	9	9	9
9	9	9	9	9	9	9	9	9	9	9	9	9	9	9
9	9	9	9	9	9	9	9	9	9	9	9	9	9	9
9	9	9	9	9	9	9	9	9	9	9	9	9	9	9
9	9	9	9	9	9	9	9	9	9	9	9	9	9	9
9	9	9	9	9	9	9	9	9	9	9	9	9	9	9

This is a 225 digit number and can be denoted by $N_4(112)$ in the notation introduced below.

The list of types of prime curios given above is by no means complete, but should give some indication of the breadth of types of primes recreational number theorists have considered. Most of the small primes of the above types have been found by repeatedly testing numbers of specific form for primality, but in general we do not have an efficient prime test when the number is large. As we saw in Chap. 11 primality testing a general number N can be very difficult, unless N has a special form and there is a concise test for it based on its form, like the Mersenne numbers, for example. Most of the numbers listed above do not have an easily specified special form and many of the ones that do have a special form for which we do not have an efficient test for primality; however, there is one surprising prime curio type that does. These are *near-repdigit palindromic primes* that, except for one digit, consist only of the digit 9; for instance, 99999199999 is a prime that satisfies these restrictions. These numbers have the special form $N_d(n) = 10^{2n+1} - d10^n - 1$ where d is in the set $A = \{1, 2, 4, 5, 7, 8\}$. Thus, if $g = 9 - d$, then

$$N_d(n) = 999\ldots 9g999\ldots 9,$$

where $999\ldots 9$ represents n consecutive nines. Since we are interested when such numbers are prime, we can immediately reject using 3 or 6 for d, as this would result in a number divisible by 3. In 1989 Dubner found several primes of this type including the 15×15 square-congruent prime $N_4(112)$ and the the titanic prime $N_7(2874)$. He further generalized the search to numbers of the form

$$N_d(n, m) = 10^{2n+m} - d10^n - 1.$$

Here d is now a number of m digits and is also palindromic. In the event that m is even, it must be the case that 11 divides $N_d(n, m)$; thus, we insist that m is odd. Further, if we wish to have only 2 distinct decimal digits, we put $d = f(10^{m-1} + 1)$, where $m \geq 3$ and f is in A. Now by setting $g = 9 - f$, $N_d(n, m)$ has the decimal representation

$$N_d(n, m) = 999\ldots 9g999\ldots 9g999\ldots 9,$$

where the first and last group of nines is of length n and the middle group is of length $m - 2$.

If we examine the Lucas-Lehmer test (Theorem 10.1) for the primality of $N = M_n$, we notice that it is characterized by a number of features that can be extended to a more general kind of test, possessing the following properties.

(1) The test is restricted to values of N given by an expression involving some base b and some exponent n, which belongs to a preselected congruence class and exceeds a given bound. (In the case of Theorem 10.1, $N = 2^n - 1, n \equiv 1 \pmod 2$ and $n \geq 3$.)

(2) The test makes use of a prespecified number l of sequences $(r_{i,k})_{k\geq 1}$, where $1 \leq i \leq l$. The values of $r_{i,1}$ $(i = 1, 2, \ldots, l)$ (the seeds) can be calculated by a simple deterministic process and the values of the successive terms of the l sequences $(r_{i,k})_{k\geq 1}$ $(i = 1, 2, \ldots, l)$ modulo N can be computed by

$$r_{i,j+1} \equiv g_i(r_{i,j}) \pmod{N},$$

where each g_i $(i = 1, 2, \ldots, l)$ is a fixed (independent of n and j) polynomial in l variables and with integer coefficients. This step terminates at a value for k determined from n.

(3) There is a closing condition which declares N a prime if and only if it divides some fixed polynomial function(s) of some of the $r_{i,j}$ values computed in (2).

(4) The entire test executes in time $O(nM(n)) = \tilde{O}(n^2)$.

We say that any primality test for N which satisfies all of the above four conditions for a particular value of l is of the L-L or Lucas-Lehmer type. Notice that the Lucas-Lehmer test itself is an instance of this more general class with $l = 1$ and $g_1(x) = x^2 - 2$. So, too, is the test of Corollary 11.1 with $l = 1$ and $g_1(x) = x(x^2 - 3)$.

What is remarkable is that it is possible to construct L-L type tests for the primality for almost all of the numbers of the form $N_d(n)$ for d an element of A, even in the difficult cases of $d = 1$ or 8. These methods can also be extended to numbers of the form $N_d(n, m)$ and $M_d(n, m)$, where

$$M_d(n, m) = 10^{2n+m-1} + d10^n + 1,$$

and d is a number of m digits. These numbers include the *Belphegor numbers* when $m = 3$ and $d = 666$. (See A232448 and A232449 of the OEIS).

Problems

12.1 (Primes based on shapes of digits) For each of the following types of primes give a 4 digit example that is different from the examples given in the chapter (warning answers are highly not unique):

(a) **[BB]** straight-digit prime
(b) curved-digit prime
(c) circular-digit prime

12.2 (Primes based on primality of digits) For each of the following types of primes give a 4 digit example that is unique from the examples given in the chapter:

(a) composite-digit prime
(b) **[BB]** prime-digit prime

12.3 (Primes based on types of digits) For each of the following types of primes give a 3 digit example that is unique from the examples given in the chapter:

(a) Yarborough prime
(b) **[BB]** anti-Yarborough prime

12.4 Notice that each of the base-10 digits occur at least once in the prime 16548071329. What name has been given to such primes?

12.5 (Woodall primes) Woodall primes have the form $n2^n - 1$.

(a) How many digits are in the Woodall prime that has $n = 30$?
(b) How many Woodall primes are known to exist?
(c) Is it known if there are infinite such primes?

12.6 (Cullen primes) Cullen primes have the form $n2^n + 1$.

(a) How many digits are in the Cullen prime that has $n = 141$?
(b) How many Cullen primes are known to exist?
(c) Is it known if there are infinite such primes?

12.7 (Cuban primes) Cuban primes have the form $(n + 1)^3 - n^3$.

(a) **[BB]** Find a 2 digit Cuban prime?
(b) Find a 3 digit Cuban prime?
(c) Is it known if there are infinite such primes?

12.8 Given a prime p, a Wagstaff prime is a prime of the form $\frac{2^p+1}{3}$. Does there exist an efficient test for primes of this form?

12.9 Recall a Fermat prime is a prime of the form $2^{2^n} + 1$. Does there exist an efficient test for primes of this form?

12.10 Given a prime p, a Mersenne prime is a prime of the form $2^p - 1$. Does there exist an efficient test for primes of this form?

12.11 Let $d = f(10^{m-1} + 1)$, where $m \geq 3$, f is in $\{1, 2, 4, 5, 7, 8\}$ and $g = 9 - f$, the number $N_d(n, m) = 10^{2n+m} - d10^n - 1$ has the decimal representation

$$N_d(n, m) = 999\ldots9g999\ldots9g999\ldots9,$$

where the first and last group of nines is of length n and the middle group is of length $m - 2$. Does there exist an efficient test for primes of this form?

Further Reading

Prime curios

Chris K. Caldwell and G.L. Honaker, *Prime Curios! The Dictionary of Prime Number Trivia*, CreatSpace, 2009. Companion website: https://primes.utm.edu/curios.

Bruce Pyne, *Prime Recreations: An Olio of Curios about Prime Numbers*, Page Publishing, NY, 2018.

David Wells, *Prime Numbers: the Most Mysterious Figures in Math*, John Wiley and Sons, Inc., NJ, 2005.

Titanic primes

Paulo Ribenboim, *The Little Book of Bigger Primes*, 2^{ed} Edition, Spring-Verlag, NY, 2004, pp. 119–122.

PrimePages, https://t5k.org/glossary/page.php?sort=TitanicPrime.

Wall-Sun-Sun primes

Jiří Klaška, Short remark on Fibonacci-Wieferich primes, *Acta Mathematica Universitatis Ostraviensis*, **15** (2007), 21–25.

Fermat's last theorem

Alf van der Poorten, *Notes on Fermat's Last Theorem*, John Wiley and Sons, NY, 1996.

Repunit primes

Albert H. Beiler, *Recreations in the Theory of Numbers–The Queen of Mathematics Entertains*, Dover Publications, New York, 1964, pp. 83–87.

H. C. Williams, *Édouard Lucas and Primality Testing*, Wiley-Interscience, John Wiley and Sons, NY, 1998, pp. 45, 99, 156–159, 181–182, 189–190, 306–314

PrimePages, PrimePages, https://t5k.org/glossary/page.php?sort=Repunit.

Lucas-Lehmer tests for $N_d(m, n)$

E. L. Roettger and H. C. Williams, Lucas-Lehmer primality tests for certain prime curios, *Integers*, **21** (2021), #A109.
E. L. Roettger and H. C. Williams, Some primality conditions for $N = 2p^n - 1$, *Integers*, **22** (2022), #A76.

Chapter 13
Linear Recurrence Sequences

This chapter is largely devoted to a question that both fascinated and perplexed Édouard Lucas. Today we know much more about this problem than he did, and we will present several pertinent results here. The reader should be warned that some of this material is fairly technical, but the basic ideas should be clear enough. We begin with a particular Lucas sequence which, as far as we know, was unknown to Lucas.

Consider the sequence (G_n), where $G_0 = 0$, $G_1 = 1$ and $G_{n+1} = G_n - 7G_{n-1}$ for all $n \geq 1$. This is the Lucas sequence (U_n) for $P = 1$ and $Q = 7$. In the table below we record some early terms of (G_n), together with their prime factorizations.

n	G_n	n	G_n
0	0	13	-93911
1	1	14	$146329 = 41 \cdot 43 \cdot 83$
2	1	15	$803706 = 2 \cdot 3 \cdot 29 \cdot 31 \cdot 149$
3	$-6 = -2 \cdot 3$	16	$-220597 = -13 \cdot 71 \cdot 239$
4	-13	17	-5846539
5	29	18	$-4302360 = -2^3 \cdot 3^2 \cdot 5 \cdot 19 \cdot 629$
6	$120 = 2^3 \cdot 3 \cdot 5$	19	$36623413 = 113 \cdot 324101$
7	-83	20	$66739933 = 13 \cdot 29 \cdot 211 \cdot 839$
8	$-923 = -13 \cdot 71$	21	$-189623958 = -2 \cdot 3 \cdot 83 \cdot 503 \cdot 757$
9	$-342 = -2 \cdot 3^2 \cdot 19$	22	$-656803489 = -8513 \cdot 77153$
10	$6119 = 29 \cdot 211$	23	$670564217 = 139 \cdot 4824203$
11	8513	24	$5268188640 = 2^5 \cdot 3 \cdot 5 \cdot 11 \cdot 13 \cdot 23 \cdot 47 \cdot 71$
12	$-34320 = -2^4 \cdot 3 \cdot 5 \cdot 11 \cdot 13$	25	$574239121 = 29 \cdot 449 \cdot 44101$

Notice that the terms with subscript divisible by 3 tend to have a lot of small prime factors belonging to no particular residue classes modulo 14, but the other terms seem only to be divisible by primes that are congruent to ± 1 modulo 14. In fact, it can be shown that any prime divisor r of G_n when $3 \nmid n$ must be a cubic residue of 7. Since the only cubic residues of 7 are ± 1, we must have $r \equiv \pm 1 \pmod{14}$. This explains

E. L. F. Roettger and H. C. Williams, *The Enchantment of Numbers*, CMS/CAIMS Books in Mathematics 16, https://doi.org/10.1007/978-3-031-76568-1_13

why terms with index a multiple of 3 will tend to be highly composite, as two-thirds of primes (those $\not\equiv \pm 1 \pmod{14}$) must wait until 3 divides n in order to divide some G_n.

The Lucas sequences are examples of linear recurrence sequences of order 2 (we will define order later). In this chapter we will show that it is possible to find extensions of Lucas' sequences of order greater than 2, some of which possess properties analogous to those that we have observed in (G_n).

In several of his works on the (U_n) and (V_n) sequences, Lucas considered the problem of extending or generalizing them. On examining what he wrote about this, we learn that he was interested in functions satisfying linear recurrence sequences; these functions should be symmetric functions of the roots of an associated polynomial with rational (in practice, usually integral) coefficients. Furthermore, it seems, just as in the case of the Lucas sequences, that he was considering more than one function in this study. Additionally, he appeared to have been particularly interested in sequences linked to polynomials of degree three or four. However, it is difficult to determine in these instances when Lucas was referring to sequences satisfying a third or fourth order linear recurrence or to sequences whose terms are symmetric functions of the roots of a third or fourth degree polynomial; he mentioned both. He emphasized the need to find addition and multiplication formulas involving these sequences, as this was what he did in order to prove the many properties of his own sequences. His suggested method of approach was to use empirical methods to attempt to elucidate what the various arithmetic laws for these extended sequences would be, and from this material he thought he should be able, as he did in the case of U_n and V_n, to derive further properties of primes, such as primality testing algorithms. However, despite this plausible program of research, he seems never to have produced any results that were analogous to his work on the Lucas sequences.

As it appears that Lucas considered linear recurrence sequences of order three (cubic recurrences) to be key to the problem of extending of the Lucas sequences, we will now provide a very brief discussion of such sequences. We consider the recurrence sequence (X_n), where P, Q, R are given integers and

$$X_{n+3} = PX_{n+2} - QX_{n+1} + RX_n, \quad \text{for } n > 0.$$

Let α, β, γ the three roots of the associated characteristic polynomial

$$f(x) = x^3 - Px^2 + Qx - R.$$

Since the recurrence relationship is a homogeneous linear difference equation of degree three with constant coefficients, it can be shown that if α, β, γ are distinct, then

$$X_n = c_1\alpha^n + c_2\beta^n + c_3\gamma^n,$$

for certain constants c_1, c_2, c_3; these constants can be determined from initial conditions on (X_n), such as the values of X_0, X_1 and X_2. This formula can be considered as an extension of the Binet formula for the Fibonacci numbers. For example, suppose $P = 1$, $Q = -1$, $R = 1$ and $X_0 = 0$, $X_1 = 0$, $X_2 = 1$. The first terms of the resulting sequence are 0, 0, 1, 1, 2, 4, 7, 13, 24, 44, 81, 149, 274, 504, 927, 1705, 3136, 5768, 10609, 19513, 35890, 66012, ... (sequence A000073 in the OEIS). After some calculation, we can find the Binet formula for this (X_n) as

$$\begin{aligned} X_n &= \frac{\alpha^{n+1}}{(\alpha-\beta)(\alpha-\gamma)} + \frac{\beta^{n+1}}{(\beta-\alpha)(\beta-\gamma)} + \frac{\gamma^{n+1}}{(\gamma-\alpha)(\gamma-\beta)} \\ &= \frac{\alpha^n}{-\alpha^2+4\alpha-1} + \frac{\beta^n}{-\beta^2+4\beta-1} + \frac{\gamma^n}{-\gamma^2+4\gamma-1}, \end{aligned}$$

where α, β, γ are the roots of the characteristic polynomial

$$x^3 - x^2 - x - 1.$$

We can compute α, β, γ here by using the formula usually attributed to the Italian polymath Gerolamo Cardano (1501–1576) to obtain

$$\begin{aligned} \alpha &= \frac{1}{3}\left(\sqrt[3]{19+3\sqrt{33}} + \sqrt[3]{19-3\sqrt{33}} + 1\right) = 1.8392\ldots \\ \beta, \gamma &= \frac{1}{6}\left(2 - \sqrt[3]{19+3\sqrt{33}} + \sqrt[3]{19-3\sqrt{33}} \pm \sqrt{3}i(\sqrt[3]{19+3\sqrt{33}} - \sqrt[3]{19-3\sqrt{33}})\right) \\ &= -0.4196\ldots \pm (0.6062\ldots)i. \end{aligned}$$

Here $i^2 = -1$.

This sequence (X_n) is today called the *Tribonacci sequence*; it was used unwittingly by Charles Darwin as an illustration for the population growth of elephants. Since $\alpha > 1$ and $\alpha\beta\gamma = 1$, we have $|\beta| = |\gamma| < 1$. Thus,

$$\alpha = \lim_{n\to\infty} \frac{X_{n+1}}{X_n},$$

which means that α here is analogous to ϕ in the case of the Fibonacci numbers. In fact, it even has a geometric interpretation analogous to that of ϕ.

For suppose we now cut a line segment into three pieces

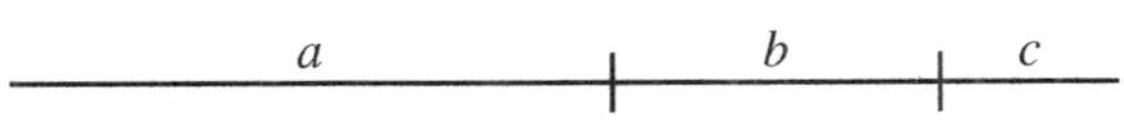

where this time

$$\frac{a+b+c}{a} = \frac{a+b}{b+c} = \frac{a}{b}.$$

If we put $\gamma = a/b$, then

$$1 + \frac{b+c}{a} = \gamma, \quad \frac{b+c}{a+b} = \frac{1}{\gamma};$$

hence,

$$\frac{b+c}{a} = \frac{a+b}{a}\frac{1}{\gamma} = \left(1 + \frac{1}{\gamma}\right)\left(\frac{1}{\gamma}\right) = \frac{1}{\gamma} + \frac{1}{\gamma^2}$$

and

$$1 + \frac{1}{\gamma} + \frac{1}{\gamma^2} = \gamma.$$

It follows that $\gamma^3 = \gamma^2 + \gamma + 1$. Since this equation has only one real root α, we see that $\gamma = \alpha$.

Next, suppose $\mathcal{C}$ is a rectangular cuboid (the three dimensional analogue of the two dimensional rectangle, i.e. a box) with edges a, $a + b$ and $a + b + c$, subject to the conditions above.

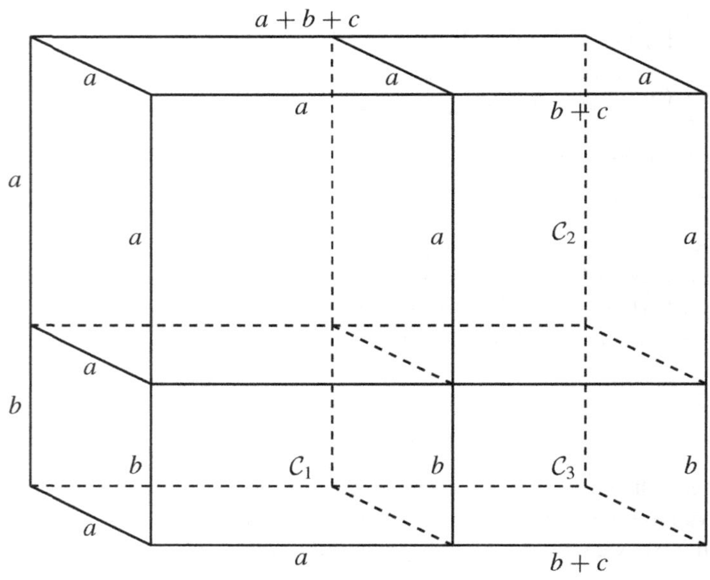

If we cut a cube of edge a from $\mathcal{C}$ by using a plane parallel to the base of $\mathcal{C}$ and another parallel to the $a \times a + b$ face, we get three cuboids: $\mathcal{C}_1$ of dimensions $a \times a \times a + b$, $\mathcal{C}_2$ of dimensions $a \times b + c \times b + c$ and $\mathcal{C}_3$ of dimensions $a \times b \times b + c$. Since

$$\frac{a}{b} = \frac{a+b}{b+c} = \frac{a+b+c}{a} = \alpha,$$

we see that $\mathcal{C}_3$ is similar to $\mathcal{C}$, analogous to the situation in the case of the golden rectangle.

Except for the Tribonacci example, much of this material was known to Lucas, but he seemed unable to find a particular set of third order recurring sequences which was satisfactory. One of the main features of the (second order) Lucas sequences is that (U_n) is a divisibility sequence (see Chap. 10), and much of the rich arithmetic theory of his sequences is dependent on this; however, it has been conjectured that the only third-order recurring sequences (X_n) which are also divisibility sequences are given by: $X_n = n^2a^n$, $X_n = nU_n$, $X_n = U_n^2$ where a is an integer and (U_n) is either a Lucas or Lehmer sequence; this long-standing conjecture has recently been proved by the talented British-born mathematician Andrew Granville. Unfortunately, none of these sequences can possess the kind of new arithmetic properties that Lucas was seeking.

We will now consider some of the properties of recurrence sequences with constant integer coefficients. As we have seen, these seemed to be the sequences that Lucas regarded as essential to solving the problem of generalizing his sequences. We begin by generalizing some of our earlier observations. Let the sequence (X_n) be determined by an initial set of k integer values

$$X_0, X_1, X_2, \ldots, X_{k-1}$$

and the kth order linear recurrence formula

$$X_{n+k} = P_1X_{n+k-1} - P_2X_{n+k-2} + \cdots + (-1)^{k-1}P_kX_n,$$

where $P_1, P_2, \ldots, P_k$ are given fixed integers. Notice that the values for X_n for all $n > 0$ are completely determined by these conditions. Such a sequence (X_n) is called *linear recurrence sequence of order k*. Notice that the Lucas sequences have $k = 2$ and the Tribonacci sequence has $k = 3$. We define the *characteristic polynomial* $f(x)$ of the linear recurrence sequence (X_n) to be

$$f(x) = x^k - P_1x^{k-1} + P_2x^{k-2} - \cdots + (-1)^kP_k.$$

As we have already mentioned, it seems, on perusing what little Lucas wrote concerning the problem of generalization, that divisibility sequences would play a prominent role. Certainly, some of his suggested possible sequences do possess this property. In view of the importance of this feature in his development of the arithmetic properties of the Lucas sequence (U_n), this is not surprising. It seems that he might have been considering linear recurrence sequences which were also divisibility sequences, i.e. linear recurrence sequences (X_n) which satisfy the condition that $X_n \mid X_{mn}$ for all positive integers m and n. Such sequences are called *linear divisibility sequences* or *linear division sequences* (LDSs). Also, for such sequences to be at all interesting, we must have $X_0 = 0$. Furthermore, if (X_n) is a LDS, we may always assume that since the first term X_1 of such a sequence must divide all other terms, we can put $X_1 = 1$ with no significant loss of generality. The old and difficult problem of classifying all such sequences has recently been solved by Andrew Granville, but we caution the reader that the mathematics required to do this is rather deep and difficult.

We now consider some examples of LDSs below. In view of what we now know, the possible LDSs with order greater than 2 must have order at least 4. Indeed, the more interesting of these must have a characteristic polynomial of the form

$$x^4 - Px^3 + (R + 2Q)x^2 - PQx + Q^2,$$

for some integers P, Q, R. We observe that if α, β, γ, δ are the roots of this polynomial, then we may assume that $Q = \alpha\beta = \gamma\delta$. Of course, since $X_0 = 0$ and $X_1 = 1$, we must also specify the values for X_2 and X_3. For example, if we put $P = 1$, $Q = 1$, $R = -7$, $X_2 = 1$ and $X_3 = 5$, we get the sequence

$$0, 1, 1, 5, 11, 36, 95, 281, 781, 2245, 6336, 18061, 51205, 145601, 413351, 1174500, 3335651, \ldots$$

(OEIS sequence A005178), satisfying the recurrence formula

$$X_{n+4} = X_{n+3} + 5X_{n+2} + X_{n+1} - X_n.$$

It turns out that each term X_n here represents the number of different ways to tile a $4 \times n$ rectangle with 2×1 tiles (dominoes) and the sequence is also a LDS. Indeed, if we put $X_2 = P$ and $X_3 = P^2 - R - 3Q$, then the resulting sequence (X_n) must be a LDS of order 4. Also, in this case

$$X_n = \frac{\alpha^n + \beta^n - \gamma^n - \delta^n}{\alpha + \beta - \gamma - \delta}.$$

It is of some interest to note that sometimes, by varying the values of the initial terms, we can get more than one LDS belonging to a given recurrence. For example, consider the case of $P = 10$, $R = 21$, $Q = 1$. When we put $X_2 = P = 10$ and $X_3 = P^2 - R - 3Q = 76$, we get the sequence

$$0, 1, 10, 76, 540, 3751, 25840, 177451, 1217160, 8344876, 57202750, 392089501, \ldots, (L_{4n} - L_{2n})/4, \ldots$$

but if we put $X_2 = 6$ and $X_3 = 38$, we get the LDS

$$0, 1, 6, 38, 252, 1705, 11628, 79547, 544824, 3733234, 25585230, 175356611, \ldots.$$

Furthermore, if we put $X_2 = 15$ and $X_3 = 128$, we get the LDS

$$0, 1, 15, 128, 945, 6655, 46080, 317057, 2176335, 14925184, 102320625, \ldots, (F_{4n} - 2F_{2n}), \ldots.$$

These three sequences, each of which satisfies the same 4th order recurrence

$$X_{n+4} = 10X_{n+3} - 23X_{n+2} + 10X_{n+1} - X_n,$$

are listed as A215465, A215466 and A127595, respectively in the OEIS. As we will discuss later, there are many LDSs of order greater than 4. For example, if $X_n = (L_n F_{2n} F_{3n})/6$, then (X_n) is a LDS of order 6.

We now return to Lucas' problem of extending his sequences. While we will use symbols like (U_n) and (V_n) to denote candidates for these extended sequences, we will try to do this in such a way as to limit any possible confusion that might arise between this notation and that of the standard Lucas sequences defined in Chap. 10.

We have seen that any such sequences should satisfy the following four basic properties:

(1) There are two sequences of integers (when $n > 0$).
(2) Both sequences satisfy the same linear recurrence relation.
(3) One of the two sequences is a divisibility sequence.
(4) There are addition and multiplication formulas for the terms of the sequences.

It has been shown that if we put

$$U_n = \lambda^{n-1} \prod_{i=1}^{k} \frac{1-\gamma_i^n}{1-\gamma_i} \quad \text{and} \quad V_n = \lambda^n \prod_{i=1}^{k} (1+\gamma_i^n),$$

where λ and $\gamma_1, \gamma_2, \ldots, \gamma_k$ are simply constrained to be non-zero algebraic numbers such that $V_1 \neq 0$ and both (U_n), (V_n) are sequences of integers for $n > 0$, then (U_n) and (V_n) possess the above four properties for $k = 1, 2, 3$. An *algebraic number* is defined to be a root of a polynomial with integer coefficients. We will denote such numbers by using letters from the Greek alphabet. Notice that, just as in the case of the Lucas sequences, we have $V_n = U_{2n}/U_n$. For any fixed value of k, conditions on λ and $\gamma_1, \gamma_2, \ldots, \gamma_k$ can be derived to ensure that both (U_n) and (V_n) are sequences of integers which satisfy the same linear recurrence. For example, one of these conditions is that $Q = \lambda^2 \gamma_1 \gamma_2 \cdots \gamma_k$ be an integer. It is also shown that under these same conditions (U_n) is a divisibility sequence.

Under the above constraints it is easy to prove that if $k = 1$, $P = \lambda(\gamma_1 + 1)$, $Q = \lambda^2\gamma_1$, then both P and Q must be integers because $V_1 = \lambda(\gamma_1 + 1)$, $V_2 = \lambda^2(1 + \gamma_1^2)$, $U_3 = \lambda^2(1 + \gamma_1 + \gamma_1^2)$ must all be integers. In this case, we see that $U_n = U_n(P, Q)$, $V_n = V_n(P, Q)$, where $\alpha = \lambda$ and $\beta = \lambda\gamma_1$. Thus, we get the standard Lucas sequences in this case.

When $k = 2$, we must have T_1, T_2 and Q integers (T_1 and T_2 are coefficients of the characteristic polynomial of (U_n)), where

$$T_1 = \lambda(1+\gamma_1)(1+\gamma_2), \;\; T_2 = \lambda^2(\gamma_1+\gamma_2)(1+\gamma_1\gamma_2) + 2\lambda^2\gamma_1\gamma_2 \;\; \text{and} \;\; Q = \lambda^2\gamma_1\gamma_2.$$

Thus, if we put $P_1 = T_1$ and $P_2 + 2Q = T_2$, we must have P_1, P_2 and Q integers. We find that both (U_n) and (V_n) above satisfy the fourth order linear recurrence:

$$X_{n+4} = P_1 X_{n+3} - (P_2 + 2Q) X_{n+2} + P_1 Q X_{n+1} - Q^2 X_n,$$

with initial conditions $U_0 = 0$, $U_1 = 1$, $U_2 = P_1$, $U_3 = P_1^2 - P_2 - 3Q$, $V_0 = 4$, $V_1 = P_1$, $V_2 = P_1^2 - 2P_2 - 4Q$, $V_3 = P_1(P_1^2 - 3P_2 - 3Q)$. That is,

$$U_n = \frac{\alpha_1^n + \alpha_2^n - \beta_1^n - \beta_2^n}{\alpha_1 + \alpha_2 - \beta_1 - \beta_2} \quad \text{and} \quad V_n = \alpha_1^n + \alpha_2^n + \beta_1^n + \beta_2^n,$$

where $\alpha_1 = \lambda\gamma_1$, $\beta_1 = \lambda$, $\alpha_2 = \lambda\gamma_2$, $\beta_2 = \lambda\gamma_1\gamma_2$ are the roots of the characteristic polynomial $x^4 - P_1 x^3 + (P_2 + 2Q)x^2 - P_1 Q x + Q^2$. Note that $\alpha_1\alpha_2 = \beta_1\beta_2$.

Suppose we put $P_1 = -89$, $P_2 = 1199$, $Q = 1331 = 11^3$. It can be shown that if r is any prime divisor of the corresponding U_n above and $5 \nmid n$, then r must be a quintic residue of 11. Thus, the behaviour of the factorizations of the U_n is similar to that of G_n, but more extreme because only two (1 and 10) of the 10 nonzero residues of 11 are quintic residues of 11.

When $k = 3$, the situation is somewhat more complicated because both sequences satisfy a linear recurrence of order 8. A difficulty then arises when we try to develop the addition and multiplication formulas, because it becomes necessary to introduce a third sequence in order to do this; thus, the sequences when $k = 3$ cannot in general satisfy condition (1). This, of course, is also likely to be the case when $k > 3$. Nevertheless, there is a special case when $k = 3$ which does allow us to have only two sequences; this is when we constrain $\gamma_1\gamma_2\gamma_3 = 1$. Under this condition we have $Q = R^2$, where $R\ (= \lambda)$ must be an integer. If we put $\alpha_i = R\gamma_i$, $\beta_i = R^2/\alpha_i$ $(i = l, 2, 3)$, we get

$$U_n = \frac{\alpha_1^n + \alpha_2^n + \alpha_3^n - \beta_1^n - \beta_2^n - \beta_3^n}{\alpha_1 + \alpha_2 + \alpha_3 - \beta_1 - \beta_2 - \beta_3} \quad \text{and} \quad V_n = \alpha_1^n + \alpha_2^n + \alpha_3^n + \beta_1^n + \beta_2^n + \beta_3^n + 2R^n.$$

We have $\alpha_1\alpha_2\alpha_3 = \beta_1\beta_2\beta_3 = R^3$ and if $W_n = V_n - 2R^n$, we find that the sequences (U_n) and (W_n) both satisfy the same sixth order linear recurrence

$$\begin{aligned} X_{n+6} = P_1 X_{n+5} - (P_2 + 3R^2) X_{n+4} + (P_3 + 2R^2 P_1) X_{n+3} - R^2(P_2 + 3R^2) X_{n+2} \\ + R^4 P_1 X_{n+1} - R^6 X_n, \end{aligned}$$

where P_1, P_2 and P_3 are integers such that

$$P_3 = R P_1^2 - 2R P_2 - 4R^3.$$

The initial values for (U_n) are $U_0 = 0$, $U_1 = 1$, $U_2 = P_1 + 2R$, $U_3 = P_1^2 + R P_1 - P_2 - 3R^2$ and those for (W_n) are $W_0 = 6$, $W_1 = P_1$, $W_2 = P_1^2 - 2P_2 - 6R^2$, $W_3 = P_1^3 - 3P_1 P_2 + 3R P_1^2 - 6R P_2 - 3R^2 P_1 - 12R^3$. It is also easy to see that $U_{-n} = -U_n/R^{2n}$ and $W_{-n} = W_n/R^{2n}$; thus, additional terms can be computed from the recurrence. Particular instances of this sequence (U_n) occur in the OEIS as A180510 with $P_1 = -1$, $P_2 = -5$, $P_3 = 7$, $R = 1$ and A005120 with $P_1 = -3$, $P_2 = 2$, $P_3 = 1$, $R = 1$. Others are: A001351, A001945 and A006235. Indeed, it is a rather enjoyable pastime to detect these kinds of sequences in the OEIS.

One particularly entertaining sequence of this type occurs when $P_1 = 14965$, $P_2 = -1017406826$, $R = 24389 = 29^3$, $P_3 = -2939567318323$ (not an example that would occur to you immediately, but one of an infinitude). In this case we get, after finding the prime decomposition of the terms of the sequence, the following table of values.

n	U_n
0	0
1	1
2	63743
3	$41^2 \cdot 105967$
4	$-63743 \cdot 120528407$
5	$18080861 \cdot 47932181359$
6	$-41^2 \cdot 1217 \cdot 63743 \cdot 105967 \cdot 548099$
7	$-2^3 \cdot 3^3 \cdot 7^3 \cdot 13 \cdot 83 \cdot 113 \cdot 127 \cdot 181 \cdot 4327 \cdot 22637 \cdot 33629$
8	$-63743 \cdot 2023951 \cdot 120528407 \cdot 154192040407$
9	$-41^2 \cdot 105967 \cdot 352043 \cdot 8608823 \cdot 41508639603521$
10	$63743 \cdot 199289 \cdot 18080861 \cdot 47932181359 \cdot 90718179089$
11	$307 \cdot 190553764987279 \cdot 1471511488046122134363179119$
12	$-41^2 \cdot 1217 \cdot 1931 \cdot 63743 \cdot 105967 \cdot 528611 \cdot 548099 \cdot 120528407$ $\cdot 799092083$
13	$911 \cdot 868829859786520299354532010846078627722993379292 13$
14	$2^6 \cdot 3^3 \cdot 7^3 \cdot 13^3 \cdot 43 \cdot 83 \cdot 113 \cdot 127 \cdot 181 \cdot 461 \cdot 827 \cdot 1093 \cdot 4327$ $\cdot 7043 \cdot 8807 \cdot 22637 \cdot 33629 \cdot 45263 \cdot 63743$
15	$-41^2 \cdot 1259 \cdot 4409 \cdot 10499 \cdot 33391 \cdot 105967 \cdot 18080861$ $\cdot 47932181359 \cdot 188657402425248539369$
16	$-63743 \cdot 2023951 \cdot 3655681 \cdot 120528407 \cdot 154192040407$ $\cdot 270791233327 \cdot 573424602581099359$
17	$13124329444423469738970728321$ $\cdot 68649190438221280509278093785106812361842 9$
18	$-41^2 \cdot 1217 \cdot 63743 \cdot 105967 \cdot 352043 \cdot 548099 \cdot 8608823$ $41508639603521 \cdot 46265077508898794765943465 1$
19	$-3191 \cdot 290737 \cdot 19018201 \cdot 500689256569970800812621056$ $3049425129045409521906956962476568 11$
20	$63743 \cdot 199289 \cdot 18080861 \cdot 120528407 \cdot 47932181359$ $\cdot 90718179089 \cdot 30443720905796435556052270866112601 9$
21	$2^3 \cdot 3^6 \cdot 5^3 \cdot 7^3 \cdot 13 \cdot 41^2 \cdot 43 \cdot 83 \cdot 113 \cdot 127 \cdot 181 \cdot 379 \cdot 4409$ $\cdot 4327 \cdot 11801 \cdot 12263 \cdot 22637 \cdot 33629 \cdot 46831 \cdot 85639 \cdot 105967$ $\cdot 116423 \cdot 887492677 \cdot 780831241$
22	$307^3 \cdot 190553764987279 \cdot 1471511488046122134363179119$ $\cdot 63743 \cdot 97789 \cdot 7841244907646133326488341710420059$
23	$25179078774401811394677769123622574703 \cdot 207253234766$ $445419497519285309564717099931568281243272800379$
24	$-41^2 \cdot 1217 \cdot 1931 \cdot 3191 \cdot 63743 \cdot 105967 \cdot 528611 \cdot 548099$ $\cdot 2023951 \cdot 2425751 \cdot 120528407 \cdot 799092083 \cdot 154192040407$ $\cdot 9798681472366641815190367 3$
25	$349^3 \cdot 2801 \cdot 18080861 \cdot 47932181359 \cdot 6174993709235093684949 01$ $\cdot 10604913411071362611895961734179080975557912513809901$

Notice that the terms with subscript divisible by 7 tend to have a lot of small prime factors belonging to no particular residue classes modulo 29, but the other terms seem only to be divisible by primes that are congruent to $\pm 1, \pm 17$ modulo 29. In fact, it can be shown that if r is any prime divisor of some U_n when $7 \nmid n$, then r must be a 7th power residue of 29, i.e.

$$r^{(29-1)/7} = r^4 \equiv 1 \pmod{29}.$$

This is similar to the situation we observed for (G_n). It explains why the terms with subscript divisible by 7 will tend to be highly composite because only

$$\frac{4}{\phi(29)} = \frac{1}{7}$$

of all primes are congruent to $\pm 1, \pm 17 \pmod{29}$; thus, the remaining $6/7$ of all primes must wait until 7 divides n in order for them to divide U_n. Hence, we expect that many primes should divide U_n when $7 \mid n$. In fact, there always exists a sequence (U_n), with similar properties to the above, for any prime $q \equiv 1 \pmod 7$. The theory of (U_n) and the corresponding (W_n) can be used to produce a necessary and sufficient primality test similar to Theorem 11.5 for integers N when $N^6 - 1$ is divisible by a large power of 7, as long as we have some prime q such that

$$N^{(q-1)/7} \not\equiv 1 \pmod{q}.$$

If we let P, Q, R be arbitrary integers and put

$$\begin{aligned}
P_1 &= PQ - 3R, \\
P_2 &= P^3R - Q^3 - 5PQR + 3R^2, \\
P_3 &= R(P^2Q^2 - 2Q^3 - 2P^3 + 4PQR - R^2),
\end{aligned}$$

then P_1, P_2, P_3 satisfy

$$P_3 = RP_1^2 - 2RP_2 - 4R^3.$$

The resulting (U_n) and (W_n) sequences turn out to be given by

$$\begin{aligned}
U_n &= \frac{(\alpha^n - \beta^n)(\beta^n - \gamma^n)(\gamma^n - \alpha^n)}{(\alpha - \beta)(\beta - \gamma)(\gamma - \alpha)}, \\
W_n &= (\alpha^n - \beta^n)(\beta^n - \gamma^n)(\gamma^n - \alpha^n) - 2\alpha^n\beta^n\gamma^n,
\end{aligned}$$

where α, β, γ are the zeros of the cubic polynomial $x^3 - Px^2 + Qx - R$. The functions U_n and W_n are symmetric functions (do not change when (α, β, γ) is replaced by any permutation of (α, β, γ)) of the zeros of a cubic polynomial, were likely known to Lucas and represent a cubic extension of the Lucas sequences.

The number theoretic properties of these pairs of sequence when $k = 2$ or $k = 3$ are very similar to those of the Lucas sequences, but allow for greater flexibility in application, particularly to the primality testing problem. Thus, Lucas' intuition was correct about the existence of fourth order extensions of his sequences. It seems that he was wrong about the existence of third order analogues which satisfy a cubic recurrence, but there are sixth order analogues, some of which involve symmetric functions of the roots of cubic polynomials. Of course, there exist further sequences for $k > 3$, but as we have observed, it seems very likely that condition (1) will be violated. This, however, should not be regarded as a significant obstacle to further advancement of the theory. It would be of some interest to discover what further results might be derived for the sequences (U_n), (V_n) when $k > 3$. Very little is known.

Problems

13.1 (4th-order LDS) Consider the 4th-order linear recurrence

$$U_{n+4} = PU_{n+3} - (R + 2Q)U_{n+2} + PQU_{n+1} - Q^2U_n,$$

where $U_0 = 0, U_1 = 1, U_2 = P$ and $U_3 = P^2 - R - 3Q$. What are the first 15 terms in the following sequences for the given P, R and Q:

(a) **[BB]** $P = R = Q = 1$
(b) $P = 1, R = 2, Q = -1$
(c) **[BB]** $P = 1, R = 2, Q = 1$

13.2 (6th-order LDS) Consider the 6th-order linear recurrence

$$\begin{aligned}U_{n+6} = P_1U_{n+5} - (P_2 + 3R^2)U_{n+4} + (P_3 + 2R^2P_1)U_{n+3} - R^2(P_2 + 3R^2)U_{n+2}\\ + R^4P_1U_{n+1} - R^6U_n,\end{aligned}$$

where P_1, P_2 and P_3 are integers such that

$$P_3 = RP_1^2 - 2RP_2 - 4R^3$$

and the initial values for (U_n) are $U_0 = 0$, $U_1 = 1$, $U_2 = P_1 + 2R$, and $U_3 = P_1^2 + RP_1 - P_2 - 3R^2$. Use the fact that $U_{-n} = -U_n/R^{2n}$ to find the below in terms of P_1, P_2, P_3 and R:

(a) **[BB]** U_4
(b) U_5
(c) U_6

13.3 (6th **-order (not LDS)**) Consider the 6th-order linear recurrence

$$\begin{aligned}W_{n+6} = P_1W_{n+5} - (P_2 + 3R^2)W_{n+4} + (P_3 + 2R^2P_1)W_{n+3} - R^2(P_2 + 3R^2)W_{n+2}\\ + R^4P_1W_{n+1} - R^6W_n,\end{aligned}$$

where P_1, P_2 and P_3 are integers such that

$$P_3 = RP_1^2 - 2RP_2 - 4R^3$$

and the initial values for (W_n) are $W_0 = 6$, $W_1 = P_1$, $W_2 = P_1^2 - 2P_2 - 6R^2$, $W_3 = P_1^3 - 3P_1P_2 + 3RP_1^2 - 6RP_2 - 3R^2P_1 - 12R^3$. Use the fact that $W_{-n} = W_n/R^{2n}$ to find the below in terms of P_1, P_2, P_3 and R:

(a) **[BB]** W_4
(b) W_5
(c) W_6

Further Reading

Extensions of the Lucas sequences (history)

H. C. Williams, *Édouard Lucas and Primality Testing*, Wiley-Interscience, John Wiley and Sons, NY, 1998, pp. 450–464.

Linear difference equations

Samuel Goldberg, *Introduction to Difference Equations*, Dover Publications, NY, 1986.

The Tribonacci numbers

W. R. Spickerman, Binet's formula for the Tribonacci sequence, *Fibonacci Quarterly*, **20** (1982), 118–120.

H. C. Williams, A generalization of the Fibonacci numbers, *Proceedings of the Louisiana Conference on Combinatorics, Graph Theory and Computing*, Utilitas Mathematica Publishing, Winnipeg, 1970, pp. 340–356.

Classification of linear divisibility sequences

Sergiy Koshkin,Non-classical linear divisibility sequences and cyclotomic polynomials, *Fibonacci Quarterly*, **57** (2019), 68–80.

Andrew Granville, *Classifying Linear Division Sequences*, https://arxiv.org/abs/2206.11823

Linear divisibility sequences of order 4

H. C. Williams and R. K. Guy, Some fourth order linear divisibility sequences, *International Journal of Number Theory*, **7** (2011), 1255–1277.

Peter Bala, *A Family of Linear Divisibility Sequences of Order Four*, unpublished MS Feb. 2014, OEIS A238536.

H. C. Williams, *Édouard Lucas and Primality Testing*, Wiley-Interscience, John Wiley and Sons, NY, 1998, pp. 257–283.

Extended Lucas sequences

Christian J.-C. Ballot and Hugh C. Williams, *The Lucas Sequences: Theory and Applications*, CMS/CAIMS Books in Mathematics, Vol. 8, Springer Nature Switzerland AG, 2023, pp. 204–218.

E. L. Roettger and H. C. Williams and R. K. Guy, Some extensions of the Lucas functions, *Number Theory and Related Fields: In Memory of Alf van der Poorten*, Springer Proceedings in Mathematics and Statistics **43**, Springer Science and Business Media, New York, 2013, pp. 271–311.

E. L. Roettger and H. C. Williams, Some arithmetic properties of certain sequences, *J. of Integer Sequences*, **18** (2015), Article 15.6.2.

E. L. Roettger and H. C. Williams, A sextic extension of the Lucas functions, *Journal of Integer Sequences*, **23** (2020), Article 20.4.8.

Cubic extensions of Lucas sequences

E. L. Roettger, A Cubic Extension of the Lucas Functions, Ph.D. Thesis, University of Calgary, 2009. Available at https://prism.ucalgary.ca/handle/1880/103553

E. L. Roettger and H. C. Williams, Some primality tests constructed from a cubic extension of the Lucas functions, *Fibonacci Quarterly*, **59** (2021), pp. 194–213.

Chapter 14
Simple Continued Fractions

In this chapter, we will introduce and briefly describe a rather peculiar looking object called a *simple continued fraction*. Suppose we use the Euclidean algorithm to find $\gcd(34711, 49500)$. We get

$$\begin{aligned}
347111 &= 7 \cdot 49500 + 611,\\
49500 &= 81 \cdot 611 + 9,\\
611 &= 67 \cdot 9 + 8,\\
9 &= 1 \cdot 8 + 1,\\
8 &= 8 \cdot 1 + 0.
\end{aligned}$$

For each of the above equations, we can divide through by the divisors to produce

$$\begin{aligned}
&\frac{347111}{49500} = 7 + \frac{611}{49500} \rightarrow \frac{347111}{49500} = 7 + \cfrac{1}{\cfrac{49500}{611}},\\
&\frac{49500}{611} = 81 + \frac{9}{611} \rightarrow \frac{49500}{611} = 81 + \cfrac{1}{\cfrac{611}{9}},\\
&\frac{611}{9} = 67 + \frac{8}{9} \rightarrow \frac{611}{9} = 67 + \cfrac{1}{\cfrac{9}{8}},\\
&\frac{9}{8} = 1 + \frac{1}{8}.
\end{aligned}$$

E. L. F. Roettger and H. C. Williams, *The Enchantment of Numbers*, CMS/CAIMS Books in Mathematics 16, https://doi.org/10.1007/978-3-031-76568-1_14

Hence, we can easily write

$$\frac{347111}{49500} = 7 + \cfrac{1}{\cfrac{49500}{611}} = 7 + \cfrac{1}{81 + \cfrac{1}{\cfrac{611}{9}}} = 7 + \cfrac{1}{81 + \cfrac{1}{67 + \cfrac{1}{\cfrac{9}{8}}}}$$

$$= 7 + \cfrac{1}{81 + \cfrac{1}{67 + \cfrac{1}{1 + \cfrac{1}{8}}}}.$$

We define a *finite simple continued fraction* to be an expression of the form

$$a_0 + \cfrac{1}{a_1 + \cfrac{1}{a_2 + \ddots + \cfrac{1}{a_{n-1} + \cfrac{1}{a_n}}}}$$

where a_i are integers and $a_i > 0$ for $i > 0$. The word "simple" here refers to our putting all the numerators in this expression equal to 1. We call the a_i $(i \geq 0)$ the *partial quotients* of the continued fraction. For ease of writing we use $[a_0; a_1, a_2, \ldots, a_n]$ to denote the continued fraction above. For instance,

$$[3; 1, 4, 5] = 3 + \cfrac{1}{1 + \cfrac{1}{4 + \cfrac{1}{5}}}.$$

Recall that a rational number is one of the form a/b where a, b are integers and $b \neq 0$. It is easy to see that every finite simple continued fraction is equal to a rational number. For instance,

$$[7; 3, 4, 2] = 7 + \cfrac{1}{3 + \cfrac{1}{4 + \cfrac{1}{2}}} = 7 + \cfrac{1}{3 + \cfrac{1}{\cfrac{8}{2} + \cfrac{1}{2}}} = 7 + \cfrac{1}{3 + \cfrac{1}{\cfrac{9}{2}}} = 7 + \cfrac{1}{3 + \cfrac{2}{9}}$$

$$= 7 + \cfrac{1}{\frac{27}{9} + \frac{2}{9}} = 7 + \cfrac{1}{\frac{29}{9}} = 7 + \frac{9}{29} = \frac{212}{29}.$$

Furthermore, as our first example illustrates, we can always use the Euclidean algorithm to find a finite simple continued fraction expression for any given rational number.

Another method of evaluating $[a_0; a_1, a_2, \ldots, a_n]$ is to put $A_{-2} = 0$, $A_{-1} = 1$, $B_{-2} = 1$, and $B_{-1} = 0$ and define

$$A_{j+1} = a_{j+1}A_j + A_{j-1}, \quad B_{j+1} = a_{j+1}B_j + B_{j-1}, \quad (j \geq -1).$$

We call $\frac{A_i}{B_i}$ the ith *convergent* of the continued fraction

$$[a_0; a_1, a_2, \ldots, a_n] = \frac{A_n}{B_n}.$$

Notice that, the terms in the sequence B_0 $(= 1)$, B_1, $B_2, \ldots,$ up to B_n strictly increase. Again, we demonstrate this with $[a_0; a_1, a_2, a_3] = [7; 3, 4, 2]$. For $j = -1$ and continuing until $j = 2$, we have

$$\begin{aligned}
A_0 &= a_0 A_{-1} + A_{-2} = a_0 \cdot 1 + 0 = a_0 = 7,\\
B_0 &= a_0 B_{-1} + B_{-2} = a_0 \cdot 0 + 1 = 1,\\
A_1 &= a_1 A_0 + A_{-1} = 3 \cdot 7 + 1 = 22,\\
B_1 &= a_1 B_0 + B_{-1} = 3 \cdot 1 + 0 = 3,\\
A_2 &= a_2 A_1 + A_0 = 4 \cdot 22 + 7 = 95,\\
B_2 &= a_2 B_1 + B_0 = 4 \cdot 3 + 1 = 13,\\
A_3 &= a_3 A_2 + A_1 = 2 \cdot 95 + 22 = 212,\\
B_3 &= a_3 B_2 + B_1 = 2 \cdot 13 + 3 = 29.
\end{aligned}$$

Thus, $\frac{A_3}{B_3} = \frac{212}{29}$.

Notice that since, $8 = 7 + \frac{1}{1}$, we have

$$\frac{347111}{49500} = [7; 81, 67, 1, 8] = [7; 81, 67, 1, 7, 1].$$

In fact, rational numbers can have two representations as simple continued fractions

$$[a_0; a_1, a_1, \ldots, a_{n-1}, a_n] \quad \text{and} \quad [a_0; a_1, a_1, \ldots, a_{n-1}, a_n - 1, 1]$$

whenever $a_n > 1$.

Formally, to find the simple continued fraction of $\frac{K}{L}$, where K and L are integers and $L > 0$. We put $R_{-2} = K$, $R_{-1} = L$ and define

$$\begin{aligned} a_j &= \lfloor R_{j-2}/R_{j-1} \rfloor, \\ R_{j-2} &= a_j R_{j-1} + R_j, \end{aligned}$$

for $j = 0, 1, 2, \ldots, n-1$. Note that $0 < R_j < R_{j-1}$. We can prove by induction that

$$(-1)^{j+1} R_j = LA_j - KB_j \text{ and } L = B_j R_{j-1} + B_{j-1} R_j \quad (j = 0, 1, 2, \ldots, n-1).$$

This process is easily recognized as the Euclidean algorithm, and accordingly, we stop when $R_n = 0$ for some n; thus, yielding

$$\frac{K}{L} = [a_0; a_1, a_2, \ldots, a_n].$$

Of course, we also have that the $\gcd(K, L) = R_{n-1}$.

A *Diophantine equation* is an indeterminate equation whose unknowns are constrained to be either integral or rational. The study of such equations goes back at least to the ancient Greeks; indeed, they are named after Diophantus of Alexandria (c. 200–284 CE) in honour of his work on them. For example, the *linear Diophantine equation in n variables* is a Diophantine equation

$$a_1 x_1 + a_2 x_2 + \cdots + a_n x_n = b,$$

where a_i, b are given integers, $a_i \neq 0$ and $x_1, \ldots, x_n$ are the unknowns. We can use the algorithm above to solve a linear Diophantine equation in 2 variables,

$$ax + by = c,$$

where a, b, and c are given integers and $a, b \neq 0$. We mentioned in Chap. 2 that for this equation to have integer solutions, it must be the case that d $(= \gcd(a, b))$ divides c. We therefore make this assumption and put $\overline{a} = a/d$, $\overline{b} = b/d$, and $\overline{c} = c/d$. Now, if we put $K = a$ and $L = b > 0$ (we can assume $b > 0$ as its sign can be absorbed by y), we see

$$\frac{\overline{a}}{\overline{b}} = \frac{a}{b} = [a_0; a_1, \ldots, a_n] = \frac{A_n}{B_n}.$$

It can be established by induction that

$$A_j B_{j-1} - B_j A_{j-1} = (-1)^{j-1};$$

hence, $\gcd(A_n, B_n) = 1$, and since $\gcd(\overline{a}, \overline{b}) = 1$, $B_n > 0$, $b > 0$, we must have $\overline{a} = A_n$ and $\overline{b} = B_n$. We then have

$$\overline{a}B_{n-1} - \overline{b}A_{n-1} = (-1)^{n-1};$$

thus,

$$\overline{a}(-1)^{n-1}\overline{c}B_{n-1} + \overline{b}(-1)^{n}\overline{c}A_{n-1} = \overline{c}$$

or

$$a(-1)^{n-1}\overline{c}B_{n-1} + b(-1)^{n}\overline{c}A_{n-1} = c.$$

Thus, one solution to $ax + by = c$ is given by

$$x_0 = (-1)^{n-1}\overline{c}B_{n-1}, \quad y_0 = (-1)^{n}\overline{c}A_{n-1}.$$

In order to express all solutions to $ax + by = c$, let x, y be any other solution. It must be the case that

$$ax + by = c = ax_0 + by_0 \rightarrow \overline{a}(x - x_0) = -\overline{b}(y - y_0).$$

Since, $\gcd(\overline{a}, \overline{b}) = 1$, we see that

$$x - x_0 = \overline{b}t \quad \text{and} \quad y - y_0 = -\overline{a}t$$

for some integer parameter t. Thus, all the solutions are represented by

$$x = x_0 + \overline{b}t \quad \text{and} \quad y = y_0 - \overline{a}t.$$

For example, we can use the above method to find all solutions to the linear Diophantine equation

$$694222x + 99000y = 60;$$

here, $a = 694222$, $b = 99000$, and $c = 60$. We can see from our earlier example that

$$\frac{694222}{99000} = \frac{347111}{49500} = [7; 81, 67, 1, 7, 1] = [a_0; a_1, a_2, a_3, a_4, a_5].$$

Thus, as the $\gcd(a, b) = 2$, we get $\overline{a} = 347111$, $\overline{b} = 49500$, $\overline{c} = 30$, and $n = 5$. Also,

$$\frac{A_{n-1}}{B_{n-1}} = \frac{A_4}{B_4} = [7; 81, 67, 1, 7] = \frac{308480}{43991}.$$

We can then set

$$x_0 = (-1)^{n-1}\overline{c}B_{n-1} = (-1)^{5-1} \cdot 30 \cdot 43991 = 1319730,$$

$$y_0 = (-1)^n \overline{c} A_{n-1} = (-1)^5 \cdot 30 \cdot 308480 = -9254400.$$

Our complete set of solutions to $694222x + 99000y = 60$ is now given by

$$x = x_0 + \overline{b}t = 1319730 + 49500t,$$
$$y = y_0 - \overline{a}t = -9254400 - 347111t.$$

We can also use continued fractions to solve the problem of representing a prime as a sum of two squares. Suppose p is a prime and

$$p = X^2 + Y^2$$

for integers X and Y. Note that since $\gcd(X, Y)$ here must be 1 and $p \nmid XY$, we must have

$$(XY^{-1})^2 \equiv -1 \pmod{p}.$$

If $p \equiv 3 \pmod 4$, this congruence has no solutions because the Legendre symbol (which was discussed in Chap. 5) $(-1|p) = (-1)^{\frac{p-1}{2}} = -1$. If $p = 2$, then $p = 1^2 + 1^2$; thus, we are left with this problem for $p \equiv 1 \pmod 4$. Since, in this case, $(-1)^{\frac{p-1}{2}} = 1$, we have $(-1|p) = 1$ and there must exist some integer c such that $0 < c < p/2$, $\gcd(p, c) = 1$ and

$$c^2 \equiv -1 \pmod{p}.$$

If we apply the Euclidean algorithm with $K = R_{-2} = c$ and $L = R_{-1} = p$, we get the sequence of remainders $R_{-2}, R_{-1}, R_0, R_1, \ldots$, where if $a_j = \lfloor R_{j-2}/R_{j-1} \rfloor$, then $R_j = R_{j-2} - a_j R_{j-1}$. We must reach the point where the denominators of the convergents A_n/B_n in the continued fraction expansion of c/p satisfy

$$1 \leq B_k < \sqrt{p} < B_{k+1}.$$

This is because $A_n = c$ and $B_n = p$ when $R_n = 0$ and (B_k) is an increasing sequence for $k \geq 0$. Since $p = L = B_{k+1}R_k + B_k R_{k+1}$, we see that $p > B_{k+1}R_k$ or $R_k < p/\sqrt{p} = \sqrt{p}$. Furthermore, since

$$(-1)^{k+1} R_k = LA_k - KB_k = pA_k - cB_k,$$

we have

$$|pA_k - cB_k| < \sqrt{p}.$$

It follows that

$$1 \leq B_k^2 + (pA_k - cB_k)^2 < 2p$$

since $p \mid c^2 + 1$, this means that since

$$B_k^2 + (pA_k - cB_k)^2 \equiv (1 + c^2)B_k^2 \equiv 0 \pmod{p},$$

we must have

$$B_k^2 + (pA_k - cB_k)^2 = p.$$

Thus, we have a constructive proof of the result, going back to Fermat, that any prime p ($\equiv 1 \pmod 4$) can always be represented as a sum of two integral squares. This construction is due to the French mathematician Charles Hermite (1822–1901), who discovered it in 1848.

For example, consider the prime $p = 2621$. Certainly $p \equiv 1 \pmod 4$, and hence, we can apply the construction above to write p as a sum of two squares. It is easily verified that the value $c = 472$ satisfies the requirements that $c < p/2$ and $c^2 \equiv -1 \pmod p$. Using methods from the beginning of the chapter, we see that

$$\frac{c}{p} = \frac{472}{2621} = [0; 5, 1, 1, 4, 4, 1, 1, 5].$$

Now, upon calculating convergents we see

$$\frac{A_5}{B_5} = [0; 5, 1, 1, 4, 4] = \frac{38}{211},$$
$$\frac{A_4}{B_4} = [0; 5, 1, 1, 4] = \frac{9}{50};$$

hence, for $k = 4$ we have $1 \le B_k < \sqrt{p} < B_{k+1}$. Substitution into

$$B_k^2 + (pA_k - cB_k)^2 = p$$

yields

$$50^2 + (-11)^2 = 2621.$$

An infinite sequence $a_0, a_1, a_2, \ldots$ of integers, all positive except possibly a_0, determines an *simple infinite continued fraction* $[a_0; a_1, a_2, \ldots]$ or SICF. The convergents C_i $(i = 0, 1, 2, \ldots)$ of this type of continued fraction are defined in the same way as those for a finite simple continued fraction. The value of

$$\alpha = [a_0; a_1, a_2, \ldots]$$

is defined as

$$\lim_{n\to\infty} C_n.$$

Such a limit always exists.

A real number which is not the ratio of two integers (i.e., rational) is said to be *irrational*. A simple example, known since antiquity, is $\sqrt{2}$. If $\sqrt{2}$ were rational, say

$\sqrt{2} = \frac{r}{s}$ where $\gcd(r, s) = 1$, then $2 = \frac{r^2}{s^2}$ or $2s^2 = r^2$. Since $s \mid r^2$ and $\gcd(r, s) = 1$, we must have $s^2 = 1$, and therefore $2 = r^2$, which is impossible because 2 is not a perfect integral square. The value of α above is also irrational and its continued fraction expression is unique.

Now suppose α is any given irrational number and define the sequence a_0, a_1, $a_2, \ldots$, using $\alpha_0 = \alpha$,

$$a_k = \lfloor \alpha_k \rfloor, \quad \text{and} \quad \alpha_{k+1} = \frac{1}{\alpha_k - a_k},$$

where $k = 0, 1, 2, \ldots$. We then have $\alpha = [a_0; a_1, a_2, a_3, \ldots]$ and also $\alpha = [a_0; a_1, a_2, \ldots, a_{n-1}, \alpha_n]$; however, here we can no longer guarantee that we have a finite simple continued fraction, as we have now introduced a partial quotient that might not be integral. It can be shown that $[a_0; a_1, a_2, \ldots, a_{n-1}, \alpha_n]$ converges to α as n goes to infinity; furthermore, this expression of α as a simple continued fraction is unique. Notice that since $0 < \lfloor \alpha_k \rfloor < \alpha_k$, we have $\alpha_i > 1$ for $i \geq 1$. We call α_k a *complete quotient* in the simple continued fraction for α. There is considerable evidence that the ancient Greeks were aware of something very like continued fractions as a consequence of their process of anthyphairesis.

It is not difficult to show that

$$|\alpha - C_i| = |\alpha - \frac{A_i}{B_i}| < \frac{1}{B_i B_{i+1}} < \frac{1}{a_{i+1} B_i^2} \quad (i \geq 0).$$

Thus, the convergents of the simple continued fraction of our irrational α provide very good rational approximations to the value of α. Indeed, a theorem of the German mathematician Adolf Hurwitz (1859–1919) asserts that of every three consecutive convergents of the simple continued fraction expansion of an irrational α, at least one satisfies

$$|\alpha - C_i| < \frac{1}{\sqrt{5} B_i^2}.$$

This means that for any irrational number α, there exists an infinitude of distinct rational numbers a/b such that

$$|\alpha - a/b| < \frac{1}{\sqrt{5} b^2}.$$

Also, the constant $\sqrt{5}$ in this result is best possible, as the above statement cannot hold if $\sqrt{5}$ is replaced by a larger value.

At this point we should remark that in 1842, Dirichlet proved, by using the pigeonhole principle discussed in Chap. 9, what is now called *Dirichlet's approximation theorem*.

Theorem 14.1 *For any real numbers α and N with $N \geq 1$, there exist integers p, q such that $1 \leq q \leq N$ and*

$$|q\alpha - p| < \frac{1}{N}.$$

From this it is easy to deduce, without resorting to continued fractions, that any real number has a sequence of good rational approximations. Indeed,

$$|\alpha - p/q| < \frac{1}{q^2}$$

holds for infinitely many pairs of integers (p, q). As we have seen, the theory of continued fractions offers us a constructive proof of this result.

On the other hand, we have the *Thue-Siegel-Roth theorem*, commonly shortened to *Roth's theorem*. This states that for any positive ϵ the inequality

$$|\alpha - p/q| < \frac{1}{q^{2+\epsilon}}$$

can have only finitely many solutions in coprime integers p and q. This theorem is the result of work by the Norwegian mathematician Axel Thue (1863–1922), the German mathematician Carl Ludwig Siegel (1896–1981) and the German-born British mathematician Klaus Roth (1925–2015). The eventual proof of it, as we have stated it here, in 1955 won Roth the Fields Medal, considered to be the Nobel Prize of mathematics.

We also have the following remarkable result concerning rational approximations to irrationals:

Theorem 14.2 *Suppose that a and b are integers and b is positive. If*

$$|\alpha - \frac{a}{b}| < \frac{1}{2b^2},$$

then $\frac{a}{b}$ must be a convergent in the simple continued fraction expression of α.

We see, then, that not only are the convergents to $\alpha = [a_0; a_1, a_2, \ldots]$ good rational approximations to α, but any good rational approximation to α must be such a convergent. As an example, consider the continued fraction expansion of

$$\begin{aligned}\pi &= 3.14159265358979323846264338 32\ldots \\ &= [3; 7, 15, 1, 292, 1, 1, 1, 2, 1, 3, 1, 14, 2, 1, 1, 2, 2, 2, 2, 1, 84, \ldots].\end{aligned}$$

We find the convergents $C_0 = 3$, $C_1 = \frac{22}{7}$, $C_2 = \frac{333}{106}$, $C_3 = \frac{355}{113}$,... . Notice that

$$C_3 = 3.14159292035398230088495575 22\ldots,$$

which is within 0.000009% of the value of π. This approximation for π is sometimes called Milü and was discovered by the Chinese mathematician Zu Chongzhi (429–500 CE) in the fifth century. It is because the convergents provide a very accurate representation of π, that those who like to recite the decimal representation of π have been encouraged instead to memorize the sequence of partial quotients in its continued fraction.

We say a real number α is a *quadratic irrational*, if α is irrational and is a root of a quadratic equation with integer coefficients, i.e. there exist integers a ($\neq 0$), b, c such that

$$a\alpha^2 + b\alpha + c = 0.$$

It follows that

$$\alpha = \frac{-b \pm \sqrt{D}}{2a},$$

where $D = b^2 - 4ac = m^2 d$, and m and d are integers with d square-free and $d \neq 1$. We denote by $\mathbb{Q}(\sqrt{d})$, the set of all numbers of the form

$$\frac{r + s\sqrt{d}}{t},$$

where r, s, t are integers and $t \neq 0$. We call d the *radicand* of $\mathbb{Q}(\sqrt{d})$. Note that $\mathbb{Q}(\sqrt{d}) = \mathbb{Q}(\sqrt{D})$. If α is any element of $\mathbb{Q}(\sqrt{d})$, we must have some integers u, v, w with $w \neq 0$ such that

$$\alpha = \frac{u + v\sqrt{d}}{w};$$

we define the *conjugate* $\bar{\alpha}$ of this α as

$$\bar{\alpha} = \frac{u - v\sqrt{d}}{w}.$$

If α and β are in $\mathbb{Q}(\sqrt{d})$, then it is easy to verify that $\overline{\alpha + \beta} = \bar{\alpha} + \bar{\beta}$ and $\overline{\alpha\beta} = \bar{\alpha}\bar{\beta}$; also, if k is any integer, then k is in $\mathbb{Q}(\sqrt{d})$ for any suitable d and $\bar{k} = k$.

We can write any quadratic irrational α with $D > 0$ as

$$\alpha = \frac{P + \sqrt{D}}{Q},$$

where P, Q are integers. We may also assume that $Q \mid D - P^2$. For if this is not the case, simply replace P by $P|Q|$, Q by $Q|Q|$ and D by $Q^2 D$. Now put

$$P_0 = P, \quad Q_0 = Q, \quad a_0 = \left\lfloor \frac{P + \sqrt{D}}{Q} \right\rfloor$$

and define

$$P_k = a_{k-1}Q_{k-1} - P_{k-1}, \quad Q_k = \frac{D - P_k^2}{Q_{k-1}}, \quad a_k = \left\lfloor \frac{P_k + \sqrt{D}}{Q_k} \right\rfloor \quad (k \geq 1).$$

It is a simple matter to establish for $\alpha_0 = \alpha$ that

$$\alpha_k = \frac{P_k + \sqrt{D}}{Q_k} \quad (k \geq 1).$$

For example, suppose $\alpha = \alpha_0 = \sqrt{85}$. Putting $P_0 = 0$, $Q_0 = 1$ and $D = 85$, we get

$$\alpha_1 = \frac{9 + \sqrt{85}}{4}, \qquad \alpha_2 = \frac{7 + \sqrt{85}}{9}, \qquad \alpha_3 = \frac{2 + \sqrt{85}}{9},$$
$$\alpha_4 = \frac{7 + \sqrt{85}}{4}, \qquad \alpha_5 = 9 + \sqrt{85}, \qquad \alpha_6 = \frac{9 + \sqrt{85}}{4} = \alpha_1.$$

Thus,

$$\sqrt{85} = [9; 4, 1, 1, 4, 18, 4, 1, 1, 4, 18, 4, 1, 1, 4, 18, 4, 1, 1, 4, 18, \ldots].$$

If α is any quadratic irrational, it can be shown that there exists some positive integer j such that $\overline{\alpha}_j < 0$. When this occurs, we must have $\overline{\alpha}_k < 0$ for all $k \geq j$. It follows, then, that if $k \geq j$, we have

$$0 < Q_k < 2\sqrt{D}, \quad |P_k| < \sqrt{D}, \quad 0 < a_k < 2\sqrt{D}.$$

Thus, for a given D the values of both Q_k and $|P_k|$ are bounded, and therefore there can only exist a finite number of distinct values for α_k whenever $k \geq j$. It follows from Dirichlet's pigeonhole principle that there must exist minimal integers j (≥ 0) and p (≥ 1) such that

$$\alpha_{j+p} = \alpha_j$$

from which it is easy to deduce that

$$\alpha_{m+p} = \alpha_m$$

for all $m \geq j$. We can then write

$$\alpha = [a_0, a_1, a_2, \ldots, a_{j-1}; \overline{a_j, a_{j+1}, \ldots, a_{j+p-1}}],$$

where the bar is placed over a repeating or periodic set of partial quotients. The ordered set of partial quotients $\{a_j, a_{j+1}, \ldots, a_{j+p-1}\}$ is called the *period* of the simple continued fraction expansion of α and the integer p is called the *period length*. Thus, $\sqrt{85} = [9; \overline{4, 1, 1, 4, 18}]$ and the period length is 5 in this case. It is an old result of the Italian born, French naturalized, mathematician and astronomer Joseph-Louis Lagrange (1736–1813), that the infinite simple continued fraction of an irrational number is periodic if and only if the number is a quadratic irrational.

If α is a quadratic irrational, we say that the SICF expansion of α is *purely periodic* when

$$\alpha = [\overline{a_0; a_1, a_2, \ldots, a_{p-1}}]$$

for some integer $p \geq 1$. In this case we have

$$\alpha = [a_0; a_1, a_2, \ldots, a_{p-1}, \alpha].$$

It is not difficult to show that the SICF of α is purely periodic if and only if

$$\alpha > 1 \quad \text{and} \quad -1 < \overline{\alpha} < 0.$$

A simple example of this phenomenon occurs when $\alpha = \phi$, the golden ratio of Chap. 8. In this case we have $\phi > 1$ and

$$-1 < \overline{\phi} = \frac{1 - \sqrt{5}}{2} < 0.$$

We find that

$$\phi = [1; 1, 1, 1, 1, \ldots] = [\overline{1}].$$

The convergents C_n $(n \geq 0)$ of this continued fraction are given by the quotients of adjacent Fibonacci numbers, i.e.

$$C_n = \frac{F_{n+2}}{F_{n+1}}.$$

We say that an irrational β is a *cubic irrational* if it is not a quadratic irrational, but is a root of the cubic equation

$$ax^3 + bx^2 + cx + d = 0,$$

where a, b, c, d are integers. For example, $\sqrt[3]{2}$ is a cubic irrationality. We see by Lagrange's result that β cannot have a periodic simple continued fraction expansion, but there exists a different type of continued fraction for which the expansion of β does exhibit periodicity. This remarkable continued fraction was discovered in 1896 by the Ukrainian mathematician Georgy Voronoi (1868–1908).

We conclude this chapter with a useful identity related to the continued fraction expansion of $\sqrt{D}$. We know that we can write

$$\alpha = [a_0; a_1, a_2, \ldots, a_{n-1}, \alpha_n].$$

Hence,

$$\alpha = \frac{\alpha_n A_{n-1} + A_{n-2}}{\alpha_n B_{n-1} + B_{n-2}}.$$

Since $\alpha_n = (P_n + \sqrt{D})/Q_n$, we get for $\alpha = \sqrt{D}$

$$Q_n A_{n-2} = DB_{n-1} - P_n A_{n-1},$$
$$Q_n B_{n-2} = A_{n-1} - P_n B_{n-1},$$

on equating rational and irrational parts. We know that $A_{n-2} = A_n - a_n A_{n-1}$, $B_{n-2} = B_n - a_n B_{n-1}$, $P_{n+1} = a_n Q_n - P_n$, and by substituting these quantities in the above identities we find that

$$Q_n A_n = P_{n+1} A_{n-1} + DB_{n-1},$$
$$Q_n B_n = P_{n+1} B_{n-1} + A_{n-1}.$$

Furthermore, since

$$A_{n-1} B_{n-2} - B_{n-1} A_{n-2} = (-1)^n,$$

we can derive the beautiful identity

$$A_{n-1}^2 - DB_{n-1}^2 = (-1)^n Q_n.$$

Problems

14.1 (Simple continued fractions) Use the Euclidean algorithm to help you write the following fractions as simple continued fractions (show all of your work):
(a) **[BB]** $\frac{789}{123}$ (b) $\frac{98912}{999}$ (c) $\frac{456}{12399}$ (d) $\frac{1444}{27}$

14.2 (Simple continued fractions) Write the following simple continued fractions as rational numbers by adding fractions (show all of your work):
(a) **[BB]** $[4; 3, 7, 5]$
(b) $[27; 1, 17, 2]$
(c) $[9; 13, 7, 15]$
(d) $[0; 13, 2, 4]$

14.3 (Simple continued fractions) Write the following simple continued fractions as rational numbers by putting $A_{-2} = 0$, $A_{-1} = 1$, $B_{-2} = 1$, and $B_{-1} = 0$ and using

$$A_{j+1} = a_{j+1} A_j + A_{j-1}, \quad B_{j+1} = a_{j+1} B_j + B_{j-1}, \quad (j \geq -1)$$

and

$$[a_0; a_1, a_2, \ldots, a_n] = \frac{A_n}{B_n}$$

(show all of your work):
(a) **[BB]** [4; 3, 7, 5]
(b) [27; 1, 17, 2]
(c) [9; 13, 7, 15]
(d) [0; 13, 2, 4]

14.4 (Diophantine equation) Find all possible solutions to the following linear Diophantine equations in two variables (show all of your work):
(a) **[BB]** $143x + 77y = 55$
(b) $72x + 42y = 30$
(c) $1919x + 4199y = 418$

14.5 (Sum of squares) Use continued fractions to express the given prime p ($p \equiv 1 \pmod 4$) and the value c such that $c^2 \equiv -1 \pmod p$ as a sum of two squares (show all of your work):
(a) **[BB]** $p = 137, c = 37$
(b) $p = 313, c = 25$
(c) $p = 853, c = 333$

14.6 (Quadratic irrationals) Express the following quadratic irrationals as infinite simple continued fractions (show all of your work):
(a) $\sqrt{17}$
(b) **[BB]** $\sqrt{83}$
(c) $\frac{2+\sqrt{32}}{7}$

Further Reading

Diophantus

Sir Thomas L. Heath, *Diophantus of Alexandria: A Study in the History of Greek Algebra*, Dover Publications, NY, 1964.

Simple continued fractions (of quadratic irrationalities)

Ivan Niven, Herbert S. Zuckerman, Hugh L. Montgomery, *An Introduction to the Theory of Numbers*, 5th Edition, John Wiley and Sons, NY, 1991, pp. 325–359.
Michael J. Jacobson, Jr. and Hugh C. Williams, *Solving the Pell Equation*, Canadian Mathematical Society Books in Mathematics, Springer Science+Business Media, LLC, NY, 2009, pp. 43–73.

Anthyphairesis

W. Knorr, *The Evolution of the Euclidean Elements*, D. Reidel, Dordrecht, Boston, 1975.

David Fowler, *The Mathematics of Plato's Academy: A New Reconstruction*, 2nd Edition, Clarendon Press, Oxford, 1999.

Continued fraction of π

Evelyn Lamb, Don't recite the digits to celebrate pi. Recite its continued fraction instead., *Roots of Unity*, Scientific American Blogs, March 11, 2015. Available at: https://blogs.scientificamerican.com/roots-of-unity/don-8217-t-recite-digits-to-celebrate-pi-recite-its-continued-fraction-instead/

Irrational numbers

Julian Havil, *The Irrationals: the Story of the Numbers You Can't count on*, Princeton University Press, Princeton, NJ, 2012.

Cubic irrationalities

B. N. Delone and D. K. Faddeev, *The Theory of Irrationalities of the Third Degree*, Translations of Mathematical Monographs, Vol. 10, American Mathematical Society, Providence, RI, 1964.

Samuel Hambleton and Hugh C. Williams, *Cubic Fields with Geometry*, CMS Books in Mathematics, Springer Nature, Cham, Switzerland, 2018.

Chapter 15
Integer Factorization

Let N be any composite integer larger than 1. The *integer factoring problem* for N is that of determining some factor f of N (Chap. 2) such that $1 < f < N$ and f divides N. Notice that if f is a factor of N, then so is N/f. For example, if $N = 85$, then 5 and 13 are both factors of N. The factoring problem has a very long history, starting in antiquity, but nothing truly significant was done until the time of Fermat. From then on progress has been steady, if not always spectacular, up to the present day. Of course, it is possible to factor a very large number if it has a small prime divisor, but we will be concerned with integers which have only large prime divisors. Richard Guy discussed the state of the factoring problem in 1975; at that point, in spite of the availability of computer technology, most of the methods used to factor integers were not very powerful; indeed, up to then the largest integer ever factored by a general purpose technique was the 39 digit Fermat number F_7. The distinguished Dutch computational number theorist, Hendrik Lenstra Jr., once observed that it seemed a new method was always needed to be devised to factor the next unfactored Fermat number. However, the main impetus in the modern invention of new factoring routines was the RSA cryptosystem that we saw in Chap. 7.

We begin our exploration into factoring by first looking at pre-computer-age factoring techniques. The earliest interesting factorization technique was given to us by Fermat in 1643. We begin by assuming that N is composite and odd; thus, $N = rs$ with both r and s being odd and without loss of generality $r \leq \sqrt{N}$. Upon putting

$$a = \frac{r+s}{2} \quad \text{and} \quad b = \frac{s-r}{2},$$

we then have $N = a^2 - b^2$ and $\sqrt{N} < a < (N+1)/2$. We can then factor N by searching values of $x = a^2 - N$, for a values $\lfloor\sqrt{N}\rfloor + 1, \lfloor\sqrt{N}\rfloor + 2, \ldots, (N+1)/2$, until one of these values yields a perfect square for x. For example, let us use this method to factor $N = 10033$, here $\lfloor\sqrt{N}\rfloor = 100$; hence:

E. L. F. Roettger and H. C. Williams, *The Enchantment of Numbers*, CMS/CAIMS Books in Mathematics 16, https://doi.org/10.1007/978-3-031-76568-1_15

$$a = \lfloor\sqrt{N}\rfloor + 1 = 101, \quad x = a^2 - N = 101^2 - 10033 = 168, \quad \text{(not helpful)}$$
$$a = \lfloor\sqrt{N}\rfloor + 2 = 102, \quad x = a^2 - N = 102^2 - 10033 = 371, \quad \text{(not helpful)}$$
$$a = \lfloor\sqrt{N}\rfloor + 3 = 103, \quad x = a^2 - N = 103^2 - 10033 = 576 = 24^2 \quad \text{(helpful)}.$$

Thus $a = 103$ and $b = 24$ and since $N = a^2 - b^2 = (a - b)(a + b)$ and so $N = (103 + 24)(103 - 24) = 127 \cdot 79$. This method of finding a difference of squares such that $N = a^2 - b^2$ is now called *Fermat factorization*.

Fermat also pointed out that if an integer is a perfect square, then its value modulo 100 must belong to the set

$$\{0, 1, 4, 9, 16, 21, 24, 25, 29, 36, 41, 44, 49, 56, 61, 64, 69, 76, 81, 84, 89, 96\}$$

or, equivalently, its last two digits must belong to the set

$$\{00, 01, 04, 09, 16, 21, 24, 25, 29, 36, 41, 44, 49, 56, 61, 64, 69, 76, 81, 84, 89, 96\}.$$

This saves time, as we notice that in the first two steps of the previous example, the last two digits are 68 and 71, and since neither of these is in our set of possible last two digits for a square, we know that neither of them can be a square without the need for further computation. This method is now called *modular exclusion*. We remark that Fermat's technique is only useful when r and s are relatively close in size. For instance, if s is much larger than r, say $s = 1433$ and $r = 7$, we then have $N = 10031$; a number comparable in size to the one previously factored. Again, $\lfloor\sqrt{N}\rfloor = 100$, so our first value to be checked for a is 101, but now we do not arrive at a perfect square for x until $a = 720$, at which point $x = 508369$.
We can use a difference of squares method in the case where we have $a^2 \equiv b^2 \pmod{N}$, as long as $a \not\equiv \pm b \pmod{N}$. If these conditions are met then it must be the case that N is composite. Moreover, $(a - b)(a + b) \equiv 0 \pmod{N}$, thus the $\gcd(a - b, N)$ and $\gcd(a + b, N)$ will yield nontrivial factors of N. The difficulty here is, of course, finding a, b such that the above is true. This can be done by initially finding a set of values of c such that

$$x^2 \equiv c \pmod{N}.$$

The idea on how to do this follows from a series of papers published between 1885 and 1887 by Paul Seelhoff (1829–1896). He began by setting $N = w^2 + r$, where $w = \lfloor\sqrt{N}\rfloor$. Recall that if p is any prime such that $(N|p) = 1$ (the Legendre symbol, which was introduced in Chap. 5), then we can solve

$$s^2 \equiv N \pmod{p^k}.$$

We can then put $m = w \pm (s + yp^k)$, where y is some integer. If we now set $t = m(2w - m) + r$, then it is easily verified that since $p^k \mid N - s^2$ and $m - w \equiv \pm s \pmod{p^k}$, we have $N = (w - m)^2 + t$ and $p^k \mid t$. Observe that if the absolute value of the size of m is near p^k, the corresponding t values will tend to be about the size $2p^k\sqrt{N}$, and since we already know that p^k divides them they should be relatively easy to factor. Doing this for a number of primes p (where $(N|p) = 1$) and various values for both y and k, we can build up a bank of equations of the form

$$x^2 \equiv c \pmod{N}.$$

Seelhoff suggested for values of N of about 15 digits, using the primes up to 97. The values for k were subject to the following conditions: for

$$\begin{aligned} p &\geq 7, \quad k \leq 2; \\ p &= 5, \quad k \leq 4; \\ p &= 3, \quad k \leq 6; \\ p &= 2, \quad k \leq 10. \end{aligned}$$

From these Seelhoff was able to select certain preliminary congruences in order to set up a congruence like

$$a^2 \equiv b^2 \pmod{N}.$$

For example, let $N = 25021$, we will use the above methods to factor N. We begin by calculating $w = \lfloor\sqrt{N}\rfloor = 158$; hence, $N = w^2 + r = 158^2 + 57$. We can find a number of congruences of the form

$$x^2 \equiv c \pmod{N},$$

by using the primes

$$\{2, 3, 13, 19, 29, 31, \ldots\}$$

until we find a set of useful congruences. For instance, after just a few trials when $p = 5$ and $k = 3$, we get

$$s^2 \equiv N \pmod{p^k} \to 39^2 \equiv N \pmod{5^3}.$$

Here, if we set $y = 1$ we can put $m = w + (s + p^k) = 158 + (39 + 5^3) = 322$ and $t = m(2w - m) + r = 322(2 \cdot 158 - 322) + 57 = -1875$. Since

$$N = (w - m)^2 + t = (158 - 322)^2 + (-1875),$$

we have

$$164^2 \equiv 1875 \pmod{N} \to 164^2 \equiv 3 \cdot 5^4 \pmod{N}.$$

In the case that $p = 13$ and $k = 2$, we have

$$s^2 \equiv N \pmod{p^k} \rightarrow 3^2 \equiv N \pmod{13^2}.$$

Again, we use $y = 1$, but this time set $m = w - (s + p^k) = 158 - (3 + 13^2) = -14$ and $t = m(2w - m) + r = -14(2 \cdot 158 - (-14)) + 57 = -4563$. Here, since

$$N = (w - m)^2 + t = (158 - (-14))^2 + (-4563),$$

we have

$$172^2 \equiv 4563 \pmod{N} \rightarrow 172^2 \equiv 3^3 \cdot 13^2 \pmod{N}.$$

At which point we can stop looking for new congruences by observing that from

$$164^2 \equiv 3 \cdot 5^4 \pmod{N} \quad \text{and} \quad 172^2 \equiv 3^3 \cdot 13^2 \pmod{N},$$

we have

$$164^2 \cdot 172^2 \equiv 3^4 \cdot 5^4 \cdot 13^2 \pmod{N},$$

or rather

$$28208^2 \equiv 2925^2 \pmod{25021} \rightarrow 3187^2 \equiv 2925^2 \pmod{25021}.$$

Thus, we have

$$N \mid (3187 - 2925) \cdot (3187 + 2925).$$

We can then use the Euclidean algorithm to compute $\gcd(3187 - 2925, 25021) = 131$, and thus we can see that $N = 25021 = 131 \cdot 191$.

We discussed in Chap. 11 some of the early attempts to factor the Fermat numbers, those numbers of the form

$$F_n = 2^{2^n} + 1.$$

Euler was able to show that if $p \mid F_n$, then p must have the form $k2^{n+1} + 1$. The justification of this result is very straightforward. For, if you assume $p \mid F_n$, we then have $2^{2^n} + 1 \equiv 0 \pmod{p} \rightarrow 2^{2^n} \equiv -1 \pmod{p}$. Squaring both sides of this congruence yields $2^{2^{n+1}} \equiv 1 \pmod{p}$. We can now see that r, the order of 2 modulo p, divides 2^{n+1}, but does not divide 2^n. It follows that $r = 2^{n+1}$; thus, $2^{n+1} \mid \phi(p) \rightarrow \phi(p) = k2^{n+1}$. Since p is a prime, we have $\phi(p) = p - 1$ and consequently $p - 1 = k2^{n+1} \rightarrow p = k2^{n+1} + 1$ for some integer k.

Lucas strengthened this result to all prime divisors of the Fermat number F_n ($n \geq 2$), must have the form $k'2^{n+2} + 1$. We already have by Euler's result that $p = k2^{n+1} + 1$ and since $n \geq 2$ we must have $8 \mid k2^{n+1}$; thus, $p \equiv 1 \pmod{8}$. By the laws of quadratic reciprocity we saw in Chap. 5 (Theorem 5.2), we know that $(2|p) = 1$, i.e. there exists some x such that $x^2 \equiv 2 \pmod{p}$. Using our earlier observations

$$2^{2^n} \equiv -1 \pmod{p} \quad \text{and} \quad 2^{2^{n+1}} \equiv 1 \pmod{p},$$

we have

$$(x^2)^{2^n} \equiv 2^{2^n} \equiv -1 \pmod{p} \rightarrow x^{2^{n+1}} \equiv -1 \pmod{p},$$
$$(x^2)^{2^{n+1}} \equiv 2^{2^{n+1}} \equiv 1 \pmod{p} \rightarrow x^{2^{n+2}} \equiv 1 \pmod{p}.$$

Thus the order of x modulo p divides 2^{n+2}, but not 2^{n+1}. Therefore there exists an integer k' such that $\phi(p) = k'2^{n+2} \rightarrow p = k'2^{n+1} + 1$.

By using his observation, Euler was able to discover that the prime 641 divides the Fermat number F_5. Interestingly, it can be shown that 641 divides F_5 without actually performing any divisions. We can do so using the facts that

$$641 = 5 \cdot 2^7 + 1 \rightarrow 5 \cdot 2^7 = 641 - 1 \quad \text{and} \quad 641 = 2^4 + 5^4 \rightarrow 2^4 = 641 - 5^4.$$

We can then see

$$\begin{aligned} F_5 = 2^{2^5} + 1 = 2^{32} + 1 = 2^4 \cdot 2^{28} + 1 = (641 - 5^4)2^{28} + 1 = 641 \cdot 2^{28} - 5^4 \cdot 2^{28} + 1 \\ = 641 \cdot 2^{28} - (5 \cdot 2^7)^4 + 1 = 641 \cdot 2^{28} - (641 - 1)^4 + 1 \\ = 641 \cdot 2^{28} - (641^4 - 4 \cdot 641^3 + 6 \cdot 641^2 - 4 \cdot 641 + 1) + 1 \\ = 641 \cdot 2^{28} - 641^4 + 4 \cdot 641^3 - 6 \cdot 641^2 + 4 \cdot 641 - 1 + 1 \\ = 641(2^{28} - 641^3 + 4 \cdot 641^2 - 6 \cdot 641 + 4). \end{aligned}$$

The next Fermat number F_6 was first factored by the Danish mathematician Thomas Clausen (1801–1885) in 1854. He announced this result in an unpublished letter to Gauss, but gave no indication of what technique he used to find the factors. It was only in 1880 that the French enthusiast Fortuné Landry (1799–1895) published the following announcement:

> I have just factored the number
>
> $$2^{64} + 1 = 18446744073709551617.$$
>
> This number is the product of two factors 274177 which is prime, and 67280421310721. I do not currently know if this last factor is a prime.

One might assume that since 274177 is not very large, that Landry, as advised by Lucas, simply performed trial division by all possible prime divisors of the form $k2^8 + 1$ that are less than or equal to $\sqrt{F_6}$; however, this was not what he did. Initially, Landry in 1867 rediscovered the techniques of Fermat and applied them to factoring numbers of the form $2^n \pm 1$. He then went on to develop new methods of factorization that aided him in factoring F_6. However, the exact details of how he did this are still somewhat speculative, as Landry never published his precise method.

As we saw in Chap. 9, the Mersenne number $M_{67} = 2^{67} - 1$ is a composite number and not a prime, contrary to Mersenne's assertion. However, showing it to be composite using a Lucas-Lehmer test does not give any indication of what its factors are. In 1903 Frank Cole (1861–1926) presented a paper to the American Mathematical society in which he demonstrated the factorization of M_{67} as $193707721 \cdot 761838257287$. This was a most remarkable accomplishment, done without the aid of a computer, and thanks to Cole's paper *On the factorization of large numbers* we are aware of his method.

In order to factor $N = M_{67}$ Cole began by finding quadratic residues of M_{67}; these he listed in his paper as follows:

$$2, -3, -7, 13, -23 \cdot 53, 37, 41, 61, -67, -71, 23 \cdot 83, 89, 97, 101, -23 \cdot 113, -127, 137,$$
$$23 \cdot 151, -23 \cdot 157, 23 \cdot 167, 173, 181, 23 \cdot 191, -23 \cdot 193, \ldots.$$

As he mentioned Seelhoff's work, he very likely used these techniques to find these quadratic residues, but provided few details.

Using ideas of Euler, Cole knew that $2^p - 1$, where p is prime, must have factors of the form $kp + 1$ and $8l \pm 1$. Here, since $p = 67$, we see that $536k + 1$ is of both forms $67k + 1$ and $8l + 1$. Also, since $536k + 135 = 8 \cdot 67k + 2 \cdot 67 + 1 = 8 \cdot 67k + 8 \cdot 17 - 1$, $536k + 135$ is of both forms $67k + 1$ and $8l - 1$. Since -3 is a quadratic residue of M_{67}, Cole was able to refine these forms to $1608k + 1$ or $1608k + 1207$. Although this work is not described in his short paper, Cole continued in this fashion using his collection of quadratic residues to eliminate (but not entirely convincingly) all possible prime factors < 16000000.

He next attempted to solve $N = a^2 - b^2$ by computing the possible residue classes for a modulo small prime powers. He was able to derive that

$$\begin{aligned}
a &\equiv 671^* \pmod{67^2},\\
a &\equiv 0^* \pmod{8},\\
a &\equiv 1, 4^* \pmod{5},\\
a &\equiv 1, 3^* \pmod{7},\\
a &\equiv 0, 1^*, 12 \pmod{13},\\
a &\equiv 10, 37, 46^*, 64 \pmod{81}.
\end{aligned}$$

We can then see that there are 48 possible residue classes here for a modulo $1323536760 = 8 \cdot 5 \cdot 7 \cdot 13 \cdot 81 \cdot 67^2$, but only one of these will lead to a factorization. After what must have been an enormous amount of work, Cole eliminated all the possibilities except those marked with the asterisks, and used the CRT to arrive at $a = 1323536760x + 1160932384$. He then proceeded to use modular exclusion, as we have seen before, to winnow down the value of x to 287, thus yielding

$$a = 381015982504 \quad \text{and} \quad b = 380822274783;$$

hence, $N = M_{67} = 193707721 \cdot 761838257287$.

There is no doubt that to factor such a number by hand calculations is an incredible achievement. Cole is quoted saying that this calculation took him "three years of Sundays".

The advent of programmable computers stimulated the development of new methods for factoring. We will discuss a few of them in what follows.

John Pollard is a British mathematician who is best known for having created a number of algorithms to perform integer factorization; in fact, many of the recent advances in this problem owe their origin to his ideas. We will discuss two of them next: Pollard's $p-1$ algorithm and Pollard's rho algorithm.

We first present Pollard's $p-1$ algorithm, which he created in 1974. Imagine N is a composite integer and let p be a prime factor of N. Recall from Fermat's little theorem in Chap. 4, that if p is a prime, then

$$a^p \equiv a \pmod{p}.$$

In the case that $\gcd(a, p) = 1$ and k is any positive integer, then we have

$$a^{k(p-1)} \equiv 1 \pmod{p}.$$

We will use this to attempt to find a number x such that $x \equiv 1$ modulo some factor of N, in which case $\gcd(x-1, N)$ will divisible by that factor. However, it may be the case that at times this factor is 1 or N; in either of these instances we need to rerun the algorithm with different initial conditions. The values of x we will use rely on selecting some value a such that $\gcd(a, N) = 1$ (generally we will just use $a = 2$, as we can easily remove all factors of 2 from the number we wish to factor before we start the algorithm) and we will set $x = a^M$, where M is the product of a set of primes (each to some power) up to a certain bound. A simple version of his algorithm can be written as follows:

Algorithm 1 Pollard's $p-1$ algorithm for factoring composite integer N.

1. Select some bound B.
2. Set
$$M = \prod_{\text{primes } q \leq B} q^{\lfloor \log_q B \rfloor}.$$
3. Select any value a coprime to N.
4. Calculate $\gcd(a^M - 1, N)$
5. If $1 < \gcd(a^M - 1, N) < N$, then return this value, and you have a factor of N.
6. If $\gcd(a^M - 1, N) = 1$, then select a larger bound B and repeat the algorithm.
7. If $\gcd(a^M - 1, N) = N$, then select a smaller bound B and repeat the algorithm.

For example, we will use Pollard's $p-1$ method to factor $N = 10961$. We can initially set the bound $B = 5$. In this case $M = 2^2 \cdot 3 \cdot 5 = 60$. Setting $a = 2$, we have $a^M - 1 \equiv 2^{60} - 1 \equiv 6908 \pmod N$; yielding $g = \gcd(a^M - 1, N) = 1$. Thus, we need to set a larger bound. If we change the bound to $B = 7$ (hence, we are using one more prime, namely 7), then $M = 2^2 \cdot 3 \cdot 5 \cdot 7 = 420$. Again, with $a = 2$, we now have $a^M - 1 \equiv 2^{420} - 1 \equiv 4633 \pmod N$; thus, we have $g = \gcd(a^M - 1, N) = 113$. Hence, upon division we see that $N = 10961 = 97 \cdot 113$.

While this idea is not very practical for hand computations when N is large, it is often very useful when computers are available.

Next we examine Pollard's rho algorithm, which he published in 1975. Again, $N = pq$, where p is some nontrivial prime factor. In this algorithm we use a polynomial, say $f(x)$, and some initial starting value x_0. Historically, $f(x) = x^2 - 1$, but more recently this algorithm has been implemented using $f(x) = x^2 + 1$; however, empirically $f(x) = x^2 + a$ suffices as long as $a \neq 0, -2$. We then calculate

$$x_{i+1} \equiv f(x_i) \pmod N,$$

to generate a pseudorandom sequence. Since the number of possible values for the sequence is finite, it must be the case that it eventually cycles. We expect by probabilistic reasoning that for some j we get $x_{2j} \equiv x_j \pmod p$, where p is a prime divisor of N and $j = O(\sqrt{p})$. Thus, we check at each step whether the $\gcd(x_{2i} - x_i, N) \neq 1$, thereby yielding a nontrivial factor of N.

For example, we can use the Pollard rho method to factor $N = 176093$ with $x_{i+1} = x_i^2 + 11$ and $x_0 = 5$. We then have (modulo N):

$$\begin{array}{llll}
x_1 = 36, & x_2 = 1307, & y_1 = x_2 - x_1 = 1271, & \gcd(y_1, N) = 1, \\
x_3 \equiv 123423, & x_4 \equiv 135882, & y_2 \equiv x_4 - x_2 \equiv 134575, & \gcd(y_2, N) = 1, \\
x_5 \equiv 38606, & x_6 \equiv 148188, & y_3 \equiv x_6 - x_3 \equiv 24765, & \gcd(y_3, N) = 1, \\
x_7 \equiv 5790, & x_8 \equiv 66441, & y_4 \equiv x_8 - x_4 \equiv 106652, & \gcd(y_4, N) = 293.
\end{array}$$

Hence, upon dividing by 293, we see the factorization of $N = 176093$ is $293 \cdot 601$.

It would be a natural question to ask, "where does the word *rho* come from in the above algorithm?" If you imagine the values from x_0 to x_i, where $x_i \equiv x_{2i}$, forming a tail, then you can draw the cyclic portion of the sequence on a circle starting and stopping at x_i. The resulting diagram forms the shape of the Greek letter rho, i.e. ρ. This is illustrated in the diagram that follows.

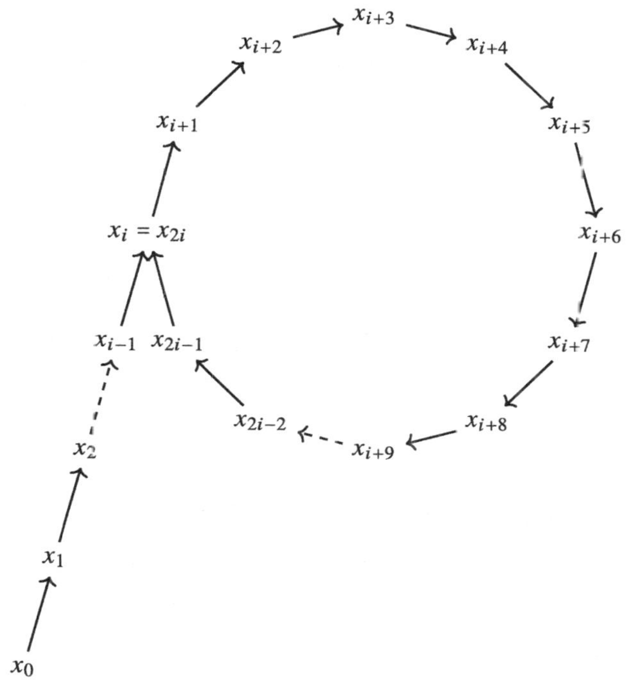

In 1980, five years after the publication of this algorithm, the Australian mathematician Richard Brent and Pollard used a modified Pollard's rho method (B-P rho) to show that 1238926361552897 is a factor of the eighth Fermat number $F_8 = 2^{2^8} + 1$. The remaining 62 digit factor,

$$93461639715357977769163558199606896584051237541638188580280321,$$

was later proved prime by Hugh Williams and J. Stephen Judd, using methods that relied on the properties of generalized Lehmer functions.

Before Brent and Pollard's computer implementation of their rho method was used to factor F_8, there was a computer implementation of a factoring method called the *continued fraction factorization method*, developed by Michael Morrison and John Brillhart (1930–2022) in 1970, but originally conceived by Lehmer and Ralph Powers (1875–1952) in 1931. In order to proceed we need first to recall some results about continued fractions mentioned in Chap. 14.

We develop the simple continued fraction expansion of $\sqrt{mN}$ for some integer multiplier $m \geq 1$, and get

$$\sqrt{mN} = [a_0; a_1, a_2, \ldots, a_{n-1}, (\sqrt{mN} + P_n)/Q_n],$$

where $\alpha_n = (\sqrt{mN} + P_n)/Q_n$. For each value of n greater or equal to 1 and less than some estimated bound, we calculate the $(n-1)$th convergent

$$\frac{A_{n-1}}{B_{n-1}} = [a_0; a_1, a_2, \ldots, a_{n-1}]$$

and exploit the identity

$$A_{n-1}^2 - mNB_{n-1}^2 = (-1)^n Q_n.$$

This identity gives us the congruence

$$A_{n-1}^2 \equiv (-1)^n Q_n \pmod{N}.$$

As we evaluate the above for different values of n, we collect pairs (A_n, Q_n). With enough of these it is possible to find a set $I = \{i_1, i_2, \ldots, i_k\}$ such that the product $\prod_{i \in I} (-1)^i Q_i$ is a perfect square. At this point we can attempt to factor N, as we have

$$A^2 \equiv \prod_{i \in I} A_i^2 \equiv \prod_{i \in I} (-1)^i Q_i = R^2 \pmod{N},$$

We note that the A value above is reduced modulo N; however, the R value is not.

For example, let $N = 126773$ and choose $m = 1$. While we convert this number to a continued fraction, we will perform the steps to produce $(A_{n-1} \pmod{N}, Q_n)$ pairs. First set $\alpha_0 = \sqrt{N}$; thus, for the $n = 1$ case we have

$$\begin{aligned} a_0 &= \lfloor \alpha_0 \rfloor = \lfloor \sqrt{126773} \rfloor = 356, \\ \alpha_1 &= \frac{1}{\alpha_0 - a_0} = \frac{1}{\sqrt{126773} - 356} = \frac{\sqrt{126773} + 356}{37} = \frac{\sqrt{N} + P_1}{Q_1}, \\ \frac{A_0}{B_0} &= 356. \end{aligned}$$

We then can see

$$A_{n-1}^2 \equiv (-1)^n Q_n \pmod{N} \rightarrow 356^2 \equiv -37 \pmod{N}.$$

For $n = 2$, we get

$$\begin{aligned} a_1 &= \lfloor \alpha_1 \rfloor = \left\lfloor \frac{\sqrt{126773} + 356}{37} \right\rfloor = 19, \\ \alpha_2 &= \frac{1}{\alpha_1 - a_1} = \frac{1}{\frac{\sqrt{126773}+356}{37} - 19} = \frac{\sqrt{126773} + 347}{172} = \frac{\sqrt{N} + P_2}{Q_2}, \end{aligned}$$

$$\frac{A_1}{B_1} = 356 + \frac{1}{19} = \frac{6765}{19}.$$

We then can see for $n = 2$

$$A_{n-1}^2 \equiv (-1)^n Q_n \pmod{N} \to 6765^2 \equiv 172 \equiv 2^2 \cdot 43 \pmod{N}.$$

Continuing from here we will show fewer details; however, it is easy to recover the values listed by performing calculations similar to those above. For $n = 3$, we have

$$a_2 = 4, \quad \alpha_3 = \frac{\sqrt{126773} + 341}{61} = \frac{\sqrt{N} + P_3}{Q_3}$$

and

$$\frac{A_2}{B_2} = 356 + \cfrac{1}{19 + \cfrac{1}{4}} = \frac{27416}{77};$$

thus,

$$A_{n-1}^2 \equiv (-1)^n Q_n \pmod{N} \to 27416^2 \equiv -61 \pmod{N}.$$

For $n = 4$, we get

$$a_3 = 11, \quad \alpha_4 = \frac{\sqrt{126773} + 330}{293} = \frac{\sqrt{N} + P_4}{Q_4}, \quad \frac{A_3}{B_3} = \frac{308341}{866};$$

Giving us,

$$308341^2 \equiv 293 \pmod{N} \to 54755^2 \equiv 293 \pmod{N}.$$

For $n = 5$,

$$a_4 = 2, \quad \alpha_5 = \frac{\sqrt{126773} + 256}{209} = \frac{\sqrt{N} + P_5}{Q_5}, \quad \frac{A_4}{B_4} = \frac{644098}{1809};$$

yielding,

$$644098^2 \equiv -209 \pmod{N} \to 10233^2 \equiv -11 \cdot 19 \pmod{N}.$$

For $n = 6$,

$$a_5 = 2, \quad \alpha_6 = \frac{\sqrt{126773} + 162}{481} = \frac{\sqrt{N} + P_6}{Q_6}, \quad \frac{A_5}{B_5} = \frac{1596537}{4484};$$

yielding,

$$1596537^2 \equiv 481 \pmod{N} \to 75241^2 \equiv 13 \cdot 37 \pmod{N}.$$

In the case of $n = 7$, we have

$$a_6 = 1, \quad \alpha_7 = \frac{\sqrt{126773} + 319}{52} = \frac{\sqrt{N} + P_7}{Q_7}, \quad \frac{A_6}{B_6} = \frac{2240635}{6293};$$

yielding,

$$2240635^2 \equiv -52 \pmod{N} \to 85494^2 \equiv -2^2 \cdot 13 \pmod{N}.$$

At this point we have enough information to factor N using the cases for $n = 1, 6$ and 7 that give us respectively:

$$\begin{aligned} 356^2 &\equiv -37 \pmod{N}, \\ 75261^2 &\equiv 13 \cdot 37 \pmod{N}, \\ 85494^2 &\equiv -2^2 \cdot 13 \pmod{N}. \end{aligned}$$

These three congruences together imply

$$(356 \cdot 75261 \cdot 85494)^2 \equiv (2 \cdot 13 \cdot 37)^2 \pmod{N};$$

hence,

$$113564^2 \equiv 962^2 \pmod{N}.$$

Both $\gcd(113564 - 962, N) = 383$ and $\gcd(113564 + 962, N) = 331$, deliver non-trivial factors of N.

In 1970 Morrison and Brillhart produced a clever implementation of the above technique, since called CFRAC, and ran it on UCLA's IBM 360/91 in order to attempt to factor the seventh Fermat number

$$F_7 = 2^{2^7} + 1 = 340282366920938463463374607431768211457.$$

An initial attempt by their method yielded only trivial factors; however, on their second attempt they found the congruence

$$\begin{aligned} &233503648380835852177232143618227956476^2 \\ &\qquad \equiv 251864781457280412973122719348520212223^2 \pmod{F_7}. \end{aligned}$$

Hence, setting

$$\begin{aligned} A &= 233503648380835852177232143618227956476, \\ R &= 251864781457280412973122719348520212223, \end{aligned}$$

the $\gcd(A - R, F_7) = 59649589127497217$; hence

$$F_7 = 59649589127497217 \cdot 5704689200685129054721.$$

CFRAC was a predecessor of and the inspiration for the factorization method called the *quadratic sieve* (QS), invented by Carl Pomerance. Today the QS is the fastest method for factoring numbers of less than 100 digits. It is the second fastest method (second to the number field sieve) for numbers of 100–150 digits. To describe how the QS works we need the following definition: we say a positive integer is *B-smooth* if none of its prime factors is greater than B. For instance, 137200 is 7-smooth as $137200 = 2^4 \cdot 5^2 \cdot 7^3$. At the heart of the QS process is a simple sieving procedure, which Pomerance realized would make QS much more efficient than the trial factorization process required by the Morrison and Brillhart technique.

Algorithm 2 QS for factoring composite integer N.

1. Select some smoothness bound B. Let $\pi(B)$ denote the number of primes less than B and set $m = \lfloor\sqrt{N}\rfloor$.
2. Find $\pi(B) + 1$ pairs (a_i, b_i) such that $b_i \equiv a_i^2 \pmod N$ and b_i is B-smooth. This is done by selecting x in $\{0, \pm1, \pm2, \dots\}$, setting $a_i = x + m$ and $b_i = a_i^2 - N$, if b_i is B-smooth this pair is kept; otherwise, the pair is discarded.
3. From the collection of (a_i, b_i) generate b, from some subset of the b_i that multiply together to form a perfect square. Calculate the product of the corresponding a_i corresponding to the b_i used in b and call this value a.
4. We then have $a^2 \equiv b^2 \pmod N$.
5. If $a \not\equiv -b \pmod N$ and $a \not\equiv b \pmod N$, then $\gcd(a - b, N)$ and $\gcd(a + b, N)$ return nontrivial factors of N.

For instance, we can use the above to factor $N = 260033$. Here $m = 509$ and we will set a bound of $B = 17$, i.e. the first seven primes. After attempting 200 trial pairs of a_i using all the values of $\pm x$ for x from 0 to 99 and keeping only those that are 17-smooth, we can produce the following table:

i	x	$a_i = x + m$	$b_i = a_i^2 - N$
1	0	509	$-2^3 \cdot 7 \cdot 17$
2	2	511	$2^6 \cdot 17$
3	4	513	$2^6 \cdot 7^2$
4	-28	481	$-2^{12} \cdot 7$
5	34	543	$2^{11} \cdot 17$
6	36	545	$2^7 \cdot 17^2$
7	-94	415	$-2^8 \cdot 7^3$
8	98	607	$2^7 \cdot 7 \cdot 11^2$

Using the cases 1, 5, and 7 from this table, we get

$$a = 509 \cdot 543 \cdot 415 = 114700605 \equiv 26052 \pmod{N}$$

and

$$b = 2^{11} \cdot 7^2 \cdot 17;$$

hence, $a^2 \equiv b^2 \pmod{N}$ and thus, $\gcd(a-b, N) = \gcd(26052 - 1705984, 260033) = 457$. Consequently, we have $N = 260033 = 457 \cdot 569$.

As mentioned above, for numbers of over 100 digits a method, called the number field sieve (NFS), outperforms QS. An early version of the NFS was created by Pollard in 1993, and has undergone considerable modification since. It relies on mathematics beyond the scope of this book and thus will not be deeply discussed herein; however, the interested reader is directed to the additional material at the end of the chapter. We would like to state that it was with the help of this method that Mark Manasse and the Dutch mathematician and cryptographer Arjen K. Lenstra (a brother of Hendrik Lenstra, Jr.) in 1993 factored F_9, a number of 155 digits. It was known that $F_9 = 2424833 \cdot c_{148}$, where c_{148} denotes a 148 digit composite number, however c_{148} resisted the methods Pollard rho, Pollard $p-1$, the continued fraction method, and the quadratic sieve, mentioned above. It was also resistant against another method that is beyond the scope of this book: Lenstra's elliptic curve method (ECM) for factoring. This method relies on properties of points on a particular type of curve, called an elliptic curve, but the strategy for factorization is not much different from that of Pollard's $p-1$ method. Again, the interested reader is directed to the further reading section at the end of the chapter.

The expected time complexity for factoring N by CFRAC is $L_N[1/2, \sqrt{2}]$, by QS is $L_N[1/2, 1]$ and by NFS is $L_N[1/3, (64/9)^{1/3}]$.

It is worth noting that while the ECM method did not work for factoring F_9 it played an important role in the factorization of F_{10} and F_{11}. Currently, the only fully factored Fermat numbers are those up to F_{11}; the table below contains some details of their factorizations.

F_n	Digits	Factors	Digits of factors	Credit
F_5	10	2	3, 7	1732 Euler
F_6	20	2	6, 14	1880 Landry
F_7	39	2	17, 22	1975 Morrison, Brillhart
F_8	78	2	16, 62	1981 Brent, Pollard
F_9	155	3	7, 49, 99	1993 Manasse, Lenstra
F_{10}	309	4	8, 10, 40, 252	1995 Brent
F_{11}	617	5	6, 6, 21, 22, 564	1988 Brent

Certainly, the challenge of factoring the Fermat numbers motivated the design of many factoring techniques as did the *Cunningham project*. The objective of this undertaking is to find the complete factorization of numbers of the form $b^n \pm 1$ for $b = 2, 3, 5, 6, 7, 10, 11, 12$. It is named after the Indian-born British mathematician and soldier Allan Cunningham (1842–1928), who along with Herbert Woodall published an initial table in 1925. Since that time this work has been carried on by a number of researchers, and the current limits on the exponents are given below:

Base	2	3	5	6	7	10	11	12
Limit	1500	900	600	550	500	450	400	400

However, the current state of factoring is best indicated by results from the *RSA factoring challenge*. In 1991, RSA Laboratories published a list of integers, each product of two large primes, along with prize money correlated to the size of the number being factored. Each number in the initial challenge was named RSA followed by the number of its decimal digits. For example, the smallest number in the challenge was

$$\begin{aligned}\text{RSA-100} = {} & 1522605027922533360535618378132637429718068114961380\,6\\ & 88657908494580122963258952897654000350692006139.\end{aligned}$$

RSA-100 carried only a small prize of US$1000 and was factored by Arjen Lenstra in the same year the challenge was issued. In order to protect the integrity of the challenge, the numbers were generated on a computer that was not connected to a network, and after the numbers were generated the computer's hard drive was destroyed. The numbers in the initial challenge were RSA-100, RSA-110, RSA-120,..., RSA-500, and then RSA-617. Later, in 2001, the challenge was updated. Unfortunately, the new numbers were labelled slightly differently, according to the number of their binary digits. The prize money was offered until 2007, at which point the challenge ended. The 2001 numbers are in the table below along with their prize values.

RSA number	Decimal digits	Binary digits	Cash prize	Credit
RSA-576	174	576	$10000	2003 Jens Franke et al.
RSA-640	193	640	$20000	2005 Jens Franke et al.
RSA-704	212	704	$30000	2012 Shi Bai, Emmanuel Thomé and Paul Zimmermann
RSA-768	232	768	$50000	2009 Thorsten Kleinjung et al.
RSA-896	270	896	$75000	
RSA-1024	309	1024	$100000	
RSA-1536	463	1536	$150000	
RSA-2048	617	2048	$200000	

We can see that the only prizes claimed before the expiry of the second challenge were for RSA-576 and RSA-640; however, even without the prize money, researchers continue to be motivated to factor these numbers, as seen by the 2009 factoring of

RSA-768 and 2012 factoring of RSA-704. Recently, F. Boudot et al. factored RSA-250 (from the initial RSA challenge, a 829 bit number), this is the largest integer ever factored by a general purpose algorithm when the factors of N are large. It may seem that since the creation of the NFS little significant progress has been made on the integer factoring problem. However, Sam Wagstaff has expressed some optimism that new factoring methods will be discovered in the future and has provided some thoughts on how these might develop.

One exciting development concerning the integer factoring problem is the possibility of enlisting a quantum computer to solve it. Quantum computers exploit the principles of quantum mechanics, such as superposition and entanglement. On a conventional computer we perform arithmetic with binary bits, which are either 0 or 1; on a quantum computer use is made of quantum bits or qubits (q-bits). A single qubit can represent a one, a zero or any quantum superposition of these two qubit states. Thus, a quantum computer with n qubits can be in any superposition of up to 2^n different states simultaneously, whereas a conventional computer can only be in one of these states at any given time. An ingenious algorithm of Peter Shor, when implemented on a quantum computer with a sufficient number of qubits, can solve the integer factoring problem in essentially polynomial time.

There are a number of obstacles to overcome if such a quantum computer is to be constructed, such as

- quantum decoherence (qubits are very fragile),
- quantum error detection and correction,
- scalability.

Nevertheless much progress has been made on these problems, and experts are predicting that a standard universal quantum computer capable of implementing Shor's algorithm might be constructed within the next two decades. Because of the power of Shor's algorithm, the likely advent of this sort of quantum computer will render currently employed cryptosystems obsolete. As this would have devastating consequences for most internet communication, past and future, work is being done on the development of post-quantum cryptographic techniques, which with any luck should produce robust, secure, and practical quantum-resistant cryptographic algorithms before too long.

Problems

15.1 (Fermat factorization) Use Fermat factorization to factor the given values for N, which are all the product of two primes, by searching values of $x = a^2 - N$, for a values $\lfloor\sqrt{N}\rfloor + 1, \lfloor\sqrt{N}\rfloor + 2, \ldots, (N+1)/2$, until one of these values yields a perfect square for x ($x = b^2$). At which point $N = a^2 - b^2 = (a-b)(a+b)$. (Hint: save time by using modular exclusion.)
(a) **[BB]** $N = 390629$
(b) $N = 341371$
(c) $N = 219341$
(d) $N = 436927$

15.2 [BB] (Seelhoff's method) Factor $N = 54253$ using Seelhoff's method. Begin by setting $N = w^2 + r$, where $w = \lfloor\sqrt{N}\rfloor$. After a search using primes p such that $(N|p) = 1$ and solving cases of

$$s^2 \equiv N \pmod{p^k},$$

we recovered

$$8^2 \equiv N \pmod{3^5} \quad \text{and} \quad 60^2 \equiv N \pmod{37^2}.$$

Use these by putting $m = w \pm (s + yp^k)$ (use $y = 1$) and $t = m(2w - m) + r$. Recall $N = (w - m)^2 + t$ and $p^k \mid t$. From the above a congruence can be found such that

$$a^2 \equiv b^2 \pmod{N},$$

and hence N can be factored using the Euclidean algorithm to find $\gcd(a - b, N)$.

15.3 (Seelhoff's method) Factor $N = 54253$ using Seelhoff's method. Begin by setting $N = w^2 + r$, where $w = \lfloor\sqrt{N}\rfloor$. After a search using primes p such that $(N|p) = 1$ and solving cases of

$$s^2 \equiv N \pmod{p^k},$$

we recovered

$$14^2 \equiv N \pmod{37}, \quad 9^2 \equiv N \pmod{47} \quad \text{and} \quad 2920^2 \equiv N \pmod{59^2}.$$

Use these by putting $m = w \pm (s + yp^k)$ (use $y = 1$) and $t = m(2w - m) + r$. Recall $N = (w - m)^2 + t$ and $p^k \mid t$. From the above a congruence can be found such that

$$a^2 \equiv b^2 \pmod{N},$$

and hence N can be factored using the Euclidean algorithm to find $\gcd(a - b, N)$.

15.4 (Pollard's $p - 1$) Factor the given N using Pollard's $p - 1$ method, Algorithm 1, using specified a and bound B:
(a) **[BB]** $N = 899$, $a = 2$, $B = 5$
(b) $N = 1517$, $a = 2$, $B = 5$
(c) $N = 2813$, $a = 2$, $B = 7$
(d) $N = 6887$, $a = 2$, $B = 7$
(e) $N = 12827$, $a = 2$, $B = 7$
(f) $N = 63083$, $a = 2$, $B = 11$
(g) $N = 121769$, $a = 2$, $B = 11$

15.5 (Pollard's rho) Factor the given N using Pollard's rho method using specified recurrence and starting value x_0:
(a) **[BB]** $N = 527$, $x_{i+1} = x_i^2 + 1$, $x_0 = 5$

(b) $N = 2173$, $x_{i+1} = x_i^2 + 11$, $x_0 = 5$
(c) $N = 3233$, $x_{i+1} = x_i^2 + 13$, $x_0 = 57$
(d) $N = 6557$, $x_{i+1} = x_i^2 + 7$, $x_0 = 11$
(e) $N = 8633$, $x_{i+1} = x_i^2 + 7$, $x_0 = 11$
(f) $N = 10379$, $x_{i+1} = x_i^2 + 7$, $x_0 = 11$
(g) $N = 88433$, $x_{i+1} = x_i^2 + 13$, $x_0 = 101$

15.6 (Continued fraction factorization method) Factor the given N using the continued fraction factorization method. Begin by setting $\alpha_0 = \sqrt{N}$:
(a) **[BB]** $N = 9167$
(b) $N = 12317$
(c) $N = 12191$
(d) $N = 72851$
(e) $N = 91123$
(f) $N = 43621$
(g) $N = 83627$

15.7 [BB] (Quadratic sieve) Factor $N = 95041$ using the quadratic sieve method and information provided. In this case $m = \lfloor\sqrt{N}\rfloor = 308$ and a bound of $B = 13$ was set, i.e. the first 6 primes. A computer search was performed and the following table was produced of pairs (a_i, b_i) such that $a_i^2 \equiv b_i \pmod{N}$:

i	x	$a_i = x + m$	$b_i = a_i^2 - N$
1	1	309	$2^3 \cdot 5 \cdot 11$
2	-1	307	$-2^3 \cdot 3^2 \cdot 11$
3	3	311	$2^4 \cdot 3 \cdot 5 \cdot 7$
4	-3	305	$-2^5 \cdot 3^2 \cdot 7$
5	-4	304	$-3 \cdot 5^3 \cdot 7$
6	-10	298	$-3^4 \cdot 7 \cdot 11$
7	11	319	$2^6 \cdot 3 \cdot 5 \cdot 7$

15.8 (Quadratic sieve) Factor $N = 135739$ using the quadratic sieve method and information provided. In this case $m = \lfloor\sqrt{N}\rfloor = 368$ and a bound of $B = 11$ was set, i.e. the first 5 primes. A computer search was performed and the following table was produced of pairs (a_i, b_i) such that $a_i^2 \equiv b_i \pmod{N}$:

i	x	$a_i = x + m$	$b_i = a_i^2 - N$
1	0	368	$-3^2 \cdot 5 \cdot 7$
2	-1	367	$-2 \cdot 3 \cdot 5^2 \cdot 7$
3	29	397	$2 \cdot 3^7 \cdot 5$
4	-36	332	$-3^6 \cdot 5 \cdot 7$
5	-126	242	$-3^2 \cdot 5^2 \cdot 7^3$
6	-301	67	$-2 \cdot 3 \cdot 5^5 \cdot 7$

Further Reading

Pre-computer-age factoring methods

H. C. Williams and J. O. Shallit, Factoring integers before computers, *Proceedings of Symposia in Applied Mathematics*, Vol. 48, Mathematics of Computation 1943–1993: A Half-Century of Computational Mathematics, American Mathematical Society, Providence RI, 1994, pp. 481–531.

Computer factoring methods and their complexity

Hans Riesel, *Prime Numbers and Computer Methods for Factorization*, 2nd Edition, Springer, NY, 1994, pp. 173–224.

Richard Crandall and Carl Pomerance, *Prime Numbers: A Computational Perspective*, Springer-Verlag, NY, 2001, pp. 191–281.

Arjen K. Lenstra, Integer factoring, *Designs, Codes and Cryptography*, **19** (2000), 101–128.

S. S. Wagstaff, Jr., *The Joy of Factoring*, Vol. 68, Student Mathematical Library, American Mathematical Society, Providence RI, 2013.

Factoring F_6

H. C. Williams, How was F_6 factored?, *Mathematics of Computation*, **61** (1993), 463–474.

Factoring F_7

Michael A. Morrison and John Brillhart, A method of factoring and the factorization of F_7, *Mathematics of Computation*, **29** (1975), 183–205.

Factoring F_8

Richard P. Brent and John M. Pollard, Factorization of the eighth Fermat number, *Mathematics of Computation*, **36** (1981), 627–630.

The number field sieve

A. K. Lenstra and H. W. Lenstra, Jr. (Eds.), *The Development of the Number Field Sieve*, Lecture Notes in Mathematics 1554, Springer-Verlag, Berlin Heidelberg, 1993.

Factoring F_9

A. K. Lenstra, H. W. Lenstra, Jr., M. S. Manasse and J. M. Pollard, Factorization of the ninth Fermat number, *Mathematics of Computation*, **61** (1993), 319–349.

Factoring F_{10}

Richard P. Brent, Factorization of the tenth Fermat number, *Mathematics of Computation*, **68** (1999), 429–451.

The Cunningham project

J. Brillhart, D. H. Lehmer, J. L. Selfridge, B. Tuckerman and S. S. Wagstaff, Jr., *Factorizations of* $b^n \pm 1, b = 2, 3, 5, 6, 7, 10, 11, 12$ *up to high powers*, Contemporary Mathematics Vol. 22, American Mathematical Society, Providence RI, 1988. Available at: https://www.ams.org/books/conm/022/
Website: https://homes.cerias.purdue.edu/~sww/cun/index.html

Future of the factoring problem

S. S. Wagstaff, Jr., *The Joy of Factoring*, Vol. 68, Student Mathematical Library, American Mathematical Society, Providence RI, 2013, pp. 262–267.

Quantum computing

Julian Brown, *The Quest for the Quantum Computer*, Touchstone, NY, 2001.

Chapter 16
Sieve Devices

In number theory a *sieve* is a process of successively deleting integers from a list, according to certain rules, until only those with desired properties remain. For example, the sieve of Eratosthenes discussed in Chap. 2 is such a procedure, where the remaining integers are the primes. We also saw sieving methods employed in the previous chapter to perform factoring; those being the quadratic and number field sieves. Modern number theoretic sieves can be much more sophisticated than these, and are frequently enlisted to estimate the size of certain sets of integers. For instance, the sieve introduced in 1915 by the Norwegian mathematician Viggo Brun (1885–1978) was used by him to study the question of the existence of infinitely many twin primes. Since then further refinements have resulted in sieve techniques having become very powerful tools for analytic number theorists. These kinds of sieves are purely intellectual constructs and have no physical manifestation; however, there are instruments that can perform a type of sieving operation, and these will form the subject of this chapter. To avoid confusion with theoretical number sieves, we call these *sieve devices*.

Suppose we consider the simple problem of trying to factor 611. We will use Fermat's idea of attempting to represent this number as a difference of squares. We write

$$611 = x^2 - y^2 = (x + y)(x - y).$$

Any perfect integral square must be one of 0, 1, 4 modulo 8 and $611 \equiv 3 \pmod 8$; thus, we can only have $x^2 \equiv 4$ and $y^2 \equiv 1 \pmod 8$, from which it follows that $x \equiv 2 \pmod 4$. In a similar fashion, because perfect squares can only be 0 or 1 modulo 3, we can deduce that $x \equiv 0 \pmod 3$. If we make use of the additional (exclusion) moduli 5 and 7, we see that because the only nonzero quadratic residues of 5 are 1 and 4 and the only nonzero quadratic residues of 7 are 1, 2, 4, we must have

E. L. F. Roettger and H. C. Williams, *The Enchantment of Numbers*, CMS/CAIMS Books in Mathematics 16, https://doi.org/10.1007/978-3-031-76568-1_16

$$x^2 \equiv 611 + 0, 1, 4 \pmod 5,$$
$$x^2 \equiv 611 + 0, 1, 2, 4 \pmod 7;$$

consequently, x can only be one of 0, 1, 4 (mod 5) and one of 2, 3, 4, 5 (mod 7). Thus, we are left with the following problem: find the nonnegative values of x such that x is less than 611 and

$$x \equiv 2 \pmod 4,$$
$$x \equiv 0 \pmod 3,$$
$$x \equiv 0, 1, 4 \pmod 5,$$
$$x \equiv 2, 3, 4, 5 \pmod 7.$$

For one value of x satisfying this system of congruences, there must be a corresponding value of y such that we can find a factorization of 611. In fact, $x = 30$ is the least solution of the above system of congruences, and we find that

$$611 = 30^2 - 17^2 = (30 - 17)(30 + 17) = 13 \cdot 47.$$

What we have here is an instance of a more general problem, called the *modular sieving problem*, in which we are given:

1. bounds A and B with B greater than A,
2. k moduli m_i ($i = 1, 2, 3, \ldots, k$), each greater than 1 and relatively prime in pairs,
3. a set of acceptable residues for each of the moduli.

We are to determine from this information all the values of x bounded between A and B such that each of these values of x is congruent modulo m_i ($i = 1, 2, 3, \ldots, k$) to an acceptable residue for m_i.

Two simple examples of the modular sieving problem are the sieve of Eratosthenes and the Chinese remainder theorem (Theorem 3.2). In the first of these we take $A = 1$ and $B = p_k^2 + 1$, where p_i, denotes the ith prime. The k moduli are p_i ($i = 1, 2, 3, \ldots, k$) and the set of acceptable residues for each p_i is the RSR modulo p_i. In the latter case, we take $A = 1$ and $B = m_1 m_2 m_3 \cdots m_k$ with the single acceptable residue a_i, for each m_i.

Even with sophisticated methods, factoring requires a great deal of computing power, and in the era before computers this generally meant a tremendous investment of time-consuming hand calculations. In order to save time performing the necessary calculations, some mathematicians set about building specialized mechanisms to help implement a more sophisticated version of Fermat's difference of squares method, namely solving

$$az^2 + 2bz + c = u^2,$$

where a, b, c are known integers and the value of z satisfies some preassigned bound.

Suppose we again consider the problem of factoring 611. We observe the tableau below.

x	0	1	2	3	4	5	6	7	8	9	10	11	12	$\cdots$	30
x (mod 4)	×	×	✓	×	×	×	✓	×	×	×	✓	×	×	$\cdots$	✓
x (mod 3)	✓	×	×	✓	×	×	✓	×	×	✓	×	×	✓	$\cdots$	✓
x (mod 5)	✓	✓	×	×	✓	✓	✓	×	×	✓	✓	✓	×	$\cdots$	✓
x (mod 7)	×	×	✓	✓	✓	✓	×	×	×	✓	✓	✓	✓	$\cdots$	✓

For each of the moduli 4, 3, 5, 7 we have placed a check mark (✓) under each acceptable residue for x and a cross (×) under the others. When, in scanning this tableau, we arrive at a value for x above a column consisting only of check marks, that value of x is one solution to our sieve problem. The difficulty with this approach is that we may have to go out a very long way on the tableau in our search for x until we find a solution, especially if k is large. We can deal with this problem by noticing that each row of the tableau exhibits a cyclic pattern of check marks and crosses, and this information can be compressed by recording these patterns only once. Any column of the original table can be generated by shifting each of the rows to the left in order to advance the appropriate set of symbols to the first column, the leftmost symbol wrapping around to the rightmost. In general, column i will appear in the first column after i of these shift operations have been performed.

We are now able to see how a device could be constructed to perform these sieving operations. We imagine each exclusion modulus m_i to be represented by a loop or *ring* which is divided into m_i, positions, with those positions representing an acceptable residue being tagged in some manner. The device examines one position from each ring at a fixed position called a window. The rings are advanced (shifted) in unison, and a solution is detected when the window is filled exclusively with tags. A trial counter records the number of shifts performed by the mechanism. Thus, this device can be set up to start searching from some point N by setting the ith ring such that the tag (or lack of one) corresponding to $N \pmod{m_i}$ is in the window when the mechanism is turned on. After s shifts, the residue of x being tested is in position $N + s$. We can call an instrument of this type a sieve device.

The earliest description of a machine that could implement this idea of using exclusion moduli was by the Englishman Frederick Lawrence (1871–1961) in his 1896 paper *Factorisations of numbers*. He mentioned the possibility of a machine that would employ movable paper strips. Another idea he put forward, is to use several gears, where the ith gear would have m_i teeth and be used for the exclusion modulus m_i (each tooth was labelled $0, 1, \ldots, m_i - 1$). With such a machine, using many gears of different diameters, each gear would, as it rotates, represent an endless strip of paper or ring. A brass stud would penetrate through the gear at the point(s) of an acceptable residue (modulo m) for which

$$az^2 + 2bz + c \equiv u^2 \pmod{m}.$$

If all the gears are penetrated at the same time, a bell would ring indicating an acceptable residue has been found modulo the least common multiple of the exclusion moduli. The user of the device would need to record the number of rotations of the drive gear in order to recover the z-value.

It seems that Lawrence never built such a machine; however, his description of how one might be constructed likely motivated others to do so when his paper was translated to French in 1910 by the French amateur mathematician André Gérardin (1879–1953). In 1912 Kraitchik constructed a similar machine with an interface that relied on light shining through holes of potential solutions.

The first fully functioning sieve device was originally envisioned by the French enthusiast Pierre Carissan (1871–1923), but was actually constructed by his brother, French Army lieutenant Eugene-Olivier Carissan (1880–1925) in 1919. This machine still exists and is in the possession (Inventory no: 43048-0000-) of the Conservatoire National des Arts et Métiers in Paris, France. There is even a fifteen minute video (in French) explaining how it works.

Subsequent to Carissan's pioneering work, several other sieves were constructed by a number of individuals, most notably by Lehmer. He constructed a photoelectric sieve in 1932 that was similar in broad design to those described by Lawrence and Kraitchik of whose work he was aware. At the time, however, he was unaware of the machines made by the Carissan brothers, and only became acquainted with their work in 1989. Lehmer also made machines for factoring that, rather than using gears, used bicycle chains of various lengths and also another using 16-mm film. Indeed, the photoelectric sieve was even exhibited at the Chicago Century of Progress Exposition or World's Fair in 1933–1934. As technology advanced, the rate of canvassing numbers improved considerably, from 35 to 40 per second to 2×10^{15} per second. The main advantage of these devices over electronic computers was the low cost of their construction and their speed. Also, they are challenging and fun to build, making them a great source of research topics for students. As many problems requiring the use of these instruments require some sophisticated software for determining the acceptable residues, it has been found convenient to connect them as peripherals to general-purpose computers.

Recently, Moore's law has resulted in rendering software sieves programmed on large computers, as originally described by Lehmer in 1953, as the winners in the race for faster sieves. Nevertheless, numerical sieving devices are much cheaper and easier to access than supercomputers and, properly programmed, can compete with these costly machines; furthermore, they are much more economical to run. Much more information on the history and construction of number sieves can be found in the reference material at the end of this chapter. In what follows we will mention several problems to which sieve technology has been applied.

We begin with the integer factoring problem. We have already described a naïve way in which a sieve device can be employed to solve this problem, but it might be instructive to examine a more refined example. The method of the difference of squares will work most efficiently when the number to be factored has two factors which are close in value, but this may not be the case. Suppose we wish to factor a composite integer N. We could write N instead as

$$mN = x^2 - y^2,$$

where m is a *multiplier* which, if selected carefully, might cause the difference of two factors of mN to be small. The sieve technique can now be applied to mN. The problem, of course, is to find a suitable value of m. Often the multiplier was put equal to a prime and then selected by some experimentation. As an example, we mention that in 1970 Lehmer succeeded in factoring a 33 digit divisor of $2^{136} + 1$; he used a 31 ring sieve called the DLS-127 and a nine digit multiplier. This process required over 2600 hours of sieving, most of the time being used to find the right multiplier. Indeed, up to 1970, well within the time of the electronic computer, sieve methods were still the fastest techniques known for integer factoring. However, the development of other factoring methods discussed in the previous chapter soon caused the use of sieve devices to be superseded in this application.

Lehmer also used his sieve devices to find the pseudosquares L_p mentioned in Chap. 11. In this case the sieve moduli are the odd primes and the corresponding acceptable residues are the quadratic residues modulo the prime. Since his work, subsequent researchers have found all the values of L_p for p up to 373. All of these, except for L_{281}, L_{367} and L_{373}, were found by making use of special purpose sieve devices. Jon Sorenson found the last two by programming a very advanced sieve algorithm, making use of the doubly focused technique of the computational number theorist Dan Bernstein, on a large supercomputer cluster, called BigDawg. All the pseudosquares currently known are tabulated below.

p	L_p	Credit
3	73	1924 Kraitchik
5	241	1924 Kraitchik
7	1009	1924 Kraitchik
11	2641	1924 Kraitchik
13	8089	1924 Kraitchik
17	18001	1924 Kraitchik
19	53881	1924 Kraitchik
23	87481	1924 Kraitchik
29	117049	1924 Kraitchik
31	515761	1924 Kraitchik
37	1083289	1924 Kraitchik
41	3206641	1924 Kraitchik
43	3818929	1924 Kraitchik
47	9257329	1924 Kraitchik
53	22000801	1928 Lehmer
59	48473881	1928 Lehmer
61	48473881	1928 Lehmer
67	175244281	1954 Lehmer
71	427733329	1954 Lehmer
73	427733329	1954 Lehmer
79	898716289	1954 Lehmer
83	2805544681	1970 Lehmer, Lehmer, Shanks
89	2805544681	1970 Lehmer, Lehmer, Shanks
97	2805544681	1970 Lehmer, Lehmer, Shanks
101	10310263441	1970 Lehmer, Lehmer, Shanks

103	23616331489	1970 Lehmer, Lehmer, Shanks
107	85157610409	1970 Lehmer, Lehmer, Shanks
109	85157610409	1970 Lehmer, Lehmer, Shanks
113	196265095009	1970 Lehmer, Lehmer, Shanks
127	196265095009	1970 Lehmer, Lehmer, Shanks
131	2871842842801	Lehmer
137	2871842842801	Lehmer
139	2871842842801	Lehmer
149	26250887023729	1973 Lehmer
151	26250887023729	1973 Lehmer
157	112434732901969	1988 Patterson, Williams
163	112434732901969	1988 Patterson, Williams
167	112434732901969	1988 Patterson, Williams
173	178936222537081	1988 Patterson, Williams
179	178936222537081	1988 Patterson, Williams
181	696161110209049	1988 Patterson, Williams
191	696161110209049	1988 Patterson, Williams
193	2854909648103881	1989 Stephens, Williams
197	6450045516630769	1989 Stephens, Williams
199	6450045516630769	1989 Stephens, Williams
211	11641399247947921	1989 Stephens, Williams
223	11641399247947921	1989 Stephens, Williams
227	190621428905186449	1991 Lukes, Williams
229	196640248121928601	1991 Lukes, Williams
233	712624335095093521	1994 Lukes, Patterson, Williams
239	1773855791877850321	1994 Lukes, Patterson, Williams
241	2327687064124474441	1994 Lukes, Patterson, Williams
251	6384991873059836689	1994 Lukes, Patterson, Williams
257	8019204661305419761	1994 Lukes, Patterson, Williams
263	10198100582046287689	1994 Lukes, Patterson, Williams
269	10198100582046287689	1994 Lukes, Patterson, Williams
271	10198100582046287689	1994 Lukes, Patterson, Williams
277	69848288320900186969	1996 Lukes, Patterson, Williams
281	208936365799044975961	2001 Bernstein
283	533552663339828203681	2003 Williams, Wooding
293	936664079266714697089	2003 Williams, Wooding
307	936664079266714697089	2003 Williams, Wooding
311	2142202860370269916129	2003 Williams, Wooding
313	2142202860370269916129	2003 Williams, Wooding
317	2142202860370269916129	2003 Williams, Wooding
331	13649154491558298803281	2003 Williams, Wooding
337	34594858801670127778801	2003 Williams, Wooding
347	99492945930479213334049	2003 Williams, Wooding
349	99492945930479213334049	2003 Williams, Wooding
353	295363487400900310880401	2003 Williams, Wooding
359	295363487400900310880401	2003 Williams, Wooding
367	3655334429477057460046489	2008 Sorenson
373	4235025223080597503519329	2008 Sorenson

For all of these pseudosquares we have $\log L_p > p \log 2/\log p$, with no indication that is likely to change for larger values of p.

Number sieves have also been employed in tabulating pseudocubes and the more complicated Eisenstein pseudocubes, which have also been applied to the problem of primality proving. They have also been used to investigate several other problems; one of these is the *Erdös-Selfridge problem*. In 1971, Erdös suggested that computer methods could be applied to the following problem due to him and Selfridge: determine or estimate the smallest integer $g(k)$ $(> k + 1)$ such that all the prime factors of $\binom{g(k)}{k}$ are greater than k. We now know that there must exist some constant c such that

$$g(k) > e^{c((\log k)^3/\log\log k)^{1/2}},$$

which implies that $g(k)$ must grow faster than any polynomial in k.

It is very difficult to compute $g(k)$, but sieving devices have been enlisted in the search for these numbers for all $k \leq 200$. The largest value of $g(k)$ known is the 23 digit

$$g(200) = 52087838892710191382732.$$

A few other applications of sieving machines are mentioned in later chapters of this book. We conclude this chapter by introducing the problem of representing an integer k as the sum of three integral cubes. This is an old problem which we will discuss in greater detail in Chap. 20. We only mention here that in 1995 a number sieve was used to solve this problem for $k = 110, 435$ and 478; this was the first time this problem had been solved for these values of k.

Further Reading

Number theoretic sieves

David Hawkins, Mathematical Sieves, *Scientific American*, **199** (1958), 105–114.

John Friedlander and Henryk Iwaniec, *Opera de Cribo*, American Mathematical Society Colloquium Publications, **57**, American Mathematical Society, Providence, RI, 2010.

Sieve devices (history)

H. C. Williams, *Édouard Lucas and Primality Testing*, Wiley-Interscience, John Wiley and Sons, NY, 1998, pp. 156–172, 180–196, 405–411. This book describes several sieve devices and includes a number of rare photographs of some of them.

S. S. Wagstaff, Jr., *The Joy of Factoring*, Vol. 68, Student Mathematical Library, American Mathematical Society, Providence RI, 2013, pp. 219–237.

Carissan's number sieve

J.O. Shallit, H.C. Williams and F. Morain, Discovery of a lost factoring machine, *Mathematical Intelligencer*, **17**, #3 (1995), 41–47.

J.O. Shallit, H.C. Williams and F. Morain, La machine à congruences, *Musée des arts et métiers La Revue*, No. 14, mars 1996, 14–19. This article, while written in French and difficult to acquire, contains some beautiful photos of Carissan's sieve.

Martine Bühler, The congruence machine of the Carissan brothers, *Let History into the Mathematics Classroom*, Springer International Publishing AG, 6330 Cham, Switzerland, 2018, pp. 75-86.

Serveur vidéo de l'IREM de Paris, Available at: https://video.irem.univ-paris-diderot.fr/videos/watch/5669ee73-e297-40ee-b11d-05e85f41f301 Photo which can be enlarged at: https://collections.arts-et-metiers.net Click on the "Recherche dans le catalogue" button; enter the inventory number (43048-0000-) in the "recherche" search box; double click on the image.

Factoring using Lehmer's sieve devices

D. N. Lehmer, Hunting big game in the theory of numbers, *Scripta Mathematica*, **1** (1932–33), 229–235.

Pseudosquares and pseudocubes

Jonathan Sorenson, Sieving for pseudosquares and pseudocubes in parallel using doubly-focused enumeration and wheel data structures, *Algorithmic Number Theory*, ANTS 2010, Lecture Notes in Computer Science 6197, Springer-Verlag, Berlin, Heidelberg, 2010, pp. 331–339.

Eisenstein pseudocubes

Kjell Wooding and H. C. Williams, Improved Primality Proving with Eisenstein Pseudocubes, *Algorithmic Number Theory*, ANTS 2010, Lecture Notes in Computer Science, 6197, Springer-Verlag, Berlin, Heidelberg, 2010, pp. 372–384.

Kjell Wooding, The Sieve Problem in One- and Two-Dimensions, PhD Thesis, University of Calgary, 2010, pp. 36-56. Available at: https://hdl.handle.net/1880/104481

Some number sieve applications

Richard F. Lukes, A Very Fast Electronic Number Sieve, PhD Thesis, University of Manitoba, 1995. Available at: https://mspace.lib.umanitoba.ca/handle/1993/12210

Erdös-Selfridge problem

Paul Erdös, Some problems in number theory, *Computers in Number Theory*, Proceedings of the Science Research Council Atlas Symposium No. 2 held at Oxford from 18–23 August 1969, Academic Press, NY, 1971, pp. 405–414.

Richard F. Lukes R. Scheidler and H. C. Williams, Further tabulation of the Erdös-Selfridge function, *Mathematics of Computation*, **66** (1997), 1709–1717.

Chapter 17
Simple Continued Fraction of $\sqrt{D}$

At the end of Chap. 14 we began to discuss the SICF for $\sqrt{D}$ for any positive and non-square value of D. In this chapter we will expand upon this discussion and in the process find that this continued fraction possesses a number of striking and beautiful properties.

We first remark that by using results concerning the floor function in Chapter 4, it is easy to see that if P and Q are integers with $Q > 0$, then

$$\left\lfloor \frac{P + \sqrt{D}}{Q} \right\rfloor = \left\lfloor \frac{P + \lfloor\sqrt{D}\rfloor}{Q} \right\rfloor.$$

In order to produce the SICF expansion of $\alpha = \sqrt{D}$ ($P_0 = 0$, $Q_0 = 1$), we first define $Q_{-1} = D$ and then let R_{i-1} ($i \geq 1$) be the remainder on dividing $P_{i-1} + \lfloor\sqrt{D}\rfloor$ by Q_{i-1}. Since

$$P_{i-1} + \lfloor\sqrt{D}\rfloor = a_{i-1}Q_{i-1} + R_{i-1} \quad (i \geq 1),$$

we get

$$P_i = a_{i-1}Q_{i-1} - P_{i-1} = \lfloor\sqrt{D}\rfloor - R_{i-1} \quad (i \geq 1)$$

and

$$\begin{aligned} Q_i &= \frac{D - P_i^2}{Q_{i-1}} = \frac{D - (a_{i-1}Q_{i-1} - P_{i-1})^2}{Q_{i-1}} = \frac{D - P_{i-1}^2}{Q_{i-1}} - a_{i-1}(a_{i-1}Q_{i-1} - 2P_{i-1}) \\ &= Q_{i-2} - a_{i-1}(P_i - P_{i-1}) \quad (i \geq 1). \end{aligned}$$

E. L. F. Roettger and H. C. Williams, *The Enchantment of Numbers*, CMS/CAIMS Books in Mathematics 16, https://doi.org/10.1007/978-3-031-76568-1_17

Thus, we can now calculate the complete quotient α_i from α_{i-1} by dividing $P_{i-1} + \lfloor\sqrt{D}\rfloor$ by Q_{i-1}, finding the quotient a_{i-1} and the remainder R_{i-1}. We get

$$P_i = \lfloor\sqrt{D}\rfloor - R_{i-1}$$

and

$$Q_i = Q_{i-2} - a_{i-1}(P_i - P_{i-1}).$$

As most computers provide both the quotient and the remainder as the result of executing an integer divide instruction, this means that (ignoring the cost of a subtraction operation) that we can replace the earlier process of finding α_i from α_{i-1}, which costs 2 divides, 1 multiplication and 1 squaring operation by a process which only requires 1 divide and 1 multiply (usually by a relatively small number). This procedure is often called *Tenner's algorithm* after the somewhat obscure German mathematician G. W. Tenner, who published part of it in 1841.

We illustrate this algorithm by computing the SICF for $\sqrt{29}$. We first put $P_0 = 0$, $Q_0 = 1$, $Q_{-1} = 29$, $\lfloor\sqrt{29}\rfloor = 5$, and then use the above formulas to produce the tableau below. Since $\alpha_1 = \alpha_6$ (recall, $\alpha_n = \frac{P_n+\sqrt{D}}{Q_n}$), we see that

i	-1	0	1	2	3	4	5	6
P_i		0	5	3	2	3	5	5
Q_i	29	1	4	5	5	4	1	4
R_i	5	0	2	3	2	0	0	
a_i		5	2	1	1	2	10	

$$\sqrt{29} = [5; \overline{2, 1, 1, 2, 10}].$$

Notice that once $\lfloor\sqrt{D}\rfloor$ has been determined, executing Tenner's algorithm requires the performance of arithmetic operations on integers only.

Suppose we put $\beta = \sqrt{D} + \lfloor\sqrt{D}\rfloor$. We have $\beta > 1$ and $-1 < \overline{\beta} = \lfloor\sqrt{D}\rfloor - \sqrt{D} < 0$. Thus, the SICF expansion of β is purely periodic with a period length p and

$$\beta = [2\lfloor\sqrt{D}\rfloor; a_1, a_2, \ldots, a_{p-1}, 2\lfloor\sqrt{D}\rfloor, a_1, a_2, \ldots, a_{p-1}, \ldots].$$

Here, the value of p is the least positive integer n such that $\beta_n = \beta_0 = \beta$. Hence, the SICF expansion of $\sqrt{D} = \beta - \lfloor\sqrt{D}\rfloor$ is periodic of period length p (≥ 1) after an initial partial quotient of $\lfloor\sqrt{D}\rfloor$, i.e.

$$\sqrt{D} = [\lfloor\sqrt{D}\rfloor; \overline{a_1, a_2, \ldots, a_{p-1}, 2\lfloor\sqrt{D}\rfloor}].$$

Since the complete quotients of this continued fraction must satisfy

$$\alpha_0 = \frac{P_0+\sqrt{D}}{Q_0} = \sqrt{D},$$
$$\alpha_n = \beta_n = \frac{P_n+\sqrt{D}}{Q_n} > 1 \quad (n \geq 1)$$

and

$$-1 < \overline{\alpha}_n < 0, \quad (n \geq 1)$$

we find that

$$0 \leq P_n \leq \lfloor\sqrt{D}\rfloor, \quad 1 \leq Q_n \leq \lfloor 2\sqrt{D}\rfloor \quad 1 \leq a_n \leq 2\lfloor\sqrt{D}\rfloor$$

for $0 \leq n \leq p$. Also, $P_p = \lfloor\sqrt{D}\rfloor$, $Q_p = 1$, $a_p = 2\lfloor\sqrt{D}\rfloor$.

Suppose we now consider the possibility that $Q_n = 1$ for some positive n. Since

$$-1 < \frac{P_n - \sqrt{D}}{Q_n} < 0,$$

we must have $P_n = \lfloor\sqrt{D}\rfloor$, and $\alpha_n = \lfloor\sqrt{D}\rfloor + \sqrt{D} = \beta_0$. If we put

$$n = qp + r \quad (0 \leq r < p),$$

then because $\alpha_{p+k} = \alpha_k$, we have

$$\beta_0 = \alpha_n = \alpha_{qp+r} = \alpha_r.$$

If $r > 0$, then since $r < p$, we have a contradiction to the definition of p; thus, it follows that $Q_n = 1$ if and only if $p \mid n$.

In the SICF expansion of $\alpha = \sqrt{D}$ we have

$$\frac{P_{n+1} + \sqrt{D}}{Q_n} = a_n + \frac{\sqrt{D} - P_n}{Q_n} = a_n - \overline{\alpha}_n.$$

Thus, since $-1 < \overline{\alpha}_n < 0$ when $n \geq 1$, we can write

$$a_n = \left\lfloor \frac{P_{n+1} + \sqrt{D}}{Q_n} \right\rfloor \quad (n \geq 1).$$

We next observe that since $P_p = \lfloor\sqrt{D}\rfloor = P_1$, $Q_p = 1$, then for $p \geq 2$ we get

$$Q_{p-1} = \frac{D - P_p^2}{Q_p} = \frac{D - P_1^2}{Q_0} = Q_1,$$

$$a_{p-1} = \left\lfloor \frac{P_p + \sqrt{D}}{Q_{p-1}} \right\rfloor = \left\lfloor \frac{P_1 + \sqrt{D}}{Q_1} \right\rfloor = a_1,$$
$$P_{p-1} = a_{p-1}Q_{p-1} - P_p = a_1 Q_1 - P_1 = P_2.$$

The following *symmetry properties* of the SICF for $\sqrt{D}$ are easily established by induction on i:

$$Q_{p-i} = Q_i, \quad a_{p-i} = a_i, \quad P_{p-i} = P_{i+1}$$

when $p \geq 2$ and $1 \leq i < p$.

Notice that if p is even ($p = 2t$), then $P_t = P_{p-t} = P_{t+1}$; and if p is odd ($p = 2t + 1$), then $Q_t = Q_{p-(t+1)} = Q_{t+1}$. Indeed, it can be shown that if t is the least positive integer such that $P_t = P_{t+1}$, then $p = 2t$. Also, if t is the least positive integer such that $Q_t = Q_{t+1}$, then $p = 2t + 1$.

Now suppose we define

$$\theta_n = A_{n-2} + \sqrt{D}B_{n-2} \quad (n \geq 0).$$

We have $\theta_0 = \sqrt{D}$ and $\theta_1 = 1$. We also see that

$$\begin{aligned}(P_n + \sqrt{D})\theta_n &= P_n A_{n-2} + DB_{n-2} + \sqrt{D}(A_{n-2} + P_n B_{n-2}) \\ &= Q_{n-1}(A_{n-1} + \sqrt{D}B_{n-1})\end{aligned}$$

by the identities near the end of Chapter 14. Hence, we get

$$Q_{n-1}\theta_{n+1} = Q_n\alpha_n\theta_n \quad (n \geq 1).$$

Putting $n = 1$, we have

$$\theta_2 = Q_1\alpha_1,$$

and if $n = 2$, we have

$$Q_1\theta_3 = Q_2\alpha_2\theta_2 = Q_1 Q_2\alpha_1\alpha_2,$$

which means that

$$\theta_3 = Q_2\alpha_1\alpha_2.$$

By induction we can prove that

$$\theta_n = Q_{n-1}\alpha_1\alpha_2\alpha_3 \cdots \alpha_{n-1}.$$

Thus, we can express θ_n multiplicatively in terms of the complete quotients up to α_{n-1}. In view of how the continued fraction for $\sqrt{D}$ is defined, this is a somewhat surprising result. Since $\alpha_k > 1$ for $k \geq 1$, we see that $\theta_n > 1$ for $n \geq 1$. Also, from the recursion formulas for A_i and B_i and the definition of θ_i, we can readily derive

$$\theta_{i+1} = a_{i+1}\theta_i + \theta_{i-1} \geq \theta_i + \theta_{i-1}.$$

Hence, it is easy to see that for $n \geq 2$, we have

$$\theta_n > F_n.$$

Since $F_n > \phi^{n-2}$, we get

$$\log \theta_n > (n-2)\log \phi.$$

From the symmetry properties we have

$$\alpha_{p-i} = \frac{P_{p-i} + \sqrt{D}}{Q_{p-i}} = \frac{P_{i+1} + \sqrt{D}}{Q_i}.$$

If we substitute this into the multiplicative equation for θ_n above, we find that we can produce the following expressions for θ_{p+1}:

$$\theta_{p+1} = A_{p-1} + \sqrt{D}B_{p-1} = Q_t(\alpha_1\alpha_2\ldots\alpha_t)^2 \quad \text{when } p = 2t,$$
$$\theta_{p+1} = (P_t + \sqrt{D})(\alpha_1\alpha_2\ldots\alpha_t)^2 \quad \text{when } p = 2t+1.$$

From these it follows that if $p = 2t$, then $P_t = P_{t+1}$ and

$$A_{p-1} = A_t B_{t-1} + A_{t-1}B_{t-2}, \quad B_{p-1} = B_{t-1}(B_t + B_{t-2}).$$

If $p = 2t+1$, then $Q_t = Q_{t+1}$ and

$$A_{p-1} = A_t B_t + A_{t-1}B_{t-1}, \quad B_{p-1} = B_t^2 + B_{t-1}^2.$$

We also observe from the last identity of Chap. 14 that

$$\begin{aligned}\theta_{p+1}\overline{\theta}_{p+1} &= (A_{p-1} + \sqrt{D}B_{p-1})(A_{p-1} - \sqrt{D}B_{p-1}) \\ &= A_{p-1}^2 - DB_{p-1}^2 = (-1)^p Q_p = (-1)^p.\end{aligned}$$

The problem of determining the parity of p (whether p is even or odd) goes back over two hundred years. In 1980 the American mathematician Jeff Lagarias produced a polynomial time algorithm for doing this when the complete factorization of D

is known, together with a quadratic nonresidue for each prime divisor of D. This technique will execute in time complexity $\tilde{O}((\log D)^5)$.

Denote by $p(D)$, the period length of the SICF expression of $\sqrt{D}$. It is easy to verify that $p(D) = 1$ whenever $D = M^2 + 1$; thus, $p(D) = 1$ infinitely often because $M^2 + 1$ can be a perfect square only for $M = 0$. The special form $M^2 + 1$ is a member of a collection of forms $D = M^2 + k$, where $k \mid 4M$. It can be shown that $p(D) \leq 12$ for any such D that is not a perfect square. This form of D is said to be of *Richaud-Degert type* after the two mathematicians (C. Richaud and G. Degert) who independently discovered it.

We can write the Richaud-Degert type as $D = A^2X^2 + 2^m A$, where $0 \leq m \leq 2$. This form can be easily generalized to $D(X) = A^2X^2 + 2BX + C$, where A, B, C are integers. In 1961 the Polish mathematician Andrzej Schinzel (1937–2021) showed that $p(D(X))$ as a function of X is bounded above if and only if

$$\Delta \mid 4 \gcd(A^2, B)^2,$$

where $\Delta = B^2 - A^2C$. We call such forms *Schinzel types*; notice that they include the Richaud-Degert types. Schinzel's methods were not constructive, and, as a consequence, he was unable to derive the pattern of the partial quotients in the period of $\sqrt{D(X)}$, nor was he able to produce an explicit upper bound on $p(D(X))$.

More than 40 years after Schinzel's discovery, Kell Cheng, as part of his Ph.D. thesis, elucidated the structure of the period of the continued fraction expansion of $\sqrt{D(X)}$ for the Schinzel types. Put $|\Delta| = \Delta_1\Delta_2^2\Delta_4^4$, where Δ_1 and Δ_2 are squarefree (it is always possible to do this for unique positive values of Δ_1 and Δ_2), $G = \gcd(A, \Delta_2\Delta_4^2)$, $A' = A/G$, $\Delta' = (\Delta_2\Delta_4^2)/G$. The period for a Schinzel type can be quite intricate and depends on the residue class of X modulo Δ'. In general, for all sufficiently large $X = \Delta' W + k$ $(1 \leq k < \Delta')$, we have

$$\sqrt{D(X)} = [q_0(W); \overline{\mathcal{S}_0, q_1(W), \mathcal{S}_1, q_2(W), \ldots, \mathcal{S}_{\kappa-1}, q_\kappa(W)}],$$

where the period consists of κ segments each made up of a string $\mathcal{S}_i$ (possibly empty, but usually composed of the partial quotients of the continued fraction expansion of a certain rational number), followed by a linear function $q_{i+1}(W)$ of W. For example, when $A = 119$, $B = 833$, $C = 98$, we get $\Delta = -7^4 \cdot 17^2$, $G = 7 \cdot 17$, $A' = 1$, $\Delta' = 7$ and

$$\sqrt{D(7W)} = [833W + 7; \overline{34W, 3, 2, 34W - 1, 1, 6, 34W, 6, 1, 34W - 1,}$$
$$\overline{2, 3, 34W, 1666W + 14}],$$

whereas

$$\sqrt{D(7W + 1)} = [833W + 126; \overline{34W + 5, 6, 1, 34W + 4, 3, 2, 34W + 4, 1, 2, 1, 1,}$$
$$\overline{34W + 4, 1, 1, 2, 1, 34W + 4, 2, 3, 34W + 4, 1, 6, 34W + 5, 1666W + 252}].$$

Cheng was also able to deduce from his results that for all sufficiently large X (the lower bound was provided),

$$p(D(X)) < \frac{3\Delta' \log(\sqrt{5}A'\Delta')}{\log \phi},$$

a bound independent of the value of X.

We have a lower bound on $p(D)$, but the question now arises of determining how large $p(D)$ can be as a function of D. It is possible to show that there exists a fixed constant c such that

$$p(D) < c\sqrt{D} \log D.$$

For example, a value of .72 for c will work, but smaller values can likely be derived. The above bound, however, seems to be much too large. There are very good reasons to believe that there exists a fixed constant c such that

$$p(D) < c\sqrt{D} \log \log D,$$

but a proof of this conjecture possibly involves a proof of the extended Riemann hypothesis (ERH), something that is currently well out of reach.

For non-square, positive integers $D \not\equiv 1 \pmod 8$, it is convenient to investigate the growth of

$$G(D) = \frac{p(D)}{\sqrt{D} \log \log 4D}.$$

(There is a similar formula for $G(D)$ when $D \equiv 1 \pmod 8$.) Numerical investigations and some heuristic reasoning suggest that when $D \not\equiv 1 \pmod 8$, $G(D)$ will be maximal when D is either of the form q or $2q$, where q is a prime and $q \equiv 3 \pmod 4$. Furthermore, we also seem to require that the Legendre symbol $(D|r) = 1$ for all the odd primes $r \leq r_{\max}$. This latter requirement is, as we have seen, best attacked by making use of a sieve device. This has been done several times and the largest value of $G(D)$ (using $r_{\max} = 157$) currently known is $G(D) = 1.093416$ for

$$D = 4826678427841846,$$

where $p(D) = 275319106$. We expect from the ERH that $G(D)$ would be bounded above by a number in the neighbourhood of $3/2$, so this numerical result is well within what would be expected. We will discuss the problem of the size of $p(D)$ from a somewhat different perspective in Chapter 19.

By our earlier results we have

$$\log \theta_{p+1} > (p-1) \log \phi;$$

hence,

$$p < \frac{1}{\log \phi} \log \theta_{p+1} + 1 < 2.078087 \log \theta_{p+1} + 1.$$

To get a better idea of how p relates to $\log \theta_{p+1}$, we need some results from the metrical theory of continued fractions. If

$$\alpha = [a_0; a_1, a_2, \ldots, a_j, \alpha_{j+1}],$$

then

$$t = \alpha - a_0 = [0; a_1, a_2, \ldots, a_j, \alpha_{j+1}].$$

Since $0 \le t < 1$, we restrict ourselves to the study of this SICF of t for $0 \le t < 1$. We use $\alpha_{j+1}(t)$ to denote the α_{j+1} above. The following result, known as *Lévy's law*, was proved by the French mathematician Paul Lévy (1886–1971) in 1936.

Theorem 17.1 *For almost all real t such that* $0 \le t < 1$, *we have*

$$\lim_{n\to\infty} \left(\prod_{i=1}^{n} \alpha_i \right)^{1/n} = e^{\lambda},$$

where $\alpha_i = \alpha_i(t)$ *and* $\lambda = \pi^2/(12 \log 2) = 1.186569\ldots$.

This means that for $t = \sqrt{D} - \lfloor \sqrt{D} \rfloor$, we would expect that $\log \theta_{p+1}$ is approximately $p\lambda$ or

$$p \approx \frac{1}{\lambda} \log \theta_{p+1} \approx .842766 \log \theta_{p+1}.$$

However, it must be emphasized that Theorem 17.1 does not hold for all real values of t. For example, if $D = M^2 + 1$, we have $\alpha_i = M = \sqrt{D}\,(i = 1, 2, 3, \ldots)$. Hence

$$\lim_{n\to\infty} \left(\prod_{i=1}^{n} \alpha_i \right)^{1/n} = M + \sqrt{D} \neq e^{\lambda}.$$

If we produce the continued fraction expansion for a particular $\sqrt{D}$, we will usually observe that the partial quotients tend to be small. Indeed, 58% of them will usually be either 1 or 2. There is another result from the metrical theory of continued fractions which explains this phenomenon. If $j \ge 0$ and $Pr(a_{j+1} = k)$ denotes the probability that the partial quotient a_{j+1} is some integer k, then it turns out that

$$Pr(a_{j+1} = k) = \frac{\log\left(1 + \frac{1}{k(k+2)}\right)}{\log 2} + O\left(\frac{q^j}{k(k+1)}\right),$$

where q is some fixed real number such that $0 < q < .76$. Some approximate values of these probabilities are given in the table below.

k	$Pr(a_{j+1} = k)$
1	.415037
2	.169925
3	.093109
4	.058894
5	.040642

Problems

17.1 (Tenner's algorithm) Find the SICF expansion of $\alpha_0 = \sqrt{D}$ ($P_0 = 0, Q_0 = 1$) using Tenner's algorithm for the values given below. Begin by setting $Q_{-1} = D$. Recover α_i from α_{i-1} by dividing $(P_{i-1} + \lfloor\sqrt{D}\rfloor)$ by Q_{i-1}; call the quotient of this a_{i-1} and the remainder R_{i-1}. Use these to find P_i and Q_i with the aid of the identities:

$$P_i = \lfloor\sqrt{D}\rfloor - R_{i-1},$$
$$Q_i = Q_{i-2} - a_{i-1}(P_i - P_{i-1}).$$

(a) **[BB]** $\alpha_0 = \sqrt{17}$
(b) $\alpha_0 = \sqrt{20}$
(c) $\alpha_0 = \sqrt{23}$

17.2 (Tenner's algorithm) Find the SICF expansion of $\alpha_0 = \frac{P_0+\sqrt{D}}{Q_0}$ using Tenner's algorithm for the values given below. Begin by setting $Q_{-1} = (D - P_0^2)/Q_0$. Recover α_i from α_{i-1} by dividing $(P_{i-1} + \lfloor\sqrt{D}\rfloor)$ by Q_{i-1}; call the quotient of this a_{i-1} and the remainder R_{i-1}. Use these to find P_i and Q_i with the aid of the identities:

$$P_i = \lfloor\sqrt{D}\rfloor - R_{i-1},$$
$$Q_i = Q_{i-2} - a_{i-1}(P_i - P_{i-1}).$$

(a) **[BB]** $\alpha_0 = \dfrac{2+\sqrt{17}}{13}$

(b) $\alpha_0 = \dfrac{3+\sqrt{20}}{11}$

(c) $\alpha_0 = \dfrac{7+\sqrt{53}}{4}$

Further Reading

Continued fraction of $\sqrt{D}$

H. Davenport, *The Higher Arithmetic: An Introduction to the Theory of Numbers*, Dover Publications, NY, 1983, pp. 79–114.

Special results and Tenner's algorithm

M. J. Jacobson, Jr. and H. C. Williams, *Solving the Pell Equation*, CMS Books in Mathematics, Springer Science+Business Media, New York, 2009, pp. 63–71.

Parity of the period

J. C. Lagarias, On the computational complexity of determining the solvability of the equation $X^2 - DY^2 = -1$, *Transactions of the American Mathematical Society*, **260** (1980), 485–508.

Richaud-Degert types

R. A. Mollin, *Quadratics* CRC Press, Inc., Boca Raton, FL, 1995, pp. 77–87.

M. J. Jacobson, Jr. and H. C. Williams, *Solving the Pell Equation*, CMS Books in Mathematics, Springer Science+Business Media, New York, 2009, pp. 125–140.

Schinzel types

Kell Cheng and H. C. Williams, Some results concerning certain periodic continued fractions, *Acta Arithmetica*, **117** (2005), 247–264.

Kell Cheng, R. K. Guy, R. Scheidler and H. C. Williams, Classification and symmetries of a family of continued fractions with bounded period length, *Journal of the Australian Mathematical Society* **93** (2012), 53–76.

Period length

H. C. Williams, A numerical investigation into the length of the period of the continued fraction of $\sqrt{D}$, *Mathematics of Computation*, **36** (1981), 593–601.

A. Stephens and H. C. Williams, An open architecture number sieve, *Lecture notes of the London Math. Soc.*, **154** (1990), 38–75.

Metric theory of continued fractions

A. Y. Khinchin, *Continued Fractions*, Dover Publications, NY, 1997, Chapter 3.
A. M. Rockett and P. Szüsz, *Continued Fractions*, World Scientific, NY, 1992, Chapter V.
D. E. Knuth, *The Art of Computer Programming*, Vol. 2, Seminumerical Algorithms, 3rd ed., Addison-Wesly, Reading, MA, 1998, Section 4.5.3.

Chapter 18
Formulas for Primes

Recall that for the nth Fibonacci number we have the useful formula

$$F_n = \frac{\phi^n - \bar{\phi}^n}{\sqrt{5}},$$

where $\phi = \frac{1+\sqrt{5}}{2}$ and $\bar{\phi} = \frac{1-\sqrt{5}}{2}$. It would be desirable to have such an explicit formula for the nth prime number.

We denote the nth prime by p_n. For example, $p_4 = 7$. An early formula for p_{n+1} ($n \geq 0$) was created in 1900 by the German mathematician Casper Isenkrahe (1844–1921). He used the expression

$$F(n,k) = \frac{k!}{P(n,k)} + \frac{P(n,k)}{(k-1)!} - \left\lfloor \frac{(k-1)!}{P(n,k)} \right\rfloor$$

where

$$P(n,k) = 2^{e_1} \cdot 3^{e_2} \cdot 5^{e_3} \cdot \cdots \cdot p_n^{e_n} \quad \text{and} \quad e_i = \lfloor k/p_i \rfloor + \lfloor k/p_i^2 \rfloor + \lfloor k/p_i^3 \rfloor + \cdots .$$

By our results at the beginning of Chap. 6, we have $P(n,k) = k!$ for $2 \leq k < p_{n+1}$; in which case,

$$\begin{aligned} F(n,k) &= \frac{k!}{P(n,k)} + \frac{P(n,k)}{(k-1)!} - \left\lfloor \frac{(k-1)!}{P(n,k)} \right\rfloor \\ &= \frac{k!}{k!} + \frac{k!}{(k-1)!} - \left\lfloor \frac{(k-1)!}{k!} \right\rfloor \\ &= 1 + k - 0 = k + 1. \end{aligned}$$

Also, for the case of $k = p_{n+1}$, we have $P(n,k) = (k-1)!$; so here,

E. L. F. Roettger and H. C. Williams, *The Enchantment of Numbers*, CMS/CAIMS Books in Mathematics 16, https://doi.org/10.1007/978-3-031-76568-1_18

$$\begin{aligned} F(n,k) &= \frac{k!}{P(n,k)} + \frac{P(n,k)}{(k-1)!} - \left\lfloor \frac{(k-1)!}{P(n,k)} \right\rfloor \\ &= \frac{k!}{(k-1)!} + \frac{(k-1)!}{(k-1)!} - \left\lfloor \frac{(k-1)!}{(k-1)!} \right\rfloor \\ &= k + 1 - 1 = k. \end{aligned}$$

From this we find that

$$p_{n+1} = F(n,k),$$

but since $k = p_{n+1}$ here, this is not very helpful.

Similarly, in 1964, Christopher Willans (1942–1971) produced the formula

$$p_n = 1 + \sum_{i=1}^{2^n} \left\lfloor \left(\frac{n}{\sum_{j=1}^{i} \left\lfloor \left(\cos \frac{(j-1)!+1}{j} \pi \right)^2 \right\rfloor} \right)^{1/n} \right\rfloor.$$

It can be shown that p_n produced by Willans' formula is simply the smallest integer m, such that the prime counting function $\pi(m) = n$. Similar methods can done with any prime counting function, $\pi(m)$ by use of

$$p_n = 2 + \sum_{i=1}^{2^n} C_n(\pi(m)),$$

where

$$C_n(a) = \begin{cases} 1, & \text{if } a < n; \\ 0, & \text{otherwise.} \end{cases}$$

Since $\pi(m) \geq n$ for $m \geq p_n$ and $\pi(m) < n$ for $m < p_n$, we have

$$C_n(\pi(m)) = \begin{cases} 1, & \text{if } m < p_n; \\ 0, & \text{otherwise.} \end{cases}$$

From this we can see Willans' formula, or any other that relies on a prime counting function, reduces to the following tautology:

$$p_n = 2 + \sum_{m=2}^{p_n-1} 1 = 2 + \sum_{m=1}^{p_n-2} 1 = 2 + (p_n - 2) = p_n.$$

What Isenkrahe's and Willans' functions have in common is that neither of these functions is actually useful for computing the nth prime. More specifically, they are not useful because neither is efficiently computable. Indeed, there are no known efficient, computable formulas for expressing p_n.

More generally, we can define a *formula for primes* to be a formula that only generates prime numbers. Say we call such a formula $f(n)$, then it must be the case that $f(n)$ is always prime for all n and ideally $f(n) \neq f(m)$ for $n \neq m$. One such simple example of the above relies on Wilson's theorem (Theorem 11.1 in Chap. 11), where we saw $n > 1$ is prime if and only if $n \mid (n-1)! + 1$. Therefore, $n+1$ $(n > 0)$ is prime if and only if $n+1 \mid n! + 1 \leftrightarrow n! + 1 \equiv 0 \pmod{n+1} \leftrightarrow n! \equiv -1 \pmod{n+1} \leftrightarrow n! \equiv n \pmod{n+1}$. We can then see that

$$f(n) = \left\lfloor \frac{n! \bmod (n+1)}{n} \right\rfloor (n-1) + 2$$

is a formula that will output 2 when $n+1$ is not prime and output $n+1$, when $n+1$ is prime, as

$$\left\lfloor \frac{n! \bmod (n+1)}{n} \right\rfloor = \begin{cases} 1, & \text{if } n+1 \text{ is prime;} \\ 0, & \text{otherwise.} \end{cases}$$

Notice that since $f(n) = 2$ for all n where $n+1$ is not prime, this function fails to satisfy the condition $f(n) \neq f(m)$ for $n \neq m$. It is perhaps obvious from the reliance on Wilson's theorem that this formula is not an efficient way to generate primes; hence, again, it is not useful. In fact, despite many efforts, no one has succeeded in producing a useful formula for generating primes.

In 1947 William Mills (1921–2007) proved there exists a real number A such that $\lfloor A^{3^n} \rfloor$ is a prime number for all integers $n > 0$. We define *Mills' constant* to be the smallest such positive real number A. To date the exact value of A is still unknown and most likely always will be. However, if we assume the Riemann hypothesis is true, then

$$A = 1.306377883863080690468614492602605712916784585156713644\ldots.$$

The first 6 known Mills primes using the Mills' constant under the assumed Riemann hypothesis are (A051254 OEIS):

n	digits	$\lfloor A^{3^n} \rfloor$
1	1	2
2	2	11
3	4	1361
4	10	2521008887
5	29	16022236204009818131831320183
6	85	41131011492151048000305295379159531704861396 23539759933135949994882770404074832568499

In 2005 Chris Caldwell provided a method to calculate A. In his paper he mentioned that he calculated A to 6850 decimal places, and using this A he calculated Mills primes b_n up to $n = 12$; however, since then, only those up to $n = 11$ have been proved to be prime (b_{11} as of 2017); b_{12} has only been shown to be a probable prime. This work has since been been extended to b_{14} and hence producing more probable primes. Using the notation $b_n = \lfloor A^{3^n} \rfloor$, and noting $b_{i+1} = b_i^3 + a_i$, where (a_i) is the sequence

$$3, 30, 6, 80, 12, 450, 894, 3636, 70756, 97220, 66768, 300840, 1623568, \ldots,$$

we can then easily represent Mills primes and Mills probable primes up to $n = 14$. We note that the largest known Mills prime, b_{11}, is an integer of 20562 digits and the largest Mills probable prime b_{14}, has a staggering 555154 digits.

Similar to Mills' work, in 1951 English mathematician Sir Edward Wright (1906–2005) showed that there exists a real number α, where if $g_0 = \alpha$ and $g_{n+1} = 2^{g_n}$ $(n \geq 0)$, then

$$\lfloor g_n \rfloor = \left\lfloor 2^{\cdots 2^{2^{\alpha}}} \right\rfloor$$

is always a prime for $n > 0$. Wright only gave α to seven decimal digits; that is, $\alpha = 1.9287800$. Using Wright's values for α, we get:

n	digits	$\lfloor g_n \rfloor$
1	1	3
2	2	13
3	5	16381

However, the value for $\lfloor g_4 \rfloor$, using Wright's α, is 1913991085 followed by 4922 zeros which is clearly even. This can be corrected by using a much more precise value for α, yielding a prime of 4932 decimal digits. It is evident from the above that both Mills' and Wright's formulas are not useful for generating primes because of the expensive need to calculate enough decimal digits for A or α , respectively, due to the exponential growth rate of the corresponding functions.

It is well known that no rational algebraic function exists that will always give primes; however, we can define a *prime representing polynomial* as one in which the set of primes equals the set of positive values of the polynomial as the variables are run through all possible nonnegative integers. Yuri Matijasevič, a Russian mathematician, in 1973 knew of the existence of a prime representing polynomial in 12 variables. Later, Martin Davis (1928–2023), Yuri Matijasevič, Hilary Putnam (1926–2016) and Julia Robinson (1919–1985) proved that every recursively enumerable set is Diophantine. We say a *set S is Diophantine* when there exists some polynomial with integer coefficients, say $f(x_1, x_2, \ldots, x_i)$, such that an integer n is in S if and only if there exist integers $a_1, a_2, \ldots, a_{i-1}$ where $f(n, a_1, \ldots, a_{i-1}) = 0$. As a consequence of this, Matijasevič's result that the set of prime numbers is Diophantine is also proved. One such example of this is given by James Jones, Daihachiro Sato, Hideo Wada and Douglas Wiens in their following theorem.

Theorem 18.1 *The set of prime numbers is identical with the set of positive values taken on by the polynomial*

$$(k+2)(1-M_1-M_2-M_3-M_4-M_5-M_6-M_7-M_8 \\ -M_9-M_{10}-M_{11}-M_{12}-M_{13}-M_{14})$$

where the polynomials M_i $(i = 1, 2, \dots, 14)$ are given by

$$\begin{aligned}
M_1 &= (wz+h+j-q)^2,\\
M_2 &= ((gk+2g+k+1)(h+j)+h-z)^2,\\
M_3 &= (2n+p+q+z-e)^2,\\
M_4 &= (16(k+1)^3(k+2)(n+1)^2+1-f^2)^2,\\
M_5 &= (e^3(e+2)(a+1)^2+1-o^2)^2,\\
M_6 &= ((a^2-1)y^2+1-x^2)^2,\\
M_7 &= (16r^2y^4(a^2-1)+1-u^2)^2,\\
M_8 &= (((a+u^2(u^2-a))^2-1)(n+4dy)^2+1-(x+cu)^2)^2,\\
M_9 &= (n+i+v-y)^2,\\
M_{10} &= ((a^2-1)l^2+1-m^2)^2,\\
M_{11} &= (ai-k+1-l-i)^2,\\
M_{12} &= (p+l(a-n-1)+b(2an+2a-n^2-2n-2)-m)^2,\\
M_{13} &= (q+y(a-p-1)+s(2ap+2a-p^2-2p-2)-x)^2,\\
M_{14} &= (z+pl(a-p)+t(2ap-p^2-1)-pm)^2,
\end{aligned}$$

where the variables $a, b, c, \dots, z$ range over the nonnegative integers.

The proof is omitted here, but it relies on using Wilson's theorem to show $k+2$ is prime if and only if $M_i = 0$ (for all $1 \le i \le 14$) is solvable in nonnegative integers.

The polynomial used in Theorem 18.1 is of degree 25. It is possible, though, to lower the degree by introducing more variables to modify the polynomial. In fact, it can be lowered to degree 5 by using 42 variables.

At a conference in Paris in 1900, the influential German mathematician David Hilbert (1862–1943) presented a list of important unsolved problems in mathematics. Two years later he published a compendium of 23 such problems, among which was his tenth problem: develop a general algorithm which, for any Diophantine equation, can decide whether the equation has integer solutions. The work of Martin Davis, Yuri Matijasevič, Hilary Putnam and Julia Robinson led to the solution of this problem by showing the answer is no, such a general algorithm cannot exist. It was this work that was used to establish that the set of prime numbers is representable by a polynomial formula. We note, however, that the polynomials created, like that in Theorem 18.1, are not actually useful for generating primes. Despite being proved to exist, it is very difficult to find values for the variables that produce $M_i = 0$ for all i.

Now consider the linear polynomial $f_{a,b}(x) = ax + b$, where a and b are coprime integers. A recent (2004) result of the number theorists Ben Green (UK) and Terence Tao (AUS) asserts that for any value of the positive integer k, there must exist integers a and b such that $f_{a,b}(x)$ is prime for all $x = 0, 1, 2, \ldots, k-1$. Since these are primes in an arithmetic progression, it has become customary to denote examples under the notation AP-k. However, it is rather difficult to find such examples, even for moderate values of k; indeed, the best current result is for $k = 27$. In this case we have

$$a = 18135696597948930, \quad b = 224584605939537911.$$

This was discovered in 2019 by Rob Gahan of Ireland, using PrimeGrid, a distributed computing project (see: http://www.primegrid.com). In 2023 a second example of an AP-27 was found by Tom Greer of the United States.

An interesting quadratic polynomial of one variable that takes on prime values for the first 40 nonnegative integers was found by Euler in 1772. It is easy to verify the following table for the first 80 nonnegative values of x in the polynomial $p(x) = x^2 + x + 41$, now called *Euler's polynomial*.

x	$p(x)$	prime	x	$p(x)$	prime	x	$p(x)$	prime	x	$p(x)$	prime
0	41	✓	20	461	✓	40	1681	×	60	3701	✓
1	43	✓	21	503	✓	41	1763	×	61	3823	✓
2	47	✓	22	547	✓	42	1847	✓	62	3947	✓
3	53	✓	23	593	✓	43	1933	✓	63	4073	✓
4	61	✓	24	641	✓	44	2021	×	64	4201	✓
5	71	✓	25	691	✓	45	2111	✓	65	4331	×
6	83	✓	26	743	✓	46	2203	✓	66	4463	✓
7	97	✓	27	797	✓	47	2297	✓	67	4597	✓
8	113	✓	28	853	✓	48	2393	✓	68	4733	✓
9	131	✓	29	911	✓	49	2491	×	69	4871	✓
10	151	✓	30	971	✓	50	2591	✓	70	5011	✓
11	173	✓	31	1033	✓	51	2693	✓	71	5153	✓
12	197	✓	32	1097	✓	52	2797	✓	72	5297	✓
13	223	✓	33	1163	✓	53	2903	✓	73	5443	✓
14	251	✓	34	1231	✓	54	3011	✓	74	5591	✓
15	281	✓	35	1301	✓	55	3121	✓	75	5741	✓
16	313	✓	36	1373	✓	56	3233	×	76	5893	×
17	347	✓	37	1447	✓	57	3347	✓	77	6047	✓
18	383	✓	38	1523	✓	58	3463	✓	78	6203	✓
19	421	✓	39	1601	✓	59	3581	✓	79	6361	✓

We can readily verify that $p(40 + x^2) = p(x-1)p(x)$ by simply evaluating both sides of the identity. Hence, it is clear that Euler's polynomial is composite when evaluated at $40 + x^2$, for all nonnegative integers x. It might appear from the table that this is the only time Euler's polynomial can be composite; however, had we extended the table just slightly further, we would have seen that $p(81) = 6683 = 41 \cdot 163$. Notice that since $p(x-1) = x^2 - x + 41$, we find that $x^2 - x + 41$ assumes prime values for all values of $x = 1, 2, \ldots, 40$. Thus, putting $q(x) = p(39 - x) =$

$x^2 - 79x + 1601$, we have $q(x)$ a prime for $x = 0, 1, 2, \ldots, 39$ and for $x = 40, 41, \ldots, 79$. However this polynomial does not assume distinct prime values for $x = 0, 1, 2, \ldots, 79$ because

$$q(79 - x) = p(-40 + x) = q(x).$$

If we insist on having a quadratic polynomial that produces distinct prime values for all x such that $x = 0, 1, 2, \ldots, m$, the current record holders are

$$q_1(x) = 47x^2 - 2247x + 21647,$$
$$q_2(x) = 103x^2 - 3945x + 34381,$$
$$q_3(x) = 36x^2 - 810x + 2753.$$

The first two of these have $m = 43$ and were discovered by Gilbert Fung in 1988; the last one has $m = 45$ and was discovered by Russell Ruby in 1989. Up to the time of writing, no quadratic polynomial which will produce distinct prime values for $x = 0, 1, 2, \ldots, m > 45$, is known, although such polynomials almost certainly exist.

Now let $f_A(x)$ denote the quadratic polynomial $x^2 + x + A$, where A is an integer. Notice that $f_A(A-1) = A^2$ and $f_A(A) = A(A+2)$; thus, no polynomial of the form $f_A(x)$ can assume only prime values for all values of x. In fact, it is easy to see that no polynomial with integer coefficients can assume only prime values.

Suppose D is any non-square integer and let $\mathbb{Q}(\sqrt{D})$ denote the set of all numbers of the form $(a + b\sqrt{D})/c$, where a, b, c ($\neq 0$) are integers. To discuss Euler's polynomial any further we need to mention the class number of $\mathbb{Q}(\sqrt{D})$. As it would take us well beyond the ambit of this work to define this class number, we will only note here that it always exists and is a positive integer for any non-square D. In 1912 the Ukranian-born, American mathematical physicist Georg Rabinowitsch (1886–1968), who later in 1923 changed his name to George Yuri Rainich, showed that for a prime $A > 0$, $f_A(x)$ represents a prime for $x = 0, 1, 2, \ldots, A-2$ if and only if $\mathbb{Q}(\sqrt{1-4A})$ has class number one. Later, in 1966/7 the English mathematician Alan Baker (1939–2018) and the American mathematician Harold Stark independently published results that showed $\mathbb{Q}(\sqrt{-D})$ always has a class number greater than 1 if $-D > 163 = 4(41) - 1$. If we define $P_A(n)$ to be the number of prime values assumed by $f_A(x)$ for $x = 0, 1, 2, \ldots, n$, we have seen that $P_A(n)$ can be $A - 2$ only for certain $A \leq 41$. In fact, the only values for these A are: 2, 3, 5, 11, 17 and 41. In 1936 Lehmer posed the question as to whether there are values of A such that $P_A(n)$ simply exceeds 40.

Suppose we define $A(n)$ to be the least positive integral value of A such that $P_A(n) \geq n + 1$. Lehmer did not find a value for $A(40)$ and no one else has found one subsequently. Nevertheless, according to a widely believed conjecture, called the *prime k-tuples conjecture* (see Chap. 21), we would expect $A(40)$ to exist. We first observe that such a value of A must be odd and if we put $K = A(40)$, we find further that

$$\begin{aligned}K &\equiv 2 \pmod 3,\\ K &\equiv 1, 2 \pmod 5,\\ K &\equiv 3, 4, 6 \pmod 7,\\ K &\equiv 1, 4, 6, 7, 8 \pmod{11}.\end{aligned}$$

If we continue this process of elimination for all possible prime moduli supported by a sieve device, we can use this device to narrow down the search for $A(40)$. This was done several years ago (1995) and it was shown that

$$A(40) > 10^{18}.$$

This is still the best bound on $A(40)$ known today, although computer technology has advanced to the point where a better result could be found. However, numerical experiments support that the actual value of $A(40)$ is likely to be very large.

If, instead of attempting to extend the run of initial primes, we try to find values of A such that $f_A(x)$ has high prime density, then we can find polynomials that out perform Euler's polynomial. For instance, in 1938 the Dutch number theorist Nicolaas George Wijnand Henri Beeger (1884–1965) discovered the polynomial $f_{27941}(x) = x^2 + x + 27941$, that has $P_{27941}(1000000) = 286128$, while Euler's polynomial only has $P_{41}(1000000) = 261080$, even though Beeger's polynomial has a more modest start with $P_{27941}(39) = 30$.

In order to investigate this question, we first put $D = 1 - 4A$ and observe that

$$4f_A(n) = (2n + 1)^2 - D.$$

Let p be any small odd prime; there exists a positive value of n $(< p)$ such that $p \mid f_A(n)$ if and only if the Legendre symbol $(D|p) = 1$. Thus, we would expect that $f_A(x)$ would represent an abundance of primes whenever $(D|p) = -1$ for as many small primes p as possible. We can make this more precise by appealing to an old (1923) conjecture of the British mathematicians Godfrey Harold Hardy (1877–1947) and John Edensor Littlewood (1885–1972) which provides an extension of the PNT to quadratic polynomials. If we apply this conjecture to $f_A(x)$, it asserts that

$$P_A(x) \sim C(D)L_A(x),$$

where

$$L_A(x) = 2\int_0^n \frac{dx}{\log f_A(x)}$$

and

$$C(D) = \prod_{p \geq 3} \left(1 - \frac{(D|p)}{p-1}\right).$$

Here we use the symbol $\prod_{p\geq 3}$ to represent the product of what follows over all the odd primes.

Thus, $C(D)$ can be considered as a measure of the density of the prime values assumed by $f_A(x)$. We caution the reader that Hardy and Littlewood's conjecture is still very far from being proved. Indeed, it has never been shown that any particular quadratic polynomial in a single variable can be prime infinitely often. Nevertheless, the heuristics that led up to this conjecture are very convincing and empirical tests have always supported it; thus, it is widely accepted by analytic number theorists.

The difficulty with $C(D)$ for any particular D is that the product formula above is very slowly convergent and therefore difficult to evaluate. In 1975, Daniel Shanks was the first to evaluate $C(-163) = 3.3197732$ and $C(-111763) = 3.6319998$. It follows that, based on the Hardy-Littlewood conjecture, one would expect that for sufficiently large values of n, $P_{27941}(n)$ would exceed $P_{41}(n)$ and this is what has been observed. Indeed, to eight significant figures we get

$$\frac{P_{41}(10^6)}{L_{41}(10^6)} = 3.3203421,$$
$$\frac{P_{27941}(10^6)}{L_{27941}(10^6)} = 3.6396821,$$

both of which are quite close to their respective $C(D)$ values. The largest $C(D)$ computed by Shanks is $C(-991027) = 4.1237067$.

In work subsequent to 1975 larger values of $C(D)$ have been identified. The process of doing this requires the application of a sieve device to find values of D for which $(D|p) = -1$ for as many primes as the device will support. In 1990, the then largest value of $C(D)$ known was computed as

$$C(D) = 5.0870883$$

for

$$A = 132874279528931.$$

For this value of A

$$\frac{P_A(10^6)}{L_A(10^6)} = 5.098.$$

Later, in 1995, a value of

$$C(D) = 5.3670819$$

was computed for

$$A = 33997146285531 18047.$$

Here

$$\frac{P_A(10^6)}{L_A(10^6)} = 5.373.$$

It is believed that $C(D) = O(\log\log|D|)$, and this is borne out by the very slow growth of $C(D)$ that we have observed. In 2002 a larger value of $C(D)$ was found for a very large D, using an unpublished idea of Lehmer and some additional techniques. The largest value of $C(D)$ currently known is 5.41351382 (this is conditional on the truth of the ERH) for a 69 digit value of

$$A = 760145000704512946750245927317675854028370375047088998413588602105311.$$

We conclude this chapter by noting that other polynomials of degree greater than 2 have been found which have a large initial run of prime values. One of these is

$$-66x^3 + 3845x^2 - 60897x + 251831,$$

which is a prime for $x = 0, 1, 2, \ldots, 45$.

Problems

18.1 (Isenkrahe's formula) Let

$$F(n,k) = \frac{k!}{P(n,k)} + \frac{P(n,k)}{(k-1)!} - \left\lfloor \frac{(k-1)!}{P(n,k)} \right\rfloor$$

where

$$P(n,k) = 2^{e_1} \cdot 3^{e_2} \cdot 5^{e_3} \cdot \cdots \cdot p_n^{e_n} \quad \text{and} \quad e_i = \lfloor k/p_i \rfloor + \lfloor k/p_i^2 \rfloor + \lfloor k/p_i^3 \rfloor + \cdots .$$

Verify $F(n,k) = k$ when $k = p_{n+1}$, where p_{n+1} denotes the $(n+1)$th prime, for:
(a) **[BB]** $5 = p_3$ $(k = 5, n = 2)$
(b) $7 = p_4$ $(k = 7, n = 3)$
(c) $11 = p_5$ $(k = 11, n = 4)$

18.2 (Isenkrahe's formula) Let

$$F(n,k) = \frac{k!}{P(n,k)} + \frac{P(n,k)}{(k-1)!} - \left\lfloor \frac{(k-1)!}{P(n,k)} \right\rfloor$$

where

$$P(n,k) = 2^{e_1} \cdot 3^{e_2} \cdot 5^{e_3} \cdot \cdots \cdot p_n^{e_n} \quad \text{and} \quad e_i = \lfloor k/p_i \rfloor + \lfloor k/p_i^2 \rfloor + \lfloor k/p_i^3 \rfloor + \cdots .$$

Verify $F(n,k) = k+1$ when $k \neq p_{n+1}$, where p_{n+1} denotes the $(n+1)$th prime, for:

(a) **[BB]** $4 \neq p_3$ $(k = 4, n = 2)$
(b) $5 \neq p_4$ $(k = 5, n = 3)$
(c) $7 \neq p_5$ $(k = 7, n = 4)$

18.3 **(Willan's formula)** Let

$$p_n = 1 + \sum_{i=1}^{2^n} \left\lfloor \left(\frac{n}{\sum_{j=1}^{i} \left\lfloor \left(\cos \frac{(j-1)!+1}{j} \pi \right)^2 \right\rfloor} \right)^{1/n} \right\rfloor.$$

Verify the above formula produces:
(a) **[BB]** $p_1 = 2$
(b) $p_2 = 3$
(c) $p_3 = 5$

18.4 The formula

$$f(n) = \left\lfloor \frac{n! \bmod (n+1)}{n} \right\rfloor (n-1) + 2$$

will output 2 when $n + 1$ is not prime and output $n + 1$, when $n + 1$ is prime. Use $f(n)$ to test the primality of the following positive integers:
(a) **[BB]** $n + 1 = 3$
(b) $n + 1 = 4$
(c) $n - 1 = 5$
(d) $n + 1 = 6$
(e) $n - 1 = 7$

18.5 **(Mill's primes)** Show that

$$A = 1.306377883$$

has a sufficient number of digits to correctly compute the second Mill's prime (11 using $n = 2$), but not enough digits to compute the third Mill's prime. Recall Mill's primes are found by calculating $\lfloor A^{3^n} \rfloor$.

18.6 **(Euler's polynomial)** Euler's polynomial is defined by

$$p(x) = x^2 + x + 41.$$

Find the value of $p(x)$ for $80 \leq x \leq 84$ and state which of these are prime and which are composite.

Further Reading

Formulas for the *n*th prime

Underwood Dudley, Formulas for Primes, *Mathematics Magazine*, **56** (1983), 17–22.
Richard K. Guy, *Unsolved Problems in Number Theory*, 3rd Edition, Springer, NY, 2004, pp. 58–65.
Eric S. Rowland, A natural prime-generating recurrence, *Journal of Integer Sequences*, **11** (2008), Article 08.2.8.

Mills' formula

Chris K. Caldwell, Determining Mills' constant and a note on Honaker's problem, *Journal of Integer Sequences*, **8** (2005), Article 05.4.1.
Simon Plouffe, A Set of Formulas for Primes, 2022. Available at: https://arxiv.org/pdf/1901.01849.pdf.

Matijasevič's formulas

James P. Jones, Daihachiro Sato, Hideo Wada and Douglas Wiens, Diophantine representation of the set of prime numbers, *American Mathematical Monthly* **83** (1976), 449–464.

Euler's polynomial

Nigel Boston and Marshall L. Greenwood, Quadratics representing primes, *The American Mathematical Monthly*, **102** (1995), 595–599.
Andrew Granville and Richard A. Mollin, Rabinowitsch revisited, *Acta Arithmetica*, XCVI.2 (2000), 139–153.
R. F. Lukes, C. D. Patterson and H. C. Williams, Numerical sieving devices: their history and some applications, *Nieuw Archief voor Wiskunde*, (4) **13** (1995),113–139.
R. A. Mollin, Prime-producing quadratics, *The American Mathematical Monthly*, **104** (1997), 529–544.

Quadratic polynomials with a large prime density

G. W. Fung and H. C. Williams, Quadratic polynomials which have a high density of prime values, *Mathematics of Computation*, **55** (1990), 345–353.
Michael J. Jacobson, Jr., Hugh C. Williams, New quadratic polynomials with high densities of prime values, *Mathematics of Computation*, **72** (2002), 499–519.

Chapter 19
The Pell Equation

The amateur English mathematician Henry Ernest Dudeney (1857–1930) was one of the most prolific mathematical puzzle creators of the past two centuries. In 1914, he posed the following problem in a column entitled "Puzzles at a Village Inn" in *The Strand Magazine*:

> A gentleman says he knows Louvain in Belgium well and used to visit a friend there, before its destruction by the German army in August. His friend's house was in a long street, numbered on his side 1, 2, 3, and so on, and that all the numbers on one side of his house added up exactly the same as all the numbers on the other side of him. There are more than fifty houses on that side of the street, but not as many as five hundred. Can you work out the number of the house where he lived?

To solve this problem, now often referred to as the *Strand Puzzle*, we let n (> 50) denote the number of houses on the Louvain street and m be the number of the particular sought-after house. According to the statement of the problem, we must have

$$1 + 2 + 3 + \cdots + (m - 1) = (m + 1) + (m + 2) + \cdots + n.$$

Since $\sum_{i=1}^{m-1} i = \frac{m(m-1)}{2}$, we can write this as

$$m(m - 1)/2 = n(n + 1)/2 - m(m + 1)/2$$

or

$$m(m - 1) + m = n(n + 1)/2.$$

On simplifying, we find that

$$2m^2 = n^2 + n;$$

hence,

$$8m^2 + 1 = 4n^2 + 4n + 1 = (2n + 1)^2.$$

E. L. F. Roettger and H. C. Williams, *The Enchantment of Numbers*, CMS/CAIMS Books in Mathematics 16, https://doi.org/10.1007/978-3-031-76568-1_19

Putting, $x = 2n + 1$ and $y = 2m$, we get

$$x^2 - 2y^2 = 1.$$

This is a particular case of the so-called Pell equation, which is more generally written as

$$x^2 - Dy^2 = 1,$$

where D is any positive non-square integer. This Diophantine equation has been studied since antiquity, but was mistakenly attributed by Euler to the English mathematician Sir John Pell (1611–1685), who had nothing to do with it. Nevertheless, it is customary today to call it the Pell equation, even if this is a complete misnomer.

One of the earliest appearances of the Pell equation is in the Cattle Problem of Archimedes, which we will briefly discuss below; for more information concerning this, the reader is advised to consult the sources in the section on further reading at the end of this chapter. This problem, which is almost certainly due to Archimedes, consists of two parts; the first of these asks us to determine 8 unknowns (numbers of bulls and cows in each of four herds), given some relatively simple linear relationships among these numbers. The 8 solutions of this can be given as

$$(10366482, 7460514, 4149387, 7358060, 7206360, 4893246, 543913, 3515820)n,$$

where n can be any positive integer. The second, far more difficult part, turns out to constrain n to be $4456749y^2$, where y is a solution of the Pell equation above with $D = 410286423278424 = (9314)^2 4729494$. Archimedes could not possibly have solved this problem precisely, because the least possible value of y is a number of more than 100000 digits; what is intriguing is that he seems to have known that the solutions to the second part of the problem existed, but were very, very difficult to find. We really don't know how he knew this.

The problem was essentially solved by the German schoolmaster Carl Ernst August Amthor (1845–1916) in 1880, but he still could not exactly compute the numbers involved. In 1965, two computers, an IBM 7040 and an IBM 1620, were used to calculate the complete decimal representation of the total number of cattle T, a number consisting of 206545 digits. This was done by three Canadian third-year university students: Gus German (1942–2022), Hugh Williams and Bob Zarnke. Later, in 1981, Harry Nelson published the complete solution (all 8 of the unknowns) in 12 pages of very fine print in *The Journal of Recreational Mathematics*. In 1998, Ilan Vardi gave the following simple formula for T:

$$T = \lfloor (25194541/184119152)(109931986732829734979866232821433543901088049 + 50549485234315033074477819735540408986340\sqrt{4729494})^{4658} \rfloor.$$

Hendrik Lenstra, Jr. provided simple representations of all the solutions of the Cattle Problem. His representation for T can be written as

$$T = 25194541(w^{4658} - w^{-4658})/184119152,$$

where $w = 300426607914281713365\sqrt{609} + 84129507677858393258\sqrt{7766}$. Observe that $4729494 = 609 \cdot 7766$.

We now return to the general Pell equation. Besides the Strand Puzzle, it has found application in a number of other Diophantine problems and puzzles. In the last fifty years, much has been learned about this equation, and in what follows we will selectively describe some of these results. Let (x, y) be a pair of integers satisfying the equation. Since D is positive and not a perfect square, we may assume with no loss of generality that $x > 0$ and $y \geq 0$. If $y = 0$, we can only have the *trivial solution* $(1, 0)$, which is not very interesting; thus, we will only concern ourselves with the nontrivial solutions of the Pell equation. From the last equation in Chap. 14 and results in Chap. 17, we know that

$$A_{np-1}^2 - DB_{np-1}^2 = (-1)^{np},$$

where A_m/B_m denotes the mth convergent C_m in the continued fraction expansion of $\sqrt{D}$ and p denotes its period length. Since we can always select n such that np is even, we see that for such an n value, (A_{np-1}, B_{np-1}) is a solution of the Pell equation. It follows that the Pell equation always has nontrivial solutions.

Notice that if (x_1, y_1) and (x_2, y_2) are both distinct solutions of the Pell equation, then we can order them according to the values of $x_i + \sqrt{D}y_i$ $(i = 1, 2)$; for example, if $x_1 + \sqrt{D}y_1 > x_2 + \sqrt{D}y_2$, we say that (x_1, y_1) is greater than (x_2, y_2). (Since D is not a perfect square, we can have $x_1 + \sqrt{D}y_1 = x_2 + \sqrt{D}y_2$ only when $x_1 = x_2$ and $y_1 = y_2$.) If (x, y) is the least nontrivial solution, we say that it is the *fundamental solution* of $x^2 - Dy^2 = 1$. Now since $(x - \sqrt{D}y)(x + \sqrt{D}y) = 1 > 0$, we must have $x > \sqrt{D}y$ for any solution (x, y). We see then that $x + \sqrt{D}y > 2\sqrt{D}y$ and

$$|x - \sqrt{D}y| = 1/(x + \sqrt{D}y) < 1/(2\sqrt{D}y).$$

Dividing the above expression by y, we get

$$|x/y - \sqrt{D}| < 1/(2\sqrt{D}y^2) < 1/2y^2.$$

By Theorem 14.2, we now know that if (x, y) is any nontrivial solution of the Pell equation, then x/y must be a convergent in the simple continued fraction expansion of $\sqrt{D}$. Hence, if (x, y) is any nontrivial solution of the Pell equation, we must have $(x, y) = (A_{np-1}, B_{np-1})$ for some positive integer n. In Table 19.1 below we give the fundamental solution of the Pell equation for all non-square values of D such that $1 < D < 100$. Notice that there is considerable variation in the values in the pairs (x, y).

Table 19.1 Fundamental solutions of the Pell equation, $x^2 - Dy^2 = 1$

D	x	y	D	x	y	D	x	y
2	3	2	37	73	12	69	7775	936
3	2	1	38	37	6	70	251	30
5	9	4	39	25	4	71	3480	413
6	5	2	40	19	3	72	17	2
7	8	3	41	2049	320	73	2281249	267000
8	3	1	42	13	2	74	3699	430
10	19	6	43	3482	531	75	26	3
11	10	3	44	199	30	76	57799	6630
12	7	2	45	161	24	77	351	40
13	649	180	46	24335	3588	78	53	6
14	15	4	47	48	7	79	80	9
15	4	1	48	7	1	80	9	1
17	33	8	50	99	14	82	163	18
18	17	4	51	50	7	83	82	9
19	170	39	52	649	90	84	55	6
20	9	2	53	66249	9100	85	285769	30996
21	55	12	54	485	66	86	10405	1122
22	197	42	55	89	12	87	28	3
23	24	5	56	15	2	88	197	21
24	5	1	57	151	20	89	500001	53000
26	51	10	58	19603	2574	90	19	2
27	26	5	59	530	69	91	1574	165
28	127	24	60	31	4	92	1151	120
29	9801	1820	61	1766319049	226153980	93	12151	1260
30	11	2	62	63	8	94	2143295	221064
31	1520	273	63	8	1	95	39	4
32	17	3	65	129	16	96	49	5
33	23	4	66	65	8	97	62809633	6377352
34	35	6	67	48842	5967	98	99	10
35	6	1	68	33	4	99	10	1

Suppose that (t, u) is the fundamental solution of the Pell equation, and put $\alpha = t + u\sqrt{D}$ (> 1), $\beta = t - u\sqrt{D}$. We have $\alpha + \beta = 2t$, $\alpha - \beta = 2u\sqrt{D}$ and $\alpha\beta = 1$; thus, from our investigations of the Lucas sequences in Chap. 10, we find that if we put $T_n = V_n(2t, 1)/2$, $U_n = uU_n(2t, 1)$ $(n \geq 0)$, then

$$\alpha^n = T_n + \sqrt{D}U_n \quad \text{and} \quad \beta^n = T_n - \sqrt{D}U_n.$$

Since $(\alpha\beta)^n = 1$, it follows that (T_n, U_n) is a nontrivial solution of the Pell equation when $n > 0$. If (x, y) is any nontrivial solution, put $\gamma = x + \sqrt{D}y$ (> 1) and $\bar{\gamma} = x - \sqrt{D}y$; since (α^n) for $n \geq 0$ is a strictly increasing sequence, there must exist some positive integer m such that

$$\alpha^{m-1} < \gamma \leq \alpha^m \quad \text{or} \quad 1 < \gamma\beta^{m-1} \leq \alpha.$$

If $1 < \gamma\beta^{m-1} < \alpha$, we get a contradiction to the fundamentality of (t, u) because $\gamma\bar{\gamma}(\alpha\beta)^{m-1} = 1$ and $\gamma\beta^{m-1} = xT_m - DyU_m + \sqrt{D}(yT_m - xU_m)$. Thus, we must have $\gamma = \alpha^m$ for some positive integer m. We have shown, then, that all the nontrivial solutions of the Pell equation can be generated from the positive powers of $t + u\sqrt{D}$, where (t, u) is the fundamental solution.

We are now able to find the solution to the problem which opened this chapter. When $D = 2$, it is easy to verify that $(3, 2)$ is the fundamental solution; hence, all the solutions can be generated by $(3 + 2\sqrt{2})^k$ for positive values of k. Thus, we have the solutions $(3, 2)$, $(17, 12)$, $(99, 70)$, $(577, 408)$, $(3363, 2378)$, etc. Since x must be larger than $2 \cdot 50 + 1 = 101$ and less than $2 \cdot 500 + 1 = 1001$, we can only have $x = 577$ and $y = 408$. Thus, $m = y/2 = 204$ and $n = (x - 1)/2 = 288$.

We have seen that once we have the fundamental solution of a Pell equation, it is easy to characterize all of its nontrivial solutions; thus, we are left with the problem of how to determine the fundamental solution (t, u). We can easily do this by making use of the continued fraction expansion of $\sqrt{D}$ because we have seen that $(t, u) = (A_{p-1}, B_{p-1})$ when the period length p is even or $(t, u) = (A_{2p-1}, B_{2p-1})$ when p is odd. In either case we see by the results in Chap. 17 that

$$p(D) = O(\log(t + u\sqrt{D})).$$

For example, consider the case of $D = 109$. We can use Tenner's algorithm to compute the period of the continued fraction for $\sqrt{109}$ and find that $Q_7 = Q_8 = 3$, $p = 15$ and

$$\sqrt{109} = [10; \overline{2, 3, 1, 2, 4, 1, 6, 6, 1, 4, 2, 1, 3, 2, 20}].$$

We compute $A_6 = 1399$, $A_7 = 9532$, $B_6 = 134$, $B_7 = 913$; hence,

$$A_{14} = A_7B_7 + A_6B_6 = 8890182, \quad B_{14} = B_7^2 + B_6^2 = 851525$$

by the formulas in Chap. 17. Since p here is odd, we have

$$A_{14}^2 - 109B_{14}^2 = -1;$$

thus, it is easy to see that $A_{29} + \sqrt{109}B_{29}$ can only be $(A_{14} + \sqrt{109}B_{14})^2$. It follows that

$$t = A_{29} = A_{14}^2 + 109B_{14}^2 = 158070671986249 \quad \text{and}$$
$$u = B_{29} = 2A_{14}B_{14} = 15140424455100.$$

The values for t and u above are much larger than what we might expect for such a small value of D, like 109; however, we observed in Chap. 17 that p can often be quite large, perhaps of the order of $\sqrt{D}$; for example, if $D = 95419$, we find that $p(D) = 750$ and t, u are integers of 376 and 374 digits, respectively. However, if $D = 49767099416053091$, then $p(D) = 278655786$ and t, u are both integers consisting of more than 143569000 decimal digits!

On the other hand, we have seen in Chap. 17 that there are special forms of D, like the Richaud-Degert or Schinzel types, for which $p(D)$ is relatively small, but most values of D seem to yield large values of $p(D)$.

Indeed, when $D = 990676090995853870156271607886$, a 30 digit number, we are unable to compute the value of $p(D)$ (it's too big to calculate exactly), but we do know that both t and u have more than 2×10^{15} decimal digits each. This means that it would require over 6,000,000 books, each of 1,000 pages, of the same small font used by Nelson to record the solution of the Cattle Problem, to write out t and u. We have gone through this calculation not only to illustrate how large the solutions of the Pell equation can get, but also to show that in some cases, even if we could in theory calculate t or u, we could not record it, at least not in decimal notation. Since the integers needed to represent P_n and Q_n are of order $\sqrt{D}$, we could use the multiplicative formulas in Chap. 17 to represent the fundamental solution; this means we would require about $p(D)\log_{10} D$ decimal digits, which is smaller than $2\log_{10} t$, but we would have to know $p(D)$, which is very hard to compute when D gets large. Thus, we are now left with two questions: is there a more efficient way than using continued fractions to compute the fundamental solution of the Pell equation and is there a better way of recording this solution other than using decimal notation? The answer to both of these questions is yes, but before we try to explain this, we need to introduce some new terms.

Suppose $D = f^2 D_0$, where D_0 has no square divisors. Put $r = 2$ when $D_0 \equiv 1 \pmod 4$ and $r = 1$, otherwise; set $\omega_0 = (r - 1 + \sqrt{D_0})/r$. Now consider the set $\mathcal{O} = i + jf\omega_0$, where i and j range over all the integers; this set is called the *maximal order* of $\mathbb{Q}(\sqrt{D})$, the set of algebraic integers in $\mathbb{Q}(\sqrt{D})$. The order $\mathcal{O}$ is made up of the elements that act like the integers of $\mathbb{Q}(\sqrt{D})$. That is, $\mathcal{O}$ is to $\mathbb{Q}(\sqrt{D})$ as $\mathbb{Z}$ is to $\mathbb{Q}$. The *discriminant* Δ $(= \Delta(\mathcal{O}))$ of this order is $(2/r)^2 D$. Notice that we can now write $\mathcal{O} = \{i + j(\Delta + \sqrt{\Delta})/2)\}$, where i and j range over all the integers. We say that an element ϵ of $\mathcal{O}$ is a *unit* of $\mathcal{O}$ if the integer $N(\epsilon) = \epsilon\bar{\epsilon}$ is a divisor of 1, i.e. $N(\epsilon) = \pm 1$. We define the *fundamental unit* ϵ_Δ of $\mathcal{O}$ to be that unit $(x + y\sqrt{\Delta})/2$ of $\mathcal{O}$ such that x and y are both positive and minimal. It can be shown that if ϵ is any unit of $\mathcal{O}$, then $\epsilon = \epsilon_\Delta^n$ for some integer n. Notice that this is similar to the case of the fundamental solution (t, u) of the Pell equation $x^2 - Dy^2 = 1$. Indeed, we can relate the two by

$$t + u\sqrt{D} = \epsilon_\Delta^\nu,$$

where ν is either 1, 2, 3 or 6 according to the following table:

r	$\Delta \pmod 2$	$y \pmod 2$	$N(\epsilon_\Delta)$	ν
1	–	–	1	1
1	–	–	-1	2
2	–	0	1	1
2	–	0	-1	2
2	1	1	1	3
2	1	1	-1	6
2	0	1	–	2

Because the value of $t + u\sqrt{D}$ can be very large, we instead deal with its logarithm, which is much smaller. Since $\log(t + u\sqrt{D}) = \nu \log \epsilon_\Delta$, we can easily relate $\log(t + u\sqrt{D})$ to R, where $R = \log \epsilon_\Delta$ is called the *regulator* of the order $\mathcal{O}$. Notice that we have $p(D) = O(R)$. The size of R as a function of Δ varies enormously, but there are values of Δ for which R can be pretty close to $\sqrt{\Delta}$. What we can say is that

$$R < \frac{\sqrt{\Delta}}{4} \log \Delta,$$

but this upper bound seems to be too large. If the ERH holds, we can prove that

$$R = O(\sqrt{\Delta} \log \log \Delta)$$

and there is good reason to believe that for any positive $\varepsilon < 1/2$,

$$R = \Omega(\Delta^{1/2-\varepsilon})$$

holds for a significant proportion of all the corresponding orders $\mathcal{O}$.

Of course, the regulator is not an integer, but algorithms exist which allow us to compute a good approximation to R, like the integer $\lfloor R \rfloor$. Unfortunately, these techniques are very technical and well beyond the scope of this book. We should point out here that Sean Hallgren has discovered another algorithm for solving Pell's equation; this procedure runs in polynomial time on a quantum computer capable of implementing Shor's algorithm. The best we can say here is that there does exist a subexponential algorithm, due to the German computer scientist Johannes Buchmann, for computing the regulator, but the value for R that it produces, while almost certainly correct, is still contingent on the truth of the generalized Riemann hypothesis (GRH), which, while accepted by many specialists, is a long way from being proved. In fact, if the GRH and ERH hold, then the expected time needed to compute R by this technique is $L_\Delta[1/2, \sqrt{2}]$. Later, Mike Jacobson, a student of Buchmann, discovered a way to verify the regulator found by this technique, using an algorithm of complexity $O(\Delta^{1/6+o(1)})$; while this method has been improved more recently, this unconditional complexity measure is still the best currently known for calculating R.

As an example of this technique, we mention that for the 65 digit prime discriminant

$$\Delta = 39286375734542594749758050835151655092118848530833398743561568481,$$

we can unconditionally compute

$$\begin{aligned} R &= \log((x + y\sqrt{\Delta})/2) \\ &= 195696290466865524253842387693615.276\ldots > 1.9 \times 10^{32}. \end{aligned}$$

Since $(x + y\sqrt{\Delta})/2 = e^R$ and $(x - y\sqrt{\Delta})/2 = \pm e^{-R}$, we easily see that both x and $\sqrt{\Delta}y$ are very close in value to e^R. Thus, the values of x and y above must both have more than 8×10^{31} decimal digits, and are therefore impossible to write down in decimal notation. Nevertheless, it is still possible to record them by making use of a *compact representation* of ϵ_Δ . This is a multiplicative representation (it is not unique) of the fundamental unit of the form

$$\epsilon_\Delta = \prod_{j=1}^{k} (\lambda_i/d_i)^{e(k,j)},$$

where $e(k, j) = 2^{k-j}$ and

(1) $k < \log_2 R + 3$,
(2) a_j, b_j, d_j are all integers and $\lambda_j = (a_j + b_j\sqrt{\Delta})/2$ is an element of $\mathcal{O}$ for all $j = 1, 2, 3, \ldots, k$,
(3) $0 < d_j < \sqrt{\Delta}, \quad |a_j| < 2\Delta^{5/2}, \quad |b_j| < 2\Delta^2 \ (1 \le j \le k)$.

Thus, the number of digits needed for such a representation is less than $8k\log_{10}\Delta$, which is roughly of the order $\log R \log \Delta$, instead of R. For example, when $D = 990676090995853870156271607886$ ($\equiv 6 \pmod 8$), the double of a prime, we see that $r = 1$ and $\Delta = 4D$; thus since $\frac{x^2-Dy^2}{4} \neq -1$, we must have $\nu = 1$ and $t + u\sqrt{D} = \epsilon_\Delta$. The value of the regulator R here is $47703729558513 43.43\ldots$. A compact representation of ϵ_Δ has $k = 46$ and can be recorded on a single page containing about 2760 digital characters, whereas the decimal representation of either x or y would require more than 2×10^{16} digits. That's a lot of digits.

There is a fast algorithm which, given Δ and $\lfloor R \rfloor$, can compute a compact representation of ϵ_Δ in $O(\log\lfloor R \rfloor \log \Delta)$ elementary operations. Also, if we are given such a compact representation, we can compute the sign of $N(\epsilon_\Delta)$ very expeditiously; furthermore, for a given m, we can compute the values of x and y (mod m) in $O(k \log m)$ arithmetic operations. Since this can be done for any m, we see that a compact representation of ϵ_Δ, in spite of being much less voluminous than the decimal representation, still contains all of the information about the coefficients x, y of ϵ_Δ.

It would seem that we could produce a compact representation requiring fewer digits if we change the $e(k, j)$ above to h^{k-j}, where $h > 2$, but while this would reduce the value of k, it also has the effect of increasing the number of digits of the coefficients of the corresponding λ_j. In fact, some analysis and extensive computations have suggested that when $\Delta > 10^{17}$, the size of the compact representation is optimally small when $h = 4$.

We next turn our attention to a problem mentioned by Joe Roberts in his charming book, *Lure of the Integers*. If we have $D = 1620$, we find that $(t, u) = (161, 4)$, but the situation changes dramatically when we look at $D = 1621 \equiv 1 \pmod 4$; in this case we have $r = 2$, $\Delta = D$ and ϵ_Δ $(= (x + y\sqrt{\Delta})/2)$ with

$$x = 4823022127875, \quad y = 119806883557, \quad N(\epsilon_\Delta) = -1;$$

thus, by the table above, we have $\nu = 6$ and $t + u\sqrt{D} = \epsilon_\Delta^6$. This means that the value of t in this case must be

$$\begin{aligned}&629810181249373234303497450009145781552994230866705141285573523\\&\qquad\qquad\qquad\qquad\qquad\qquad\qquad\qquad 10169665125001,\end{aligned}$$

a number of 76 digits! (This was also pointed out in the footnote of Carmichael discussed in Chap. 1.) Roberts asked, "is it mysterious that neighbouring integers can act so very differently?"

To answer this question, we let D be a positive integer such that neither D nor $D - 1$ is a perfect square. Let (t_0, u_0), (t_1, u_1) be the fundamental solutions of the Pell equations $x^2 - (D - 1)y^2 = 1$ and $x^2 - Dy^2 = 1$, respectively. In view of the above observation, it seems appropriate to define $\rho(D) = \log t_1 / \log t_0$. When $D = 1621$, we get $\rho(D) = 34.35$, but when $D = 118681$, we have $\rho(D) = 633.84$. Indeed, it can be shown that there are infinite families of values of D for which $\rho(D) = \Omega(D^{1/6}/\log D)$. Also, under certain very reasonable hypotheses, it is possible to show that $\rho(D)$ must exceed some infinite set of values of D for which $\rho(D) = \Omega(\sqrt{D}\log\log D/\log D)$. Furthermore, under the ERH it can be shown that $\rho(D) = O(\sqrt{D}\log\log/D\log D)$; thus, $\rho(D) = \Omega(\sqrt{D}\log\log D/\log D)$ seems to be best possible.

In an attempt to find values of D for which this latter estimate might be the case, values of D of the form $M^2 + M + 1$ were selected as it is easy to see that $D - 1 = M^2 + M$ is of Richaud-Degert type, and $t_0 = 2M + 1$ $(u_0 = 2)$, making $\log t_0$ relatively small. A sieving device was then used in much the same way as mentioned in Chap. 17 to find prime values of D for which the fundamental solution is likely to be large. A computer search produced the 80 digit

$$\begin{aligned}D = {}&12779403100260586715025492824657916044067403863724697039719\\&\qquad\qquad\qquad\qquad\qquad\qquad 777303886059655053681,\end{aligned}$$

where by Buchmann's subexponential method the value of

$$R = 182871089219957536671992302657711426769 45.486\ldots$$

was computed, and this was used to evaluate

$$\rho(D) = 3.98 \times 10^{38} > \sqrt{D} \log \log D / \log D.$$

It seems, then, that there is really nothing mysterious about the existence of values of D for which $\rho(D)$ is large. The mystery resides in whether it can be proved that there exist infinite families of D for which $\rho(D) = \Omega(\sqrt{D} \log \log D / \log D)$.

There exist higher degree analogues of the Pell equation; one of these is the cubic analogue

$$x^3 + Dy^2 + D^2 Z^3 - 3Dxyz = 1,$$

where D is any integer which is not a perfect cube. The behaviour of the solutions of this Diophantine equation is very similar to that of the solutions of the Pell equation; in that there always exists a fundamental solution (t, u, v) from which all other solutions can be derived by computing $(t + u\sqrt[3]{D} + v\sqrt[3]{D})^n$ $(n \geq 1)$. The determination of the fundamental solution is a rather intricate process for most values of D, but it can be solved by making use of Voronoi's continued fraction, mentioned in Chap. 14. The application of this technique to $\delta = \sqrt[3]{D}$ will result in a periodic algorithm of period length $r(D)$ which can be used to compute (t, u, v) in a manner analogous to the use of the simple continued fraction of $\sqrt{D}$ to calculate the fundamental solution (t, u) of the Pell equation.

However, this algorithm, as originally described, requires that we essentially be able to determine when certain algebraic irrationals exceed 0. As a computer can only store an approximate (rational) value A of an algebraic irrational α, we need to know when $A > 0$ implies that $\alpha > 0$. Fortunately, this problem can be efficiently solved in the case of Voronoi's algorithm. This necessitates that a number of technical modifications be made in order for it, like the simple continued fraction expansion of $\sqrt{D}$, to execute by using integer arithmetic only. In the case of the simple continued fraction expansion of $\sqrt{D}$, we first need to compute the floor of $\sqrt{D}$. In the case of Voronoi's algorithm, we need to precompute the floors of a few fixed integral multiples of δ, δ^2, $D^2\delta$ and $D^2\delta^2$ in order to begin using only integer arithmetic. Once this has been done the algorithm works very effectively with minimal precision. As examples, we mention $D = 2$, where $r(D) = 1$; $D = 23$, $r(D) = 21$; $D = 239$, $r(D) = 390$; $D = 8429$, $r(D) = 15481$; $D = 96797$, $r(D) = 197114$.

Just as in the case of the Pell equation, it is also possible to have very small values of $r(D)$ for certain forms of D. For example, if $D = M^3 + 1$, then $r(D) = 1$, and if $D = M^3 - M$, then $r(D) = 5$. Ron Rudman showed that whenever D has the parametric form $D = M^3 + k$, where $k \mid 3M^2$, then it is a simple matter to find the fundamental solution of the cubic Pell equation. For example, if we apply his results in the particular case of $k \mid 3M$, where $|k|$ is not 1, we get $(t, u, v) =$

$(9M^6 + 9M^3k + k^2, 9M^5 + 6M^2k, 9M^4 + 3Mk)k^{-2}$. Also, if $D = M^3 + \epsilon$, where $\epsilon = \pm 1$, then, the Norwegian mathematician Trygve Nagell (1895–1988) showed in 1923 that $t = M^2$, $u = M$, $v = 1$, but if $D = M^3 + 3\epsilon$, then $t = M^6 + 3\epsilon M^3 + 1$, $u = M^5 + 2\epsilon M^2$, $v = M^4 + \epsilon M$. Thus, it appears that the form $D = M^3 + k$, where $k \mid 3M^2$, is analogous to the Richaud-Degert form for the Pell equation; however, apart from the fact that it must be small, very little is known about the exact value of $r(D)$ when D has such a form.

Problems

19.1 (Fundamental solution) Find the fundamental solution of $x^2 - Dy^2 = 1$ using the given continued fraction expansion of $\sqrt{D}$ (found using Tenner's algorithm) and recalling that $(t, u) = (A_{p-1}, B_{p-1})$ when the period length p is even or $(t, u) = (A_{2p-1}, B_{2p-1})$ when p is odd.
(a) **[BB]** $D = 111$, $\sqrt{111} = [10; \overline{1, 1, 6, 1, 1, 20}]$
(b) $D = 113$, $\sqrt{113} = [10; \overline{1, 1, 1, 2, 2, 1, 1, 1, 20}]$
(c) $D = 202$, $\sqrt{202} = [14; \overline{4, 1, 2, 2, 1, 4, 28}]$
(d) $D = 216$, $\sqrt{216} = [14; \overline{1, 2, 3, 2, 1, 28}]$

Further Reading

Pell equation puzzles and problems

Edward J. Barbeau, *Pell's Equation*, Springer, NY, 2003, pp 16–31.

The Cattle Problem of Archimedes

Albert H. Beiler, *Recreations in the Theory of Numbers—the Queen of Mathematics Entertains*, Dover Publications, NY, 1964, pp. 248–268.
Sir Thomas Heath, Ed., *The Works of Archimedes with The Method of Archimedes*, Dover Publications, NY, 2002, pp. 319–326.
André Weil, *Number theory: An Approach through History from Hammurapi to Legendre*, Birkhäuser, Boston, 1984, pp. 17–19.
H. L. Nelson, A solution to Archimedes' Cattle Problem, *Journal of Recreational Mathematics*, **13** (1981), 162–176.
Ilan Vardi, Archimedes' Cattle Problem, *American Mathematical Monthly*, **105** (4) (1998), 305–319.
H. W. Lenstra, Jr., Solving the Pell equation, *Notices of the American Mathematical Society*, **49** (2002), 182–192.
H. C. Williams, Solving the Pell Equation, *Surveys in Number Theory: Papers from the Millennial Conference on Number Theory*, A. K. Peters, Natick, MA, 2003, pp. 328–330.

Solution of the Pell equation

H. W. Lenstra, Jr., Solving the Pell equation, *Notices of the American Mathematical Society*, **49** (2002), 182–192.

M. J. Jacobson, Jr. and H. C. Williams, *Solving the Pell Equation*, CMS Books in Mathematics, Springer Science + Business Media, New York, 2009.

Quantum methods

Sean Hallgren, Polynomial-time quantum algorithms for Pell's equation and the principal ideal problem, *STOC '02: Proceedings of the Thirty-Fourth Annual ACM Symposium on Theory of Computing*, 2002, pp. 653–658.

Size of the regulator

M. J. Jacobson, Jr. and H. C. Williams, *Solving the Pell Equation*, CMS Books in Mathematics, Springer Science + Business Media, New York, 2009, pp. 226-231.

Compact Representations

A. K. Sylvester, M. J. Jacobson, Jr., H. C. Williams, Shorter compact representations in real quadratic fields, *Number Theory and Cryptography*, Papers in honor of Johannes Buchmann on the occasion of his 60th birthday, Lecture Notes in Computer Science, LCNS 8260, Springer-Verlag, Berlin, 2013, pp. 50–72.

Roberts' question

Joe Roberts, *Lure of the Integers*, The Mathematical Association of America, Providence, RI, 1992, pp. 260–262.

M. J. Jacobson, Jr. and H. C. Williams, The size of fundamental solutions of consecutive Pell equations, *Experimental Mathematics*, **9** (2000), 631–640.

The cubic Pell equation

Edward J. Barbeau, *Pell's Equation*, Springer, NY, 2003, pp. 92–112.

Voronoi's algorithm

H. C. Williams, G. Cormack and E. Seah, Calculation of the regulator of a pure cubic field, *Mathematics of Computation*, **34** (1980), 567–611.

Samuel A. Hambleton and Hugh C. Williams, *Cubic Fields with Geometry*, CMS Books in Mathematics, Springer Nature, Cham, Switzerland, 2018, pp. 277–425.

H. C. Williams, The period length of Voronoi's algorithm for certain cubic orders, *Publicationes Mathematicae Debrecen*, **37** (1990), 245–265.

Ronald J. Rudman, On the fundamental unit of a purely cubic field, *Pacific Journal of Mathematics*, **46** (1973), 253–256.

Chapter 20
Some Diophantine Equations

Recall that in Chap. 14 we defined a Diophantine equation to be an indeterminate equation whose solutions are constrained to be either integers or rational numbers. We also provided there an algorithm for solving the general linear Diophantine equation in 2 variables

$$ax + by = c,$$

where a, b, c are given integers and $ab \neq 0$. Later, in Chap. 19, we discussed another Diophantine equation, the Pell equation and its cubic analogue. In this chapter, we intend to initiate a broader investigation of Diophantine equations by examining a few more examples.

We emphasize here that the topic of Diophantine equations is vast. Some idea of how extensive the subject is can be gleaned by examining the second volume of Dickson's *History of the Theory of Numbers*. This is an 800 page book devoted to such equations. This book was written over 100 years ago, and much has been learned since. More importantly, very powerful and sophisticated methods have been developed which allow for a much more general approach to these objects. As any one of these ideas would require a great deal of discussion, we have elected here to present instead a rather superficial examination of certain Diophantine equations on which there has been recent progress. We have selected these with an eye to their importance and/or history. We warn the reader that this subject is highly addictive and has been known to seduce its practitioners, both professional and novice, into devoting much of their time to the pleasure of its study. To quote Louis Mordell

> It is well known that for many centuries, no other topic has engaged the attention of so many mathematicians, both professional and amateur, or has resulted in so many published papers.

We begin by defining the *general quadratic Diophantine equation in 2 variables* or *general binary quadratic Diophantine equation* as an equation of the form

$$ax^2 + bxy + cy^2 + dx + ey + f = 0,$$

E. L. F. Roettger and H. C. Williams, *The Enchantment of Numbers*, CMS/CAIMS Books in Mathematics 16, https://doi.org/10.1007/978-3-031-76568-1_20

where it is required to find integer solutions for x and y for given integers a ($\neq 0$), b, c ($\neq 0$), d, e and f. The first comprehensive method for solving such equations was developed by the Italian-born, French mathematician Joseph-Louis Lagrange (1736–1813). In the 200 years that followed Lagrange's investigation, there have not been many significant improvements to his method; this is unfortunate, as this technique becomes less efficient once the coefficients become large. Moreover, the solutions to a general binary quadratic Diophantine equation can be rather large despite the coefficients being relatively small. Indeed, if we put $H = \max\{|a|, |b|, |c|, |d|, |e|, |f|\}$, it has been shown that there is an infinite collection of such equations with integer solutions where none of the solutions (x, y) satisfies $\max\{|x|, |y|\} \leq 2^{H/5}$.

In order to solve

$$ax^2 + bxy + cy^2 + dx + ey + f = 0,$$

for integers x and y using Lagrange's method we begin by putting $D = b^2 - 4ac$, $E = bd - 2ae$, $F = d^2 - 4af$; hence, the equation above is equivalent to writing

$$DY^2 = (Dy + E)^2 + DF - E^2,$$

where $Y = 2ax + by + d$. Further, if we put $N = E^2 - DF$, we can then write this as

$$X^2 - DY^2 = N,$$

where $X = Dy + E$. Thus, if we have any solution X and Y in the above such that there exist integers x, y for which

$$X = Dy + E \quad \text{and} \quad Y = 2ax + by + d,$$

then we have a solution to

$$ax^2 + bxy + cy^2 + dx + ey + f = 0.$$

For example, consider the quadratic Diophantine equation

$$x^2 - 9y^2 + 3x + 9 = 0.$$

Clearly, here $a = 1, b = 0, c = -9, d = 3, e = 0$ and $f = 9$. Thus, using Lagrange's method we have

$$\begin{aligned}
D &= b^2 - 4ac = 0^2 - 4 \cdot 1 \cdot (-9) = 36,\\
E &= bd - 2ae = 0 \cdot 3 - 2 \cdot 1 \cdot 0 = 0,\\
F &= d^2 - 4af = 3^2 - 4 \cdot 1 \cdot 9 = -27,\\
N &= E^2 - DF = 0^2 - 36 \cdot (-27) = 972.
\end{aligned}$$

Hence, $X^2 - DY^2 = N$, in this case $X^2 - 36Y^2 = 972$. Note that since D is a perfect square, we can write this as $X^2 - (6Y)^2 = 972 = 2^2 \cdot 3^5$ and we know we can have only a finite number of solutions. The problem of solving this equation now reduces to factoring N. Fortunately, in this case N is easily factored due to it small size. Since we have a difference of squares, we may write this as

$$(X + 6Y)(X - 6Y) = 972 = 2^2 \cdot 3^5;$$

we observe that both $X + 6Y \mid 972$ and $X - 6Y \mid 972$. Thus, $X + 6Y$ must belong to

$$K = \{1, 2, 3, 4, 6, 9, 12, 18, 27, 36, 54, 81, 108, 162, 243, 324, 486, 972\}$$

and have integer solutions for $X + 6Y = k$ and $X - 6Y = 972/k$ where $k \in K$. By inspection of the possible cases, we get

$$\begin{aligned} X &= \pm 84 \quad \text{and} \quad Y = \pm 13, \\ X &= \pm 36 \quad \text{and} \quad Y = \pm 3. \end{aligned}$$

To illustrate this, imagine we check to see if $X + 6Y = 4$ has the necessary integer solutions. If $X + 6Y = 4$, then $X - 6Y = 243$, but in this case $X + 6Y$ and $X - 6Y$ must have different parity, which is impossible. However, if $X + 6Y = 18$, then $X - 6Y = 54$ and we can see

$$\begin{array}{rrl} & \mathrm{X} + 6\mathrm{Y} & = 18 \\ + & \mathrm{X} - 6\mathrm{Y} & = 54 \\ \hline & 2\mathrm{X} & = 72 \end{array}$$

and hence $X = 36$. Now since $X^2 - (6Y)^2 = 972$ we have $36^2 - (6Y)^2 = 972 \rightarrow (6Y)^2 = 324 \rightarrow Y^2 = 9 \rightarrow Y = \pm 3$. Upon checking all cases, the only solutions to $X^2 - (6Y)^2 = 972$ are, in fact,

$$\begin{aligned} X &= \pm 84 \quad \text{and} \quad Y = \pm 13, \\ X &= \pm 36 \quad \text{and} \quad Y = \pm 3. \end{aligned}$$

To find all solutions x, y, we now merely need to check the solutions X, Y, to see if there exist integers that satisfy

$$X = Dy + E = 36y + 0 \quad \text{and} \quad Y = 2ax + by + d = 2x + 0y + 3.$$

Notice that if $X = \pm 84$, then $X = 36y$ does not have integer solutions. However, $X = \pm 36$ has solutions $y = \pm 1$, respectively, for $X = 36y$. Similarly, using $Y = 2x + 3$, if $Y = 3$, then it must be that $x = 0$ and if $Y = -3$, then $x = -3$. We can now see all of the integer solutions to

$$x^2 - 9y^2 + 3x + 9 = 0$$

are (0, 1), (−3, 1), (0, −1) and (−3, −1).

We next provide a second example in which $D < 0$; consider the quadratic Diophantine equation

$$x^2 - xy + y^2 - x - y = 0.$$

In this example $a = 1$, $b = -1$, $c = 1$, $d = -1$, $e = -1$ and $f = 0$. Again, using Lagrange's method we get

$$\begin{aligned}
D &= b^2 - 4ac = (-1)^2 - 4 \cdot 1 \cdot 1 = -3, \\
E &= bd - 2ae = (-1) \cdot (-1) - 2 \cdot 1 \cdot (-1) = 3, \\
F &= d^2 - 4af = (-1)^2 - 4 \cdot 1 \cdot 0 = 1, \\
N &= E^2 - DF = 3^2 - (-3) \cdot 1 = 12.
\end{aligned}$$

Hence, we are now interested in finding integer solutions to $X^2 + 3Y^2 = 12$. Note that since $D < 0$ there are certainly a finite number of solutions, and we can see that we are limited by size $X^2 \in \{0, 1, 4, 9\}$ and $Y^2 \in \{0, 1, 4\}$. Thus, we can easily see by inspection that our integer solutions to $X^2 + 3Y^2 = 12$ are

$$\begin{aligned}
X &= \pm 3 \quad \text{and} \quad Y = \pm 1, \\
X &= 0 \quad \text{and} \quad Y = \pm 2.
\end{aligned}$$

Using the same method as above, we see that each pair of X, Y yields integer solutions x, y to our original Diophantine equation; these are provided in the table below

X	Y	x	y	X	Y	x	y	X	Y	x	y
−3	1	2	2	3	1	1	0	0	2	2	1
−3	−1	1	2	3	−1	0	0	0	−2	0	1

We can classify the equation

$$X^2 - DY^2 = N$$

into two main cases. We will assume in what follows that we have the complete factorization of N and a quadratic nonresidue for each prime divisor of N.

(1) If $D < 0$, there can be only a finite number of solutions (X, Y). For large values of N, these can be computed by using an old (1908) algorithm of the otherwise little-known Italian mathematician Giuseppe Cornacchia . This technique, a generalization of Hermite's procedure in Chap. 14, makes use of the Euclidean algorithm for N and solutions W $(0 < W \leq N/2)$ of

$$W^2 \equiv D \pmod{N}.$$

(In most cases such a congruence must be solvable if the equation is to hold.) Under our initial assumption, this process will find all the solutions in time complexity $O((\log N)^3)$.

(2) If $D > 0$ and D is not a perfect square, there can either be no solutions (X, Y) or an infinitude of them. The process of characterizing these for large D and/or N has recently been improved over that of Lagrange, and makes use of continued fractions and new techniques for solving the Pell equation mentioned in Chap. 19.

Another class of interesting Diophantine problems arises from the work of Diophantus himself. In problem 20 of Book IV of his *Arithmetica*, he raises the question of whether there can exist four rational numbers such that the product of any two increased by 1 can be a perfect square. He answers this positively by mentioning that

$$\left(\frac{1}{16}, \frac{33}{16}, \frac{68}{16}, \frac{105}{16}\right)$$

is such a quadruple. The first quadruple of integers with this property was found much later by Fermat:

$$(1, 3, 8, 120).$$

We now define a *Diophantine m-tuple* as a set of m positive integers with the property that the product of any two distinct elements from this set increased by 1 is a perfect square; that is, if $(a_1, a_2, \ldots, a_m)$ is a Diophantine m-tuple, then $a_i a_j + 1$ is a perfect square for all i, j such that $1 \le i < j \le m$. In 1969, Alan Baker and the English mathematician Harold Davenport (1907–1969) showed that the 3-tuple $\{1, 3, 8\}$ was only capable of being extended to a 4-tuple in one way; consequently, Fermat's set cannot be extended to a 5-tuple or quintuple. Indeed, it was an open problem until relatively recently as to whether quintuples even existed. Significant progress towards solving this open problem was first made by the Croatian mathematician Andrej Dujella, when in 2004 he was able to prove that there are at most finitely many Diophantine quintuples. The result that there are, in fact, no quintuples was proved by three authors in a paper which appeared in 2019. These three investigators are Bo He (PRC), Alain Togbé (Benin-born, USA) and Volker Ziegler (Austria).

It is well known that there exist infinitely many Diophantine quadruples. This is easily seen by considering the family of sets of the form

$$(k-1, k+1, 4k, 16k^3 - 4k) \quad \text{for } k \ge 2$$

and noting

$$(k-1)(k+1) + 1 = k^2,$$
$$(k-1)4k + 1 = 4k^2 - 4k + 1 = (2k-1)^2,$$
$$(k-1)(16k^3 - 4k) + 1 = 16k^4 - 16k^3 - 4k^2 + 4k + 1 = (4k^2 - 2k - 1)^2.$$

$$(k+1)4k+1 = 4k^2+4k+1 = (2k+1)^2,$$
$$(k+1)(16k^3-4k)+1 = 16k^4+16k^3-4k^2-4k+1 = (4k^2+2k-1)^2, \quad \text{and}$$
$$4k(16k^3-4k)+1 = 64k^4-16k^2+1 = (8k^2-1)^2.$$

Indeed, setting $k = 2$ in this family of sets returns Fermat's quadruple.

Just as the identity $(k-1)(k+1)+1 = k^2$ was used as a starting off point to construct the previous family of quadruples, we can also employ Cassini's identity

$$F_{n-1}^2 - F_n F_{n-2} = (-1)^n$$

to find an infinitude of Diophantine quadruples involving Fibonacci numbers. On setting $n = m+2$ in Cassini's identity and rearranging, we have

$$F_m F_{m+2} + (-1)^m = F_{m+1}^2.$$

We will also use a generalization of Cassini's identity called Vajda's identity, that being

$$F_{n+i} F_{n+j} - F_n F_{n+i+j} = (-1)^n F_i F_j.$$

With these identities in 1977 Verner Emil Hoggatt Jr. (1921–1980) and Gerald E. Bergum (1931–2021) showed that for any $k > 0$, the set

$$(F_{2k}, F_{2k+2}, F_{2k+4}, 4F_{2k+1}F_{2k+2}F_{2k+3})$$

generates a Diophantine quadruple. This is also easily verified as

$$F_{2k}F_{2k+2} + 1 = F_{2k+1}^2$$

follows from

$$F_m F_{m+2} + (-1)^m = F_{m+1}^2$$

where $m = 2k$. Similarly,

$$F_{2k}F_{2k+4} + 1 = F_{2k+2}^2$$

is simply Vajda's identity above with $n = 2k$ and $i = j = 2$. In what follows, we will use Vajda's identity as well as Cassini's identity in the form

$$F_n^2 - F_{n-1}F_{n+1} = (-1)^{n+1}.$$

We begin by rearranging and using Vajda's identity twice and expanding to yield

$$\begin{aligned}
F_{2k} \cdot 4F_{2k+1}F_{2k+2}F_{2k+3} + 1 &= 4(F_{2k}F_{2k+2})(F_{2k+1}F_{2k+3}) + 1 \\
&= 4(F_{2k+1}^2 - 1)(F_{2k+2}^2 + 1) + 1 \\
&= 4(F_{2k+1}^2 F_{2k+2}^2 + F_{2k+1}^2 - F_{2k+2}^2 - 1) + 1.
\end{aligned}$$

We continue working on the right and now we apply

$$F_n^2 - F_{n-1}F_{n+1} = (-1)^{n+1}$$

twice, for both $n = 2k + 1$ and $n = 2k + 2$, to see

$$4(F_{2k+1}^2 F_{2k+2}^2 + F_{2k+1}^2 - F_{2k+2}^2 - 1) + 1$$

is equal to

$$\begin{aligned}
&4(F_{2k+1}^2 F_{2k+2}^2 + (F_{2k}F_{2k+2} + 1) - (F_{2k+1}F_{2k+3} - 1) - 1) + 1 \\
&= 4(F_{2k+1}^2 F_{2k+2}^2 + F_{2k}F_{2k+2} - F_{2k+1}F_{2k+3} + 1) + 1 \\
&= 4(F_{2k+1}^2 F_{2k+2}^2 + F_{2k}F_{2k+2} - F_{2k+1}(F_{2k+1} + F_{2k+2}) + 1) + 1 \\
&= 4(F_{2k+1}^2 F_{2k+2}^2 + (F_{2k}F_{2k+2} + 1) - F_{2k+1}^2 - F_{2k+1}F_{2k+2}) + 1 \\
&= 4(F_{2k+1}^2 F_{2k+2}^2 + (F_{2k+1}^2) - F_{2k+1}^2 - F_{2k+1}F_{2k+2}) + 1 \\
&= 4(F_{2k+1}^2 F_{2k+2}^2 - F_{2k+1}F_{2k+2}) + 1 \\
&= 4F_{2k+1}^2 F_{2k+2}^2 - 4F_{2k+1}F_{2k+2} + 1 \\
&= (2F_{2k+1}F_{2k+2} - 1)^2.
\end{aligned}$$

It is a simple exercise to use the methods above to verify the remaining:

$$\begin{aligned}
&F_{2k+2}F_{2k+4} + 1 = F_{2k+3}^2, \\
&F_{2k+2} \cdot 4F_{2k+1}F_{2k+2}F_{2k+3} + 1 = (2F_{2k+2}^2 - 1)^2, \quad \text{and} \\
&F_{2k+4} \cdot 4F_{2k+1}F_{2k+2}F_{2k+3} + 1 = (2F_{2k+2}F_{2k+3} - 1)^2.
\end{aligned}$$

Another Diophantine problem, originally inspired by Diophantus, is Problem 17 in Book VI of the *Arithmetica*. Here we are asked to "find a right-angled triangle such that the area added to the hypotenuse gives a square, while the perimeter is a cube." Diophantus considered the triangle with short sides 2 and a with area a. He let the hypotenuse be $y^2 - a$ so that the area added to the hypotenuse is a square. The second condition requires that the perimeter $2 + a + y^2 - a$ be a cube, or $y^2 = x^3 - 2$, which he solved with $y = 5$, $x = 3$, giving $a = 621/50$.

In 1621, the French aristocrat Claude-Gaspard Bachet Sieur de Méziriac (1581–1638) produced a Latin translation of the original Greek version of the *Arithmetica*. Fermat asserted in the margin of his copy of this book that the only solution of

$$y^2 = x^3 - 2$$

in positive integers is $y = 5$ and $x = 3$. However, he did not provide any details of his proof. Euler produced an incomplete proof of this statement, and Pepin finally provided a complete proof in 1875. *Fermat's sandwich theorem* is the modern assertion that 26 is the only integer that is sandwiched between both a perfect square and a perfect cube.

We define a *Mordell equation* or *Mordell curve* by

$$Y^2 = X^3 + k,$$

where k is any fixed nonzero integer. These curves are named after the British number theorist Louis Joel Mordell (1888–1972), due to his inspired and seminal work on them. If we were to graph the Mordell curve where $k = 8$, we would have the following plot.

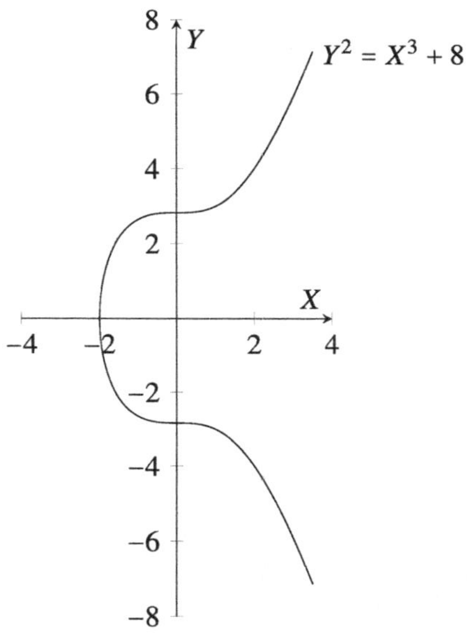

They are special cases of the elliptic curves introduced in Chap. 7. If we consider only integer coordinates for points on such curves, we have then introduced a new type of Diophantine equation. Because of the appearance of

$$Y^2 = X^3 - 2$$

in Bachet's version of the *Arithmetica*, Mordell equations are also sometimes called *Bachet's equations*. Mordell proved the significant result that any curve of the form

$$Y^2 = X^3 + k,$$

where k is a given nonzero integer, has only finitely many integer solutions. The complete set of such solutions is given in the table below for $|k| \leq 15$:

k	(X, Y)	k	(X, Y)
1	$(-1, 0), (0, \pm 1), (2, \pm 3)$	-1	$(1, 0)$
2	$(-1, \pm 1)$	-2	$(3, \pm 5)$
3	$(1, \pm 2)$	-3	
4	$(0, \pm 2)$	-4	$(5, \pm 11), (2, \pm 2)$
5	$(-1, \pm 2)$	-5	
6		-6	
7		-7	$(2, \pm 1), (32, \pm 181)$
8	$(-2, 0), (1, \pm 3), (2, \pm 4), (46, \pm 312)$	-8	$(2, 0)$
9	$(-2, \pm 1), (0, \pm 3), (3, \pm 6), (6, \pm 15), (40, \pm 253)$	-9	
10	$(-1, \pm 3)$	-10	
11		-11	$(3, \pm 4), (15, \pm 58)$
12	$(-2, \pm 2), (13, \pm 47)$	-12	
13		-13	$(17, \pm 70)$
14		-14	
15	$(1, \pm 4), (109, \pm 1138)$	-15	$(4, \pm 7)$

Remarkably, in 2015 the Canadian number theorist Michael Bennett and his then Ph.D. student Amir Ghadermarzi calculated the number of integer points for all Mordell curves such that $0 < |k| \leq 10^7$. It is worth noting that, for both positive and negative values of k, the most likely number of integer points on the curve is zero. Indeed, the ordered positive values of k for which the Mordell curve has no integer solutions is

$$6, 7, 11, 13, 14, 20, 21, 23, 29, 32, 34, 39, 42, \ldots,$$

the sequence A054504 in the OEIS. Also, the ordered negative values of k for which the Mordell curve has no integer solutions is

$$-3, -5, -6, -9, -10, -12, -14, -16, -17, -21, -22, \ldots,$$

sequence A081121 in the OEIS. We conclude this all-too-brief discussion by mentioning that when $k = -4090263$ the only positive solution of Mordell's equation is

$$(X, Y) = (16544006443618, 67291628068556097113).$$

In Chap. 14, we learned that any prime $p \equiv 1 \pmod 4$ can be expressed as a sum of two integral squares. In fact, these two squares are unique to p. For example, $p = 61 = 6^2 + 5^2$, and there is no other way of writing 61 as sum of two squares. By making use of the old identity (known, it seems, to Diophantus in Book III problem 19 of the *Arithmetica*):

$$(a^2 + b^2)(c^2 + d^2) = (ac + bd)^2 + (ad - bc)^2 = (ac - bd)^2 + (ad + bc)^2,$$

we can easily establish that an integer $k > 1$ can be written as the sum of two squares if and only if its prime power representation contains no factor of the form p^m where $p \equiv 3 \pmod 4$ and m is odd. Between 1797 and 1798, Legendre showed that

$$k = X^2 + Y^2 + Z^2$$

if and only if $k \neq 4^m(8n + 7)$; earlier in 1770 Lagrange had proved that any positive k can be written as a sum of at most four squares.

To solve

$$k = X^3 + Y^3$$

for a given integer k is relatively easy because $X^3 + Y^3 = (X + Y)(X^2 - XY + Y^2)$; thus, $X + Y$ must be a divisor of k. However, solving

$$k = X^3 + Y^3 + Z^3$$

such that $XYZ \neq 0$ is much more difficult. In 1956 Mordell wrote,

> I do not know anything about the integer solutions of $X^3 + Y^3 + Z^3 = 3$ beyond the existence of the four sets $(1, 1, 1)$, $(4, 4, -5)$, etc.; and it must be very difficult indeed to find out anything about any other solutions. One may wonder if the problem of finding other solutions is comparable in difficulty with that of finding when an assigned sequence, e.g. 123456789, occurs in the decimal expansion of π.

This claim of difficulty was certainly prescient, as a new solution was only found as recently as 2021 when Andrew Booker and Andrew Sutherland found that

$$569936821221962380720^3 + (-569936821113563493509)^3 + (-472715493453327032)^3 = 3.$$

In the more general case of

$$X^3 + Y^3 + Z^3 = k,$$

we note that since any integer cube must be 0 or ± 1 modulo 9, it is easily seen that for $k \equiv 4, 5 \pmod 9$ the equation cannot have integer solutions. Notice that if

$$X^3 + Y^3 + Z^3 = k,$$

then

$$(-X)^3 + (-Y)^3 + (-Z)^3 = -k;$$

thus, we need only consider $k \geq 0$. Also, as we will see later, there can be no solution except the trivial ones ($XYZ = 0$) when $k = 0$.

Prior to 1955, Mordell suggested to two pioneers of English computing, J. C. P. Miller and M. F. C. Woollett, that they use the computing facilities of the

Cambridge Mathematical Laboratory (in particular, they made use of the computer EDSAC) to search for solutions of

$$X^3 + Y^3 + Z^3 = k$$

for $k = 3$. However, as the search yielded no new results in their search range, they decided to extend their investigation to include all integers k such that $0 < k \leq 100$. No solutions were produced for $k = 30, 33, 39, 42, 52, 74, 75, 84$ and 87. Subsequently, other researchers extended this search to integers k such that $0 < k \leq 1000$. By the end of 2007, solutions had still not been found for 3 values of $k \leq 100$ $(33, 42, 74)$ and for 24 values of k such that $100 < k \leq 1000$. A solution for $k = 74$ was found by Sander G. Huisman in 2016, and a solution was found for $k = 33$ by Andrew Booker in 2019.

In 2021, Andrew Booker and Andrew Sutherland considerably improved Booker's earlier search algorithm and finally found a solution for $k = 42$. Indeed, they found the solutions

$$
\begin{aligned}
&(-80538738812075974)^3 + 80435758145817515^3 + 12602123297335631^3 = 42,\\
&(-385495523231271884)^3 + 383344975542639445^3 + 98422560467622814^3\\
&\qquad\qquad\qquad\qquad\qquad\qquad\qquad\qquad\qquad\qquad\qquad\qquad = 165,\\
&143075750505019222645^3 + (-143070303858622169975)^3\\
&\qquad\qquad\qquad\qquad + (-6941531883806363291)^3 = 579,\\
&(-74924259395610397)^3 + 72054089679353378^3 + 35961979615356503^3 = 906.
\end{aligned}
$$

At the end of their search only seven candidates for $k \leq 1000$ remained:

$$114, 390, 627, 633, 732, 921, 975.$$

Before rashly embarking on a search for a solution for any of these values of k, the reader is advised to study carefully Booker and Sutherland's paper.

The difficulty in finding solutions to

$$X^3 + Y^3 + Z^3 = k$$

has inspired some investigators to believe that there may be values of $k \not\equiv 4, 5 \pmod 9$ for which there are either no or only a finite number of solutions, but for very good mathematical reasons, the distinguished British mathematician David Rodney "Roger" Heath-Brown has conjectured otherwise. If k is fixed, B is some positive bound and we let $N(B)$ denote the number of solutions in the box

$$\max\{|X|, |Y|, |Z|\} \leq B,$$

then Heath-Brown conjectures that there exists a positive constant C_k dependent only on k such that

$$N(B) \sim C_k \log B.$$

Thus, as long as any fixed $k \not\equiv 4, 5 \pmod 9$, we would expect

$$X^3 + Y^3 + Z^3 = k$$

to have an infinitude of solutions. We know this is true for $k = 1, 2$, but as we have seen solutions are hard to find when $k \geq 3$.

A more familiar Diophantine equation is the *Pythagorean equation*

$$x^2 + y^2 = z^2.$$

We call a *Pythagorean triple* (x, y, z) a set of three positive integers which satisfy the Pythagorean equation. Pythagorean triples go back to antiquity. Indeed, they have been found on ancient Babylonian tablets that predate Pythagoras by over 1000 years. We say a Pythagorean triple (x, y, z) is *primitive* in the case that x, y and z have no common divisor greater than 1. It is easy to verify that both $(3, 4, 5)$ and $(9, 12, 15)$ are Pythagorean triples; yet only $(3, 4, 5)$ is primitive, as 9, 12, 15 have the common divisor of 3. Certainly, any integer multiple of k of a Pythagorean triple yields a new Pythagorean triple. Hence, there is clearly an infinitude of integer solutions to

$$x^2 + y^2 = z^2.$$

Moreover, using what is called *Euclid's formula*, it can be easily shown that there exists an infinitude of primitive Pythagorean triples. This works as follows: given two integers m, n such that $m > n > 0$, then it is easy to verify that

$$x = m^2 - n^2, \quad y = 2mn, \quad z = m^2 + n^2 \quad \text{(Euclid's formula)}$$

yields a Pythagorean triple; moreover, it is primitive if and only if $\gcd(m, n) = 1$ and one of m or n is even. In the case that both $\gcd(m, n) = 1$ and both m and n are odd, then all of x, y and z are even and a new primitive Pythagorean triple can be recovered upon halving each of x, y and z. For example, consider $m = 5, n = 3$; we have

$$x = 5^2 - 3^2 = 16, \quad y = 2 \cdot 5 \cdot 3, \quad z = 5^2 + 3^2$$

and halving these values yields the primitive Pythagorean triple $(8, 15, 17)$. Of course, we can get this same triple by putting $m = 4, n = 1$ in Euclid's formula.

Interestingly, Euclid's formula will generate all possible primitive Pythagorean triples, but it does not generate all possible Pythagorean triples. For example, there do not exist m, n as above to generate $(9, 12, 15)$. This is easily remedied by multiplying by a constant k, as we saw above, i.e.

$$x = k(m^2 - n^2), \quad y = 2kmn, \quad z = k(m^2 + n^2),$$

where m, n and k are positive integers and $m > n$, $\gcd(m, n) = 1$ and at least one of m or n is even. We can then see (9, 12, 15) follows from the above when $m = 2$, $n = 1$ and $k = 3$.

In the *Arithmetica*, Diophantus posed as his eighth problem in Book II that of expressing the square of an integer as the sum of two squares. Next to this problem, in 1634, Fermat wrote in his copy of the *Arithmetica*, mentioned earlier, the following now famous remark (translated from original Latin):

> It is impossible to separate a cube into two cubes or a fourth power into two fourth powers or in general any power higher than the second into two like powers. I have discovered a truly marvellous proof of this, which this margin is too narrow to contain.

It is highly unlikely that Fermat was actually in possession of such a proof; nevertheless, because of his assertion this became to be known as *Fermat's last theorem*. Today this theorem is usually stated as follows.

Theorem 20.1 *No three positive integers x, y, z satisfy the equation*

$$x^n + y^n = z^n$$

for any integer value of n greater than 2.

Establishing Fermat's last theorem proved to be an incredibly tough nut to crack. Fortunately, failed attempts to prove it over the last 350 years generated many interesting ideas in mathematics; however, most are well beyond the purview of this book. We will only mention here that a proof was finally developed by the English mathematician Sir Andrew John Wiles, with some assistance from his former student Richard Taylor, in 1995.

Work in this area continues on a yet more comprehensive equation called a *generalized Fermat equation*; that is, find relatively prime integers x, y, and z which satisfy

$$x^p + y^q = z^r,$$

where p, q and r are integers. Let $\sigma(p, q, r) = \frac{1}{p} + \frac{1}{q} + \frac{1}{r}$; we call this the *signature* of the equation. When $\sigma(p, q, r) > 1$, then (p, q, r) can only be $(2, 2, r)$, $(2, q, 2)$, $(2, 3, 3)$, $(2, 3, 4)$, $(2, 4, 3)$ or $(2, 3, 5)$. In each of these cases, just as in the Pythagorean case $(2, 2, 2)$, there are infinitely many solutions, all of which have been completely characterized. When $\sigma(p, q, r) = 1$, then (p, q, r) (up to reordering) can only be $(2, 6, 3)$, $(2, 4, 4)$, $(4, 4, 2)$, $(3, 3, 3)$ or $(2, 3, 6)$. Only one of these, $(2, 3, 6)$, produces a nontrivial solution, which is

$$1^6 = 3^2 + (-2)^3.$$

Because the situation has been completely resolved when $\sigma(p, q, r) \geq 1$, research has focused on the case of $\sigma(p, q, r) < 1$. The only known solutions for this case are

$$1^p + 2^3 = 3^2 \quad (p \geq 6)$$

and

$$2^5 + 7^2 = 3^4, 7^3 + 13^2 = 2^9, 2^7 + 17^3 = 71^2, 3^5 + 11^4 = 122^2,$$
$$17^7 + 76271^3 = 21063928^2, 1414^3 + 2213459^2 = 65^7, 9262^3 + 15312283^2 = 113^7,$$
$$43^8 + 96222^3 = 30042907^2, \quad \text{and} \quad 33^8 + 1549034^2 = 15613^3.$$

Notice that in all these solutions the $\min\{p, q, r\} = 2$. In 1993 the Texan businessman, philanthropist and amateur mathematician Daniel Andrew Beal observed this and made the assertion that for all integer solutions to

$$x^p + y^q = z^r,$$

where p, q and r are integers such that $\frac{1}{p} + \frac{1}{q} + \frac{1}{r} \leq 1$, and it must be that $\min\{p, q, r\} = 2$; this is now known as *Beal's conjecture*. In 1997, Beal offered a monetary prize for a peer-reviewed proof of his conjecture or a counterexample. It is still unclaimed and is now valued at one million U.S. dollars.

What we do know when the signature is less than 1 is that the number of solutions is at most finite. This is a consequence of a remarkable theorem of Andrew Granville and the eminent Canadian mathematician Henri Darmon.

Theorem 20.2 *If A, B, C, p, q, r are fixed integers with $\frac{1}{p} + \frac{1}{q} + \frac{1}{r} < 1$, then the equation*

$$Ax^p + By^q = Cz^r$$

has at most finitely many solutions in nonzero integers x, y, z such that $\gcd(x, y, z) = 1$.

Solutions of the form $1^p + 2^3 = 3^2$ for these generalized Fermat equations are related to an old Diophantine problem called *Catalan's conjecture*, which is that 8 and 9 are the only consecutive perfect powers. We can also state this as

$$x^u - y^v = 1,$$

where x, y are positive integers and u, v are integers greater than 1, and has no solutions other than $x^u = 3^2$ and $y^v = 2^3$. Catalan published his conjecture in 1844, but well before he made his assertion, Euler had already shown that 8 and 9 are the only consecutive integers for squares and cubes. However, the more general case is much more difficult to establish, eluding proof until 2002. At this time a complete proof of Catalan's conjecture was provided by the Romanian mathematician Preda Mihăilescu, who had no prior experience in dealing with it. Surprisingly, his proof had very little to do with computation, but rather he employed deep results from the theory of cyclotomic fields, a topic which is (yes) also beyond the scope of this book. In connection with this we mention that there is a conjecture, due to the Indian mathematician Subbayya Sivasankaranarayana Pillai (1901–1951), which essentially states that the Diophantine equation

$$x^u - y^v = k,$$

where k is a positive integer, has only a finite number of solutions (x, y, u, v) with $\min\{x, y, u, v\} \geq 2$. We have seen that this conjecture is true for $k = 1$, but there is no known value of k (≥ 2) for which it has been shown that Pillai's equation has only a finite number of solutions.

In 1769 Euler made the following statement:

> It has seemed to many Geometers that this theorem [Fermat's last theorem] may be generalized. Just as there do not exist two cubes whose sum or difference is a cube, it is certain that it is impossible to exhibit three biquadrates whose sum is a biquadrate, but that at least four biquadrates are needed if their sum is to be a biquadrate, although no one has been able up to the present to assign four such biquadrates. In the same manner it would seem to be impossible to exhibit four fifth powers whose sum is a fifth power, and similarly for higher powers.

This remark is now known as *Euler's conjecture*. It was shown to be wrong in 1969 when L. J. Lander and T. R. Parkin announced that a computer search on a CDC 6600 yielded

$$27^5 + 84^5 + 110^5 + 133^5 = 144^5.$$

Later, in 1988, the American mathematician Noam Elkies discovered the counterexample

$$2682440^4 + 15365639^4 + 18796760^4 = 20615673^4.$$

Indeed, he went considerably further by showing that

$$x^4 + y^4 + z^4 = w^4$$

has an infinitude of integer solutions and how to characterize them. He also reported that Roger Frye's example, found by computer search,

$$95800^4 + 217519^4 + 414560^4 = 422481^4$$

is the smallest solution.

We hope that, even in this brief discussion, we have amply illustrated the fascination and charm of Diophantine equations. They and their solutions have exerted an enormous influence on the direction of number theory in particular, but their study has also advanced mathematics in general.

Problems

20.1 (Lagrange's method) Use Lagrange's method to find all integer solutions to the following general quadratic Diophantine equations in 2 variables:
(a) **[BB]** $x^2 - 4y^2 + 2x + 4 = 0$

(b) $x^2 - xy + y^2 - 2x - y = 0$
(c) $3x^2 + xy - 2y^2 + 3 = 0$
(d) $2x^2 - xy + y^2 - 3x + 2y = 0$.

20.2 (Diophantine quadruple) Construct Diophantine quadruples of the form

$$(k-1, k+1, 4k, 16k^3 - 4k)$$

for the given value of k:
(a) $k = 3$
(b) **[BB]** $k = 10$
(c) $k = 15$
(d) $k = 215$.

20.3 (Diophantine quadruple) Construct Diophantine quadruples of the form

$$(F_{2k}, F_{2k+2}, F_{2k+4}, 4F_{2k+1}F_{2k+2}F_{2k+3})$$

for the given value of k:
(a) $k = 1$
(b) **[BB]** $k = 3$
(c) $k = 5$
(d) $k = 7$.

20.4 (Sum of two cubes) Find values X and Y such that

$$k = X^3 + Y^3$$

for the given value of k:
(a) $k = 1358$
(b) **[BB]** $k = 1241$
(c) $k = 407$
(d) $k = 793$.

20.5 (Euclid's formula) Construct a primitive Pythagorean triple using Euclid's formula

$$x = m^2 - n^2, \quad y = 2mn, \quad z = m^2 + n^2$$

for the given value of m and n:
(a) **[BB]** $m = 3, n = 2$
(b) $m = 4, n = 1$
(c) $m = 5, n = 2$
(d) $m = 7, n = 3$.

Further Reading

Diophantine equations in general

L. E. Dickson, *History of the Theory of Numbers*, Vol. II: Diophantine Analysis, Dover Publications, NY, 2013.
L. J. Mordell, *Diophantine Equations*, Academic Press, NY, 1969.
Richard K. Guy, *Unsolved Problems in Number Theory*, 3rd Edition, Springer, NY, 2004, Section D, particularly D1 and D2.

Some methods of solving Diophantine equations

Nigel P. Smart, *The Algorithmic Resolution of Diophantine Equations*, London Mathematical Society Student Texts 41, Cambridge University Press, Cambridge, 1998.

General quadratic Diophantine equations

R. E. Sawilla, A. K. Silvester and H. C. Williams, A new look at an old equation, *Proceedings of the* 8th *Algorithmic Number Theory Symposium, ANTS-VIII*, LNCS 5011, Springer-Verlag, Berlin, 2008, pp. 37-59.

The algorithm of Cornacchia

Kenneth Hardy, Joseph B. Muskat and Kenneth S. Williams, A deterministic algorithm for solving $n = fu^2 + gv^2$ in coprime integers u and v, *Mathematics of Computation*, **55** (1990), 327–343.
Kjell Wooding, The Sieve Problem in One- and Two-Dimensions, PhD Thesis, University of Calgary, 2010, pp. 69-83. Available at: https://hdl.handle.net/1880/104481

Diophantine quadruples

A. Dujella, Diophantine quadruples and Fibonacci numbers, *Bulletin of Kerala Mathematics Association*, **1** (2004), 133–147. Available at: https://web.math.pmf.unizg.hr/~duje/pdf/pondi.pdf.

Diophantine quintuples

B. He, A. Togbé and V. Ziegler, There is no Diophantine quintuple, *Transactions of the American Mathematical Society*, **371** (2019), 6665-6709.

Diophantine m-tuples

A. Dujella, *Diophantine m-tuples and Elliptic Curves*, Developments in Mathematics, Springer Nature Switzerland AG, 2024.

Elliptic curves and cryptography

Darrel Hankerson, Alfred J. Menezes and Scott Vanstone, *Guide to Elliptic Curve Cryptography*, Springer-Verlag, NY, 2004.

Mordell equations

Michael A. Bennett and Amir Ghadermarzi, Mordell's equation: a classical approach, *London Mathematical Society Journal of Computation and Mathematics*, **18** (2015), 633–646.

Sums of two squares

Daniel Shanks, *Solved and Unsolved Problems in Number Theory*, 3[d] ed., Chelsea Publishing Co., NY, 1985, pp. 159–160.

Sums of two cubes

Kevin A. Broughan, Characterizing the sum of two cubes, *Journal of Integer Sequences*, **6** (2003), Article 03.4.6.

Sums of three integer cubes

L. J. Mordell, On the integer solutions of the equation $x^2 + y^2 + z^2 + 2xyz = n$, *Journal of the London Mathematical Society*, **s1–28** (1953), 500–510.

Michael Beck, Eric Pine, Wayne Tarrant and Kim Yarbrough Jensen, New integer representations as the sum of three cubes, *Mathematics of Computation*, **76** (2007), 1683–1690.

S. Siksek, Sums of integer cubes, *Proceedings of the National Academy of Sciences (PNAS)*, **118** (16) (2021), 1–3.

Fermat's equation

Simon Singh, *Fermat's Enigma: the Epic Quest to Solve the World's Greatest Mathematical Problem*, Penguin Random House, NY, 1998.

Paulo Ribenboim, *13 Lectures on Fermat's Last Theorem*, Springer-Verlag, NY, 1979.

Alf van der Poorten, *Notes on Fermat's Last Theorem*, John Wiley and Sons Inc., NY, 1996.

M. Bennet, P. Mihăilescu and S. Siksek, The generalized Fermat equation, *Open Problems in Mathematics*, Springer International publishing, Switzerland, 2016, pp. 173–205.

Catalan's equation

T. Metsänkylä, Catalan's conjecture: another old Diophantine problem solved, *Bulletin of the American Mathematical Society*, **1** (2003), 43–57.

Euler's conjecture

Noam Elkies, On $A^4 + B^4 + C^4 = D^4$, *Mathematics of Computation*, **51** (1988), 825–835.

Chapter 21
Conclusion

We seem to have covered a great deal of ground in this work, but we have left considerably more unsaid. Much of the delight of number theory derives from its boundless breadth and depth, allowing even non-professionals the scope to make discoveries and, more importantly, to enjoy its endless fascination. This, and a natural human desire to solve puzzles, explains why so many people are willing to devote so much time studying some aspect of it, even though it may seem to have little or no practical application. However, despite many years of very significant progress, there are still several important questions whose answers elude us. In this final chapter we will discuss a few of these.

The Nobel Laureate, Steven Weinberg, famously remarked that "the more the universe becomes comprehensible, the more it also seems to be pointless". This comment, coming from such an acclaimed physicist, continues to generate a great deal of controversy among not only physicists, but also theologians, philosophers and people in general. We should emphasize here that, apart from adding to the abundance of human knowledge, number theory doesn't really have a point; it simply is. Whether it exists only as a manifestation of human consciousness or independently is a philosophical point on which the authors do not feel competent to comment. To many mathematicians, the answer to the question of whether mathematics is invented or discovered is irrelevant. We seem to get along perfectly well without knowing the answer. Indeed, perhaps the very question itself is questionable.

Number theory, indeed mathematics in general, advances through the posing of problems concerning what is known. Attempts to discover the answers invariably inspire more questions and so on. In this final chapter, we will discuss a few of the many unanswered questions that arise from the material we have presented in earlier chapters. Some of these questions have already been asked there, but, of course, there are many other problems that the reader could easily invent—such is the nature and beauty of number theory. However, our space is limited and, as the authors, we get to choose.

E. L. F. Roettger and H. C. Williams, *The Enchantment of Numbers*, CMS/CAIMS Books in Mathematics 16, https://doi.org/10.1007/978-3-031-76568-1_21

We begin with what is called the *Ankeny-Artin-Chowla* or *AAC conjecture*. We hasten to remark that neither of these three distinguished mathematicians actually made this conjecture; what they did was ask a question in a joint paper written in 1952. The conjecture came about later when it was assumed that the answer to the question must be no. We should mention that Nesmith Cornett Ankeny (1927–1993) was American, Emil Artin (1898–1962) was Austrian and Sarvadaman Chowla (1907–1995) was Indian American. Collaboration in mathematics is a great deal more common than many might assume.

In Chap. 19 we defined the fundamental unit ϵ_Δ of the quadratic order $\mathcal{O}$ in $\mathbb{Q}(\sqrt{D})$. For $\epsilon_\Delta = (x + y\sqrt{\Delta})/2$, where x, y are integers, the AAC conjecture states that when $\Delta = D$ is a prime p and $p \equiv 1 \pmod 4$, then $p \nmid y$. In their original paper concerning this matter, the authors only asked if this were possible. The question was converted into a conjecture by Mordell in 1960. It is necessary that D be a prime because if $\Delta = D = 1817 = 23 \cdot 79 \equiv 1 \pmod 4$, we find that $D \mid y$, although this is the only example like this for all $D \equiv 1 \pmod 4$ and $D < 10^7$. More such numbers can be found in the OEIS list in A135735. Neither Ankeny, Artin, Chowla nor Mordell provided any convincing justification for believing the conjecture. It is possible that it is true because of considerations that stem far from our current understanding. In this respect it might be similar to the Wall-Sun-Sun conjecture in Chap. 12.

By making use of modern research into the Pell equation, alluded to in Chap. 19, AAC has been tested for all primes p ($\equiv 1 \pmod 4$) less than 2×10^{11}, and no counterexample has been found. However, while this may seem like compelling evidence for its truth, the "log log argument" may suggest otherwise. The reasoning is based on two seemingly plausible assumptions: i) the probability that a prime p will divide a given y is $1/p$ ii) for two different primes p_1 and p_2 the events p_1 divides y and p_2 divides y are independent. It then follows that we might expect the number of exceptions to the conjecture in the interval between a and b to be given by

$$\sum_{\substack{a \le p \le b \\ p \equiv 1 \pmod 4}} \frac{1}{p}.$$

Since it can be shown that

$$\sum_{\substack{p \le x \\ p \equiv 1 \pmod 4}} \frac{1}{p} \sim \frac{1}{2} \log\log x,$$

the above sum can be fairly well approximated by

$$N(a, b) = \frac{1}{2} \log\left(\frac{\log b}{\log a}\right).$$

If $a = 5$ and $b = 2 \times 10^{11}$, we would only get $N(a, b) \approx 1.39$. Thus, even if the AAC conjecture is false, we cannot be too surprised that there are no counterexamples up to 2×10^{11}.

In connection to this, we mention that in 1961, Mordell conjectured that if D is a prime $p \equiv -1 \pmod 4$, then $p \nmid y$. Not long ago this conjecture was disproved by the Austrian mathematician Andreas Reinhart, who discovered that if $p = 39028039587479$, then $p \mid y$. As a hurriedly added postscript to this discussion we mention that Reinhart has very recently produced a counterexample to the AAC conjecture: if D is the prime 331914313984493, then the conjecture is false.

We next visit a question that arises from the Hardy-Littlewood conjecture mentioned in Chap. 18. This conjecture was generalized by Paul Trevier Bateman (1919–2012) and Roger Alan Horn in 1962. The Bateman-Horn conjecture provides a presumed density for the positive integers for which a given set of polynomials all have prime values. To state it, we first require some additional notation.

We denote the set of all polynomials in x with integer coefficients by $\mathbb{Z}[x]$. Suppose we let $f_1(x), f_2(x), \ldots, f_k(x)$ be k distinct, irreducible polynomials in $\mathbb{Z}[x]$, each with positive leading coefficients. By *irreducible* here we mean that each of the polynomials $f_i(x)$ $(1 \le i \le k)$ cannot be factored into polynomials in $\mathbb{Z}[x]$ of lesser degree. For example, since

$$x^4 + x^2 + 1 = (x^2 + x + 1)(x^2 - x + 1),$$

we see that $x^4 + x^2 + 1$ is not irreducible, but $x^4 + 1$ is irreducible. In short, the irreducible polynomials with positive leading coefficients in $\mathbb{Z}[x]$ are to $\mathbb{Z}[x]$ as the primes in $\mathbb{Z}$ are to $\mathbb{Z}$. We say that an integer n is *prime-generating* if every value of $f_i(n)$ $(i = 1, 2, \ldots, k)$ is a prime. If we put

$$F(x) = \prod_{i=1}^{k} f_i(x),$$

we see that for such an n to exist, we require that there not be any prime p such that p divides $F(m)$ for every value of the positive integer m. If no such prime exists, we say that the set of polynomials is *suitable*. For example, if $k = 2$, $f_1(x) = x$, $f_2(x) = x + 1$, then $\{f_1(x), f_2(x)\}$ is not suitable because $2 \mid m(m + 1)$ for all integers m.

If $P(x)$ is the number of prime-generating integers among the positive integers less than x for a suitable set S of k polynomials, the Bateman-Horn conjecture asserts that

$$P(x) \sim \frac{C}{E} \int_2^x \frac{dt}{(\log t)^k},$$

where E is the product of the degrees of all the polynomials in S and

$$C = \prod_p \frac{1 - N(p)/p}{(1 - 1/p)^k}$$

with $N(p)$ representing the number of possible solutions of $F(x) \equiv 0 \pmod{p}$. Note that $N(p) < p$ because S is suitable.

In the case of $S = \{f(x)\}$, where $f(x) = x^2 + x + A$ is a quadratic polynomial of discriminant $D = 1 - 4A$, we have $k = 1$, $N(2) = 0$, $N(p) = 1 + (D|p)$ $(p \geq 3)$, $E = 2$, and

$$\frac{C}{E} = \prod_{p \geq 3} \left(1 - \frac{(D|p)}{p-1}\right).$$

Thus, the Bateman-Horn conjecture includes the Hardy-Littlewood conjecture in Chap. 18. By putting $S = \{x, x+2\}$, we see that it also includes the twin-prime conjecture mentioned in Chap. 12. In this case $k = 2$, $N(2) = 1$, $N(p) = 2$ $(p \geq 3)$, $E = 1$ and

$$\frac{C}{E} = 2 \prod_{p \geq 3} \frac{p(p-2)}{(p-1)^2} \approx 1.3203236.$$

In the more general case of $S = \{x + r_1, x + r_2, \ldots, x + r_k\}$, where $r_1, r_2, \ldots, r_k$ are k integers such that S is suitable, the Bateman-Horn conjecture includes the prime k-tuple conjecture.

If $S = \{x, x^2 + x + 1\}$, we have $k = 2$, $E = 2$, $N(2) = 1$, $N(p) = 2 + (-3|p)$ and

$$\frac{C}{E} = \prod_{p \geq 3} \frac{p(p - 2 + (-3|p))}{(p-1)^2} \approx 0.761.$$

Thus, since

$$\int_2^x \frac{1}{(\log t)^k} \sim \frac{x}{(\log x)^k},$$

the conjecture tells us that the number of primes $p \leq x$ such that $p^2 + p + 1$ is also a prime is about

$$0.761 \frac{x}{(\log x)^2},$$

which has been verified up to certain values of x by computer counts.

We see, then, that the Bateman-Horn conjecture is very wide ranging and, as the originators provided a heuristic proof of it, convincing. It has also passed a number of tests by computer. However, it seems to be a long, long way from being proved. This remains one of the great challenges of analytic number theory.

A great challenge of Diophantine analysis is the *abc*-conjecture, which has been called its most important unsolved problem. We begin by defining the *radical* ($\mathrm{rad}(n)$) or *square-free kernel* ($\mathrm{sfk}(n)$) of an integer n as the product of all the distinct primes which divide n. For example, if $n = 360$, we have $n = 2^3 \cdot 3^2 \cdot 5$ and $\mathrm{rad}(n) = 2 \cdot 3 \cdot 5 = 30$. There is no known method, short of factoring n, of determining $\mathrm{rad}(n)$ from n. Note that if $\gcd(m, n) = 1$, then $\mathrm{rad}(mn) = \mathrm{rad}(m)\,\mathrm{rad}(n)$, i.e. rad is a multiplicative function of the integers. Also, it is easy to see that $\mathrm{rad}(n) \leq n$.

The *abc-conjecture* is the result of a discussion by the French mathematician Joseph Oesterlé and the British mathematician David Masser in 1985. It states that for every positive real number ϵ, there can only exist a finite number of triples (a, b, c) of coprime ($\gcd(a, b, c) = 1$) positive integers with $a + b = c$ such that

$$c > \operatorname{rad}(abc)^{1+\epsilon}.$$

The condition that $\epsilon > 0$ is necessary, as there exist infinitely many triples (a, b, c) such that $a + b = c$ and

$$c > \operatorname{rad}(abc).$$

For example, consider $a = 1$, $b = 2^{6n} - 1$, $c = 2^{6n}$. We note that $2^{6n} - 1 = 64^n - 1 \equiv 0 \pmod 9$. Since

$$\operatorname{rad}(abc) = \operatorname{rad}(a)\operatorname{rad}(b)\operatorname{rad}(c) = 1 \cdot \operatorname{rad}(2^{6n} - 1) \cdot 2,$$

we have

$$\operatorname{rad}(abc) = 2\operatorname{rad}(2^{6n} - 1) = 2\operatorname{rad}(9 \cdot b/9) \leq 2 \cdot 3 \cdot b/9 \leq 2b/3 < 2c/3.$$

If we define the *quality* of (a, b, c) by

$$q(a, b, c) = \frac{\log c}{\log(\operatorname{rad}(abc))},$$

we can produce another version of the abc-conjecture as

> For any positive real number ϵ, there exist only finitely many triples (a, b, c) of coprime integers with $c = a + b$ such that $q(a, b, c) > 1 + \epsilon$.

Computer searches, particularly at Leiden University in the Netherlands, have been run over many triples of integers in order to find large values for q, but so far, the largest value of q known is $q = 1.6299$ for $a = 2$, $b = 3^{10} \cdot 109$, $c = 23^5$, which was discovered in 1987. Much more information concerning this can be found at https://pub.math.leidenuniv.nl/~smitbde/abc/. If we examine the abc-conjecture from a different point of view, we see that with only a finite number of exceptions, the conjecture asserts that if a, b, c are relatively prime integers which satisfy $a + b = c$, then for any $\epsilon > 1$, we have

$$c < \operatorname{rad}(abc)^{1+\epsilon}.$$

Thus, there is an equivalent formulation of the conjecture as

> For every positive real number ϵ, there exists a constant $\kappa(\epsilon)$, depending only on ϵ, such that for all triples (a, b, c) of coprime positive integers with $c = a + b$,
>
> $$c < \kappa(\epsilon)\operatorname{rad}(abc)^{1+\epsilon}.$$

While this version of the conjecture, on first glance, may appear to be quite innocuous, it has a number of very significant consequences. For example, suppose we consider the Fermat equation

$$x^n + y^n = z^n,$$

with x, y, z positive integers and $\gcd(x, y, z) = 1$; if we put $a = x^n, b = y^n, c = z^n$, we get

$$\mathrm{rad}(abc) = \mathrm{rad}(x^n y^n z^n) = \mathrm{rad}(xyz) \leq xyz < z^3.$$

According to the *abc*-conjecture we must have

$$z^n < \kappa(\epsilon) z^{3(1+\epsilon)}$$

for any $\epsilon > 0$. Putting $\epsilon = 1/6$ and assuming $n \geq 4$, we get $n - 3(1+\epsilon) \geq n/8$; hence,

$$z^{n/8} \leq z^{n-3(1+\epsilon)} < \kappa(1/6),$$

and therefore

$$z^n < \kappa(1/6)^8.$$

As we have now shown that x^n, y^n, z^n must each be less than some constant, which is independent of the value of n, there can only be a finite number of solutions for any $n \geq 4$.

It has been suggested that the *abc*-conjecture might be valid for $\kappa(\epsilon) = 1$. If this is the case, then Fermat's last theorem follows immediately for $n \geq 6$. As the case for $n = 3, 4, 5$ has been known for almost 200 years, this would provide us with another proof of Fermat's last theorem for $n \geq 3$.

The *abc*-conjecture can also be used to establish that there can only be a finite number of exceptions to Beal's conjecture and to Pillai's conjecture. Indeed, there are many diverse problems in Diophantine analysis and other areas to which it can be applied. Thus, resolving it would have an enormous impact on number theory. Of course, this means that many number theorists have tried to prove it, but in spite of occasional reports of success, the current consensus of the mathematical community is that the *abc*-conjecture has not yet been proved. In connection with this we mention that a very highly regarded Japanese mathematician, Shinichi Mochizuki, has announced a proof, but it is extremely lengthy and intricate, involving very deep mathematics, which only a very few individuals understand. At the time of writing, Mochizuki's proof was controversial and not fully accepted by the number theory community. In order to resolve this issue a Japanese industrialist, Nobuo Kawkami, has offered a one million US dollar prize to the first individual to find a serious flaw in Mochizuchi's proof.

The last substantive chapter of Beiler's book, which we cited in the Preface, is entitled, "The Stone Wall", referring to the fact that in the 330 years prior to the book's publication, no acceptable proof had been devised for Fermat's last theorem. We have seen in the previous chapter that this theorem has finally been proved, so we would like to offer up another candidate conjecture for which there is currently a stone wall. This is the celebrated *Riemann hypothesis* (RH) a supposition of the German mathematician Bernhard Riemann (1826–1866).

In an 8 page paper written in 1859, Riemann defined a function which he denoted by $\zeta(s)$, which today is called the *Riemann zeta function*. A difficulty which arises in any attempt to discuss the RH is that it is rather technical. Unlike, the functions that have been discussed so far in this work, this is a function of a complex variable s. A *complex number* is a number consisting of a *real* part x and an imaginary part y and written as $x + iy$, where x and y are real numbers and $i^2 = -1$. Yes, there is no real number equal to i; numbers of the form iy for real values of y are said to be *imaginary*, although they are no more imaginary than any other number. All numbers are in a sense imaginary in that they have no physical presence. Have you ever touched a 3?

We call the *complex plane* that plane which is comprised of all complex numbers and expressed by a Cartesian coordinate system with the x-axis (the real axis) representing the real numbers and the y-axis (the imaginary axis) representing the imaginary numbers (Fig. 21.1).

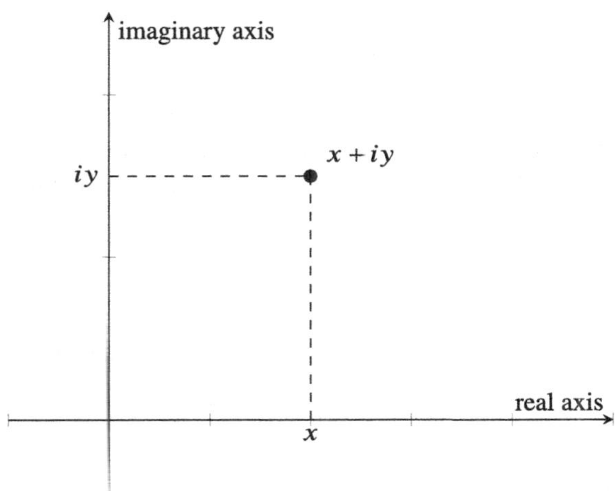

Fig. 21.1 Complex number example $x + iy$

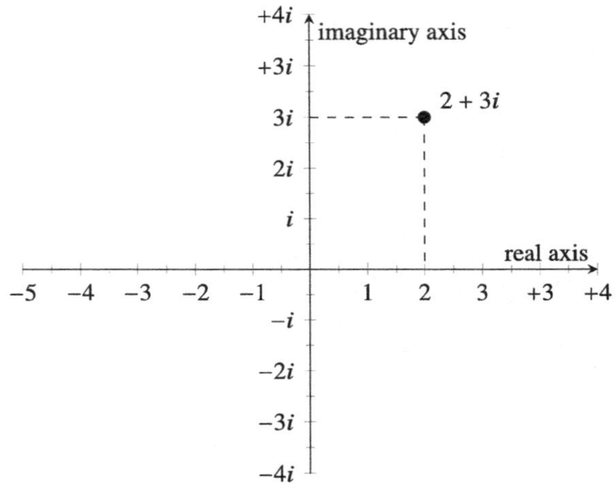

Fig. 21.2 Complex number example $2 + 3i$

For example, if $x = 2$ and $y = 3$, then we would have the complex number $2 + 3i$ which can be plotted on the Cartesian plane as in Fig. 21.2.

The Riemann zeta function is a function whose argument s is any complex number, except 1, and whose values are complex. It is defined as

$$\zeta(s) = \sum_{n=1}^{\infty} \frac{1}{n^s}.$$

When s is real (has no imaginary part) and $s > 1$, we know that the infinite series represented by $\zeta(s)$ will always converge, i.e. the value of $\zeta(s)$ will always exist and in this case be real. For example, in 1734 Euler evaluated

$$\zeta(2) = \pi^2/6.$$

More generally, we can evaluate $\zeta(2m)$ for positive integers m as

$$\zeta(2m) = \frac{(-1)^{m+1}(2\pi)^{2m} B_{2m}}{2(2m)!},$$

where B_m is the *Bernoulli number* defined as part of the coefficient of t^n in the Taylor series expansion of $\frac{t}{e^t-1}$, given by

$$\frac{t}{e^t - 1} = \sum_{n=0}^{\infty} \frac{B_n t^n}{n!} \quad (|t| < 2\pi).$$

From this we see that $B_n = 0$ when n is an odd integer, except for $n = 1$ when $B_1 = -1/2$ (see OEIS A027641). The first few values of B_n are given below

n	0	1	2	3	4	5	6	7	8	9	10
B_n	1	$-1/2$	1/6	0	$-1/30$	0	1/42	0	$-1/30$	0	5/66

B_n $(n \geq 0)$ is a rational number which we can write as $B_n = N_n/D_n$, where $\gcd(N_n, D_n) = 1$, $D_n > 0$. We call N_n and D_n the numerator and denominator of B_n, respectively. It can be shown that if $n > 0$ and $2 \mid n$, then D_n is the product of those primes p such that $p - 1 \mid n$. There is no known simple formulation for N_n.

The Bernoulli numbers have been and remain objects of much research. One of the important occurrences of these numbers is in the famous formula

$$\sum_{k=1}^{n} k^m = \frac{1}{m+1} \sum_{k=0}^{m} (-1)^k \binom{m+1}{k} B_k n^{m+1-k},$$

discovered by the early seventeenth century German mathematician Johann Faulhaber (1580–1635). We hope the reader will forgive this lengthy digression when we point out that the AAC conjecture is equivalent to the conjecture that p divides the numerator of B_{p-1}, whenever p is a prime of the form $1 + 4k$. Also, in 1850 the German mathematician Ernst Eduard Kummer (1818–1893) famously showed that if the odd prime p does not divide the numerator of any of the Bernoulli numbers B_k for $k = 2, 4, 6, \ldots, p - 3$, then there is no integer solution of the Fermat equation

$$x^p + y^p = z^p$$

such that $p \nmid xyz$. This was the first major result concerning Fermat's last theorem and was much and rightly celebrated.

When s is complex, the reader might well wonder what n^s represents in the formula for $\zeta(s)$. In complex analysis, the study of functions of a complex variable, this has the meaning

$$e^{s \log n}.$$

We know by a remarkable formula of Euler that if y is real, then

$$e^{iy} = \cos y + i \sin y,$$

where cos and sin are the trigonometric sine and cosine functions, respectively. Thus when $s = \sigma + it$, where σ and t are real, we can write

$$n^s = e^{\sigma \log n}(\cos(t \log n) + i \sin(t \log n)),$$

which is what we mean by n^s.

We use the term *zeros of a function* f to designate those values of x such that $f(x) = 0$. Riemann was interested in the zeros of $\zeta(s)$. We have seen that $\zeta(s)$ is well defined when s ($= \sigma$) is real and $\sigma > 1$. It is possible to extend the region in the complex plane for which $\zeta(s)$ is defined by making use of what is called *analytic continuation*. When this is done, we find a form of $\zeta(s)$ which is valid for all complex numbers $s \neq 1$.

Riemann knew that for this $\zeta(s)$, we get $\zeta(s) = 0$ whenever s is an even negative integer. In fact, we now know that if n is an integer, then

$$\zeta(-n) = \frac{(-1)^n B_{n+1}}{n+1}.$$

We call the zeros $-2, -4, -6, \ldots$, the *trivial zeros* of $\zeta(s)$. Riemann also knew that all the nontrivial zeros of $\zeta(s)$ must lie within the *critical strip* of the complex plane, that region for which $0 < x < 1$. Furthermore, they are symmetric with respect to the *critical line* $x = 1/2$. While he did not express this in the exact terminology we are using here, he declared his belief that all the nontrivial zeros of $\zeta(s)$ must have $\sigma = 1/2$; this is the Riemann hypothesis (RH).

The RH is of enormous importance in number theory, particularly for investigating questions concerning the distribution of primes. As we saw in Chap. 11, it allows us to refine considerably the prime number theorem. It also permits us to say something about the size of gaps between successive primes. In its more general forms like the ERH and the GRH, it can be applied to a host of other problems such as the Artin conjecture in Chap. 4 and to those mentioned in Chaps. 5, 17 and 19. To quote Harold Edwards,

> "Hilbert included the problem of proving the Riemann hypothesis in his list of the most important unsolved problems which confronted mathematics in 1900, and the attempt to solve this problem has occupied the best efforts of many of the best mathematicians of the twentieth century. It is now unquestionably the most celebrated problem in mathematics and it continues to attract the attention of the best mathematicians, not only because it has gone unsolved for so long but also because it appears tantalizingly vulnerable and because its solution would probably bring to light new techniques of far reaching importance."

Because of this importance, there have been several attempts to test it by using computers. Today the first 103800788359 nontrivial zeros of the zeta function have been tabulated (https://www.lmfdb.org/zeros/zeta/), and each has its real part equal to $1/2$. All the zeros have also been found in other large regions within the critical strip, but these too have real part equal to $1/2$. In view of this seemingly overwhelming evidence for the truth of the RH, there are still specialists who doubt that it is true. They have good reasons for doing so, but most investigators today are believers.

The Clay Mathematics Institute (CMI) of Cambridge, Massachusetts, has offered as one of 7 Millennium Prizes, the sum of \$1 million (USD) for a solution to this problem. The award of this prize is subject to three conditions:

(1) The proposed solution must be published in an approved venue;
(2) At least two years must have passed since publication;
(3) The proposed solution must have received general acceptance in the global mathematics community.

It is very likely that the CMI's money with regard to the RH is safe for the foreseeable future.

The questions raised here and throughout this book represent only a small sample of many queries that have been raised by numerous investigators. It is difficult to know at this point which may be resolved soon and which may have to wait many years or even centuries for a verdict, but the mere existence of these questions, taken together with the relentlessness of human curiosity, ingenuity and stamina, bode well for the health of our discipline. With respect to this, we should mention another less famous quote from Weinberg:

> "The effort to understand the universe is one of the very few things that lifts human life a little above the level of farce, and gives it some of the grace of tragedy."

Apart from acknowledging the rather depressing tone of this statement, we also point out that physics is not the only human endeavour; surely much the same, without the gloomy crosscurrent, could be said for the effort to understand the universe of the integers.

In closing, we remark that we are mindful of a beautiful simile attributed to the great mathematician and physicist, Sir Isaac Newton (1642–1727):

> "I do not know what I may appear to the world, but to myself I seem to have been only like a boy playing on the seashore, and diverting myself in now and then finding a smoother pebble or a prettier shell than ordinary, whilst the great ocean of truth lay all undiscovered before me."

We admit that in our research we have often been just like that boy, greatly enjoying the immediate investigation, but remaining entirely ignorant of its greater significance, if any. Without being too trite, we know the truth is out there, but it may be very difficult to uncover. Please don't forget to have fun finding whatever small piece of it you are inclined to explore; after all, as any biker knows, the journey is much more important than the destination. Have a good ride!

Further Reading

Weinberg quotes

Steven Weinberg, *The First Three Minutes: A Modern View of the Origin of the Universe*, Basic Books, Inc., NY, 1977, pp. 154–155.

The AAC conjecture

A. J. van der Poorten, H. J. J. te Riele and H. C. Williams, Computer verification of the Ankeny-Artin-Chowla conjecture for all primes less than 100000000000, *Mathematics of Computation*, **70** (2000), 1311–1328, Corrigendum, **72** (2002), 521-523.

Takashi Agoh, Congruences related to the Ankeny-Artin-Chowla conjecture, *Integers* **16** (2016), #A12.
A. Reinhart, A counterexample to the Pellian conjecture of Mordell, *Acta Arithmetica*, **215** (2024), 85–95.
A. Reinhart, A counterexample to the conjecture of Ankeny, Artin and Chowla. Available at: https://arxiv.org/html/2410.21864v1

The Bateman-Horn conjecture

Soren Laing Aletheia-Zomlefer, Lenny Fukshansky and Stephen Ramon Garcia, The Bateman-Horn conjecture: heuristic, history, and applications, *Expositiones Mathematicae*, **38** (2020), 430–479.

The *abc*-conjecture

Andrew Granville and Thomas J. Tucker, It's as easy as abc, *Notices of the American Mathematical Society*, **49** (2002), 1224–1231.
Michel Waldschmidt, Lecture on the abc conjecture and some of its consequences, *Springer Proceedings in Mathematics and Statistics*, **98** (2015), 211–230.
Manon Bischoff, $1 Million will go the mathematician who busts the "ABC Conjecture" theory, *Scientific American*, July 28, 2023. Available at: https://www.scientificamerican.com/article/1-million-will-go-to-the-mathematician-who-busts-the-abc-conjecture-theory/

The Riemann hypothesis

Marcus du Sautoy, *The Music of the Primes: Searching to Solve the Greatest Mystery in Mathematics*, HarperCollins Publishers Inc., NY, 2003.
Peter Borwein, Stephen Choi, Brendan Rooney, Andrea Weirathmueller, Eds., *The Riemann Hypothesis: A Resource for the Afficionado and Virtuoso Alike*, CMS Books in Mathematics, Springer, Science+Business Media, LLC, NY 2008.
H. M. Edwards, *Riemann's Zeta Function*, Dover Publications, NY, 2001.
Manon Bischoff, The biggest problem in mathematics is finally a step closer to being solved, *Scientific American*, July 1, 2024. Available at: https://www.scientificamerican.com/article/the-riemann-hypothesis-the-biggest-problem-in-mathematics-is-a-step-closer/

Solutions

Problems of Chap. 2

2.1 (a) $100 = 4 \times 21 + 16$.
2.2 (c) $\{1, 2, 3, 5, 6, 10, 15\}$.
2.3 (c) 42.
2.4 (c) $s_0(8) = 8$, $s_1(8) = 7$, $s_2 = s(7) = 1$.
2.5 (a) $2 \cdot 3 \cdot 5 \cdot 7 \cdot 11 + 1 = 2311$ and 2311 is a prime that is not in S.
2.6 (b) $2^3 \cdot 17 \cdot 43$.
2.7 (a) $\#p_3 = p_1 \cdot p_2 \cdot p_3 = 2 \cdot 3 \cdot 5 = 30$.
2.8 (a) $\sqrt{211} \approx 14.52583$ so check primes up to 13 to see if they divide 211, if none do, then 211 must be a prime. We see that $2 \nmid 211$, $3 \nmid 211$, $5 \nmid 211$, $7 \nmid 211$, $11 \nmid 211$ and $13 \nmid 211$.
2.10 (a) $\gcd(321, 241) = 1$.
2.11 (a) $s = 5$, $t = -9$, $\gcd(435, 240) = 15$.
2.13 (a) $x = 1$.

Problems of Chap. 3

3.1 (b) 𝍸𝍸𝍸𝍸||.
3.2 (a) □.
3.4 (b) $745_8 = 485$.
3.5 (a) $176 = 10110000_2$.
3.7 (a) ˏ ɤ ⩽ $= 2015$.
3.8 (b) $4352 =$ > ⩽̷ ⩾̷.
3.9 (a) $< \vee \quad << \vee\vee\vee\vee\vee\vee\vee = 687$.
3.10 (b) $10623 = \vee\vee \quad <<<<< \vee\vee\vee\vee\vee\vee\vee \quad \vee\vee\vee$.
3.11 (a) $\begin{matrix}\dot{\overline{\equiv}}\\ \ddot{=}\end{matrix} = 372$.
3.12 (a) $201 = \begin{matrix}=\\ \cdot\end{matrix}$.

E. L. F. Roettger and H. C. Williams, *The Enchantment of Numbers*, CMS/CAIMS Books in Mathematics 16, https://doi.org/10.1007/978-3-031-76568-1

3.13 (a) $T'\Pi'B' = 982$.
3.14 (b) $6789 =, F\Psi'\Pi'\Theta'$.
3.15 (b) $DCXXIX = 629$.
3.16 (b) $955 = CMLV$.
3.17 (b)  $= 954$.
3.18 (b) $851 =$ .
3.19 (a) $(10-1)_{\text{NAF}} = 3$.
3.20 (a) $7 = (100-1)_{\text{NAF}}$.
3.21 (a) $2^0 \cdot 3^3 + 2^0 \cdot 3^0 = 28$.
3.22 (b) $100 = 2^5 \cdot 3^1 + 2^2 \cdot 3^0$.
3.23 $\{a_1, a_2, a_3, a_4\} = \{3, 15, 17, 46\}$.
3.25 $x = 211$.

Problems of Chap. 4

4.1 (a) 5.
4.2 (b) 2.
4.3 (c) 10.
4.4 (a) $|f(x)| = |9x^2 + 2x + 6| < 9x^2 + |2x| + 6 \leq 9x^2 + 2x^2 + 6x^2 = 17x^2 = 17g(x)$.
4.5 (a) $\{1, 5, 7, 11\}$.
4.6 (b) $\phi(12) = 4$.
4.7 (b)

$$\begin{aligned}
2^1 &\equiv 2 \pmod{11}, & 2^6 &\equiv 9 \pmod{11} \\
2^2 &\equiv 4 \pmod{11}, & 2^7 &\equiv 7 \pmod{11} \\
2^3 &\equiv 8 \pmod{11}, & 2^8 &\equiv 3 \pmod{11} \\
2^4 &\equiv 5 \pmod{11}, & 2^9 &\equiv 6 \pmod{11} \\
2^5 &\equiv 10 \pmod{11}, & 2^{10} &\equiv 1 \pmod{11}.
\end{aligned}$$

4.8 (a) $x^7 \equiv 9 \pmod{50} \rightarrow 7\,\text{ind}_3\, x \equiv \text{ind}_3\, 920\text{m}$; note that from table $\text{ind}_3\, 9$ is 2 and by EEA we can see 7^{-1} modulo 20 is 3. Thus, $\text{ind}_3\, x \equiv 7^{-1} \cdot 2 \equiv 3 \cdot 2 \equiv 6 \pmod{20} \rightarrow x \equiv 3^6 \equiv 29 \pmod{50}$.

Problems of Chap. 5

5.1 (a) No solutions, as $a^{\phi(m)/d} \equiv 6^{20/5} \equiv 6^4 \equiv 21 \pmod{25}$.
5.2 (a) Solve $a^{(q-1)/2}$ modulo q for $a = 1, 2, 3, 4$:

$$\begin{aligned}
1^2 &\equiv 1 \pmod 5), && \text{1 is a quadratic residue,} \\
2^2 &\equiv -1 \pmod 5, && \text{2 is a quadratic nonresidue,} \\
3^2 &\equiv -1 \pmod 5, && \text{3 is a quadratic nonresidue,} \\
4^2 &\equiv 1 \pmod 5, && \text{4 is a quadratic residue.}
\end{aligned}$$

5.3 (b) Solve k^2 modulo q for $k = 1, 2, 3, 4, 5, 6$:

$$\begin{aligned} 1^2 &\equiv 1 \pmod 7, & 4^2 &\equiv 2 \pmod 7,\\ 2^2 &\equiv 4 \pmod 7, & 5^2 &\equiv 4 \pmod 7,\\ 3^2 &\equiv 2 \pmod 7, & 6^2 &\equiv 1 \pmod 7; \end{aligned}$$

thus, the quadratic residues are 1, 2, 4.
5.4 (a)

$$\begin{aligned}(35|431) &= (5|431)(7|431) = (-1)^{(5-1)(431-1)/4}(431|5)(-1)^{(7-1)(431-1)/4}(431|7)\\ &= 1\cdot(431|5)\cdot(-1)\cdot(431|7) = -(1|5)(4|7) = -(1)\cdot(1) = -1;\end{aligned}$$

thus, there are no solutions.
5.5 (a) $(13|223) = (-1)^{(13-1)(223-1)/4}(223|13) = 1\cdot(223|13) = (2|13) = -1$; thus, there are no solutions.

Problems of Chap. 6

6.1 (a) $\sum_{i=1}^{5} i = 1+2+3+4+5 = 15$.
6.2 (a) $\prod_{i=1}^{5} i = 1\cdot 2\cdot 3\cdot 4\cdot 5 = 120$.
6.3 (b) 1.02101.
6.4 (a) $\nu_3(459) = 3$.
6.5 (a) $\nu_7(47!) = \lfloor \frac{47}{7} \rfloor = 6$.
6.6 (a) $(a+b)^5 = a^5 + 5a^4b + 10a^3b^2 + 10a^2b^3 + 5ab^4 + b^5$.
6.7 (a) For $m = 3$ and $n = 2$, we have

$$\sum_{k=0}^{m}\binom{n+k}{k} = \sum_{k=0}^{3}\binom{2+k}{k} = \binom{2}{0}+\binom{3}{1}+\binom{4}{2}+\binom{5}{3} = 20$$

and

$$\binom{m+n+1}{m} = \binom{6}{3} = 20.$$

Upon plotting these on Pascal's triangle, we see the hockey-stick emerge

$$
\begin{array}{ccccccccccccccccccccc}
&&&&&&&&&&1&&&&&&&&&& \\
&&&&&&&&&1&&1&&&&&&&&& \\
&&&&&&&&\boxed{1}&&2&&1&&&&&&&& \\
&&&&&&&1&&\boxed{3}&&3&&1&&&&&&& \\
&&&&&&1&&4&&\boxed{6}&&4&&1&&&&&& \\
&&&&&1&&5&&10&&\boxed{10}&&5&&1&&&&& \\
&&&&1&&6&&15&&\boxed{20}&&15&&6&&1&&&& \\
&&&1&&7&&21&&35&&35&&21&&7&&1&&& \\
&&1&&8&&28&&56&&70&&56&&28&&8&&1&& \\
&1&&9&&36&&84&&126&&126&&84&&36&&9&&1& \\
1&&10&&45&&120&&210&&252&&210&&120&&45&&10&&1
\end{array}
$$

6.8 (b) $C_n = \frac{1}{n+1}\binom{2n}{n} = \frac{1}{4+1}\binom{2\cdot 4}{4} = 70/5 = 14$.
6.9 (a) $\binom{n}{k} = \binom{25}{2} = 300$, $\binom{\lfloor n/p \rfloor}{\lfloor k/p \rfloor}\binom{r}{s} = \binom{3}{0}\binom{4}{2} = 6$ and $300 \equiv 6 \pmod 7$.

Problems of Chap. 7

7.1 (a) Alice sends $A \equiv 50 \pmod{73}$ to Bob and the shared secret is 3.
7.2 (a) Note entries in respective tables match at $i = 11$ and $j = 2$ and $N = \lfloor \sqrt{239 - 1} \rfloor = 16$; thus, $x = i + Nj = 11 + 16 \cdot 2 = 43$.
7.4 (a) Solve $g^x \equiv c \pmod p$ for the case $2^x \equiv 29 \pmod{37}$. Note $37 - 1 = 36 = 2^2 \cdot 3^3$. We will do the $q = 2$ case first by solving $x \equiv x_0 + 2x_1 \pmod{2^2}$.

- $c^{(p-1)/q} \equiv 29^{36/2} \equiv 36 \pmod{37}$ and $g^{(p-1)/q} \equiv 2^{36/2} \equiv 36 \pmod{37}$. Since $c^{(p-1)/q} \equiv g^{x_0(p-1)/q} \rightarrow x_0 = 1$.
- $c_1 \equiv c \cdot g^{-x_0} \pmod p \rightarrow c_1 \equiv 29 \cdot 2^{-1} \equiv 29 \cdot 19 \equiv 33 \pmod{37}$ and so $c_1^{(p-1)/q^2} \equiv 33^{36/4} \equiv 1 \pmod{37}$. Since $c_1^{(p-1)/q^2} \equiv g^{x_1(p-1)/q} \pmod{37} \rightarrow 1 \equiv (-1)^{x_1} \pmod{37} \rightarrow x_1 = 0$. Thus, $x \equiv x_0 + 2x_1 \equiv 1 + 2 \cdot 0 \equiv 1 \pmod 4$.

Second, we will solve the $q = 3$ case, by solving $x \equiv x_0 + 3x_1 \pmod{3^2}$.

- $c^{(p-1)/q} \equiv 29^{36/3} \equiv 1 \pmod{37}$ and $g^{(p-1)/q} \equiv 2^{36/3} \equiv 26 \pmod{37}$. Since $c^{(p-1)/q} \equiv g^{x_0(p-1)/q} \rightarrow x_0 = 0$.
- $c_1 \equiv c \cdot g^{-x_0} \pmod p \rightarrow c_1 \equiv 29 \cdot 2^0 \equiv 29 \pmod{37}$ and so $c_1^{(p-1)/q^2} \equiv 29^{36/9} \equiv 26 \pmod{37}$. Since $c_1^{(p-1)/q^2} \equiv g^{x_1(p-1)/q} \pmod{37} \rightarrow 26 \equiv 26^{x_1} \pmod{37} \rightarrow x_1 = 1$. Thus, $x \equiv x_0 + 3x_1 \equiv 0 + 3 \cdot 1 \equiv 3 \pmod 9$.

Now we recover the solution using the CRT for

$$
\begin{aligned}
x &\equiv 1 \pmod 4 \quad (x \equiv a_1 \pmod{m_1}),\\
x &\equiv 3 \pmod 9 \quad (x \equiv a_2 \pmod{m_2}).
\end{aligned}
$$

Let $M = m_1 m_2 = 4 \cdot 9 = 36$, $M_1 = M/m_1 = 36/4 = 9$ and $M_2 = M/m_2 = 36/9 = 4$. Use EEA to find $y_1 = 1$, $y_2 = 7$ in

$$y_1 M_1 \equiv 1 \pmod{m_1} \rightarrow y_1 9 \equiv 1 \pmod{4},$$
$$y_2 M_2 \equiv 1 \pmod{m_2} \rightarrow y_1 4 \equiv 1 \pmod{9}.$$

Thus, $x \equiv y_1 a_1 M_1 + y_2 a_2 M_2 \equiv 1 \cdot 1 \cdot 9 + 7 \cdot 3 \cdot 4 \equiv 93 \equiv 21 \pmod{36}$, i.e. $x = 21$.

7.5 (a) From the table given, we see $\text{ind}_{19} 2 = 44$ and $\text{ind}_{19} 5 = 50$. Also, after taking indices of $29 \cdot 19^3 \equiv 2^4 \cdot 5 \pmod{191}$, we see $\text{ind}_{19} 29 \equiv -3 + 4\,\text{ind}_{19} 2 + \text{ind}_{19} 5 \equiv -3 + 4 \cdot 44 + 50 \equiv 223 \equiv 33 \pmod{190}$, i.e. $x = 33$.

7.7 (a) Let $p = 17$, $q = 19$, then $N = pq = 323$ and $\phi(N) = (p-1)(q-1) = 288$. Run EEA on $\phi(N)$ and e to yield $-3\phi(N) + 5e = 1 \rightarrow 5e \equiv 1 \pmod{\phi(N)}$. Thus, $d = 5$ and $m \equiv c^d \equiv 268^5 \equiv 13 \pmod{323}$.

7.8 (a) $p = \frac{(n-\phi(n)+1) \pm \sqrt{(n-\phi(n)+1)^2 - 4n}}{2} = 461, 439.$

Problems of Chap. 8

8.2 (a) 8 ways.

8.3 (a) $\frac{F_6}{F_5} = \frac{8}{5} = 1.6$.

8.4 (b) 89 miles is approximately 987 km.

8.7 (b) $L_5^2 - 5F_5^2 = 11^2 = 5 \cdot 5^2 = 121 - 125 = -4 = 4(-1)^5$.

8.8 (b) $F_{2 \cdot 5} = F_{10} = 55 = 5 \cdot 11 = F_5 \cdot L_5$.

8.9 (a) $L_{2 \cdot 2} = L_4 = 7 = 9 - 2 = 3^2 - 2(-1)^2 = L_2^2 - 2(-1)^2$.

8.10 (a) $F_{3+2} = F_5 = 5 = 2 \cdot 2 + 1 \cdot 1 = F_3 F_3 + F_2 F_2 = F_3 F_{2+1} + F_2 F_{3-1}$ and $L_{3+2} = L_5 = 11 = 8 + 3 = 4 \cdot 2 + 1 \cdot 3 = L_3 F_3 + F_2 L_2 = L_3 F_{2+1} + F_2 L_{3-1}$.

8.11 (a) $\sum_{i=0}^{2} F_i = F_0 + F_1 + F_2 = 0 + 1 + 1 = 2 = 3 - 1 = F_4 - 1 = F_{2+2} - 1$.

8.12 (a) $\sum_{i=0}^{2} F_i^2 = F_0^2 + F_1^2 + F_2^2 = 0^2 + 1^2 + 1^2 = 2 = 1 \cdot 2 = F_2 F_3 = F_2 F_{2+1}$.

8.13 (a) $\sum_{i=1}^{2 \cdot 2 - 1} F_i F_{i+1} = \sum_{i=1}^{3} F_i F_{i+1} = 1 \cdot 1 + 1 \cdot 2 + 2 \cdot 3 = 9 = 3^2 = F_4^2 = F_{2 \cdot 2}^2$.

8.14 (b) $\sum_{k=0}^{\lfloor 5/2 \rfloor} \binom{n-k}{k} = \sum_{k=0}^{2} \binom{5-k}{k} = \binom{5}{0} + \binom{4}{1} + \binom{3}{2} = 1 + 4 + 3 = 8 = F_6 = F_{5+1}$

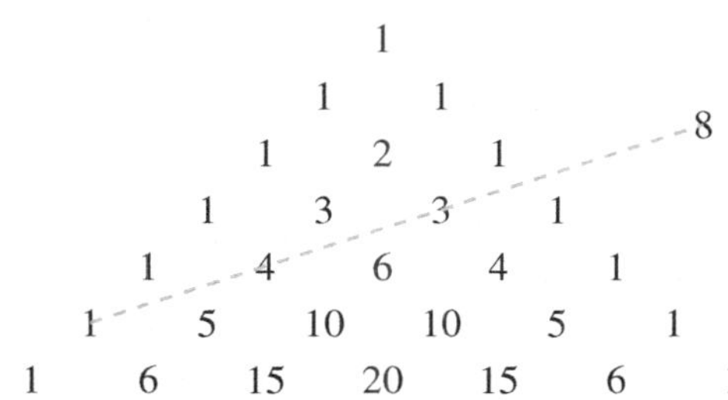

8.15 (b) $\sum_{k=0}^{\lfloor 5/2 \rfloor} \frac{5}{5-k} \binom{5-k}{k} = \sum_{k=0}^{2} \frac{5}{5-k} \binom{5-k}{k} \frac{5}{5}\binom{5}{0} + \frac{5}{4}\binom{4}{1} + \frac{5}{3}\binom{3}{2} = 11 = L_5$.

8.16 (a) $\sum_{i=0}^{2} \binom{2}{i} F_i = \binom{2}{0} F_0 + \binom{2}{1} F_1 + \binom{2}{2} F_2 = 1 \cdot 0 + 2 \cdot 1 + 1 \cdot 1 = 3 = F_4 = F_{2 \cdot 2}$.

8.17 (a) $\sum_{i=0}^{n} \binom{2}{i} L_i \binom{2}{0} L_0 + \binom{2}{1} L_1 + \binom{2}{2} L_2 = 1 \cdot 2 + 2 \cdot 1 + 1 \cdot 3 = 7 = L_4 = L_{2 \cdot 2}$.

8.18 (a) $\sum_{i=0}^{2} (-1)^{i+1} \binom{2}{i} F_i = (-1)^1 \binom{2}{0} F_0 + (-1)^2 \binom{2}{1} F_1 + (-1)^3 \binom{2}{2} F_2 = (-1)(1)(0) + (1)(2)(1) + (-1)(1)(1) = 1 = F_2$.

8.19 (a) $\sum_{i=0}^{2} (-1)^i \binom{n}{i} L_i = (-1)^0 \binom{2}{0} L_0 + (-1)^1 \binom{2}{1} L_1 + (-1)^2 \binom{2}{2} = (1)(1)(2) + (-1)(2)(1) + (1)(1)(3) = 3 = L_2$.

8.21 (a) $2 \mid 4$, $F_2 = 1$ and $F_4 = 3$ and $1 \mid 3$.
8.22 (a) $\epsilon(3) = (3|5) = (-1)^{(3-1)(5-1)/4}(5|3) = (2|3) = -1$, $F_{3-\epsilon(3)} = F_4 = 3$; thus, $3 \mid F_{3-\epsilon(3)}$ as $3 \mid 3$.
8.23 (c) $\epsilon(7) = (7|5) = (2|5) = -1, L_{7-\epsilon(3)} = L_8 = 47$; thus, $L_{7-\epsilon(3)} \equiv 5 \equiv -2 \equiv 2\epsilon(7) \pmod 7$.

Problems of Chap. 9

9.1 (a) $\sigma(68962) = \frac{2^2-1}{2-1}\frac{29^3-1}{29-1}\frac{41^2-1}{41-1} = 109746$.
9.6 (a) $s_0(95) = 95, s_1(95) = 25, s_2(95) = s(25) = 6, s_i(95) = 6$ for all $i \geq 3$, as 6 is a perfect number.
9.7 (a) $s(6) = 1 + 2 + 3 = 6$.
9.8 (a) $(2^{13-1})(2^{13} - 1) = 33550336$.
9.9 $r_1 = 3$, $r_2 \equiv 7 \pmod{M_{19}}$, $r_3 \equiv 47 \pmod{M_{19}}$, $r_4 \equiv 2207 \pmod{M_{19}}$, $r_5 \equiv 152264 \pmod{M_{19}}$, $r_6 \equiv 354554 \pmod{M_{19}}$, $r_7 \equiv 244924 \pmod{M_{19}}$, $r_8 \equiv 420095 \pmod{M_{19}}$, $r_9 \equiv 86240 \pmod{M_{19}}$, $r_{10} \equiv 326503 \pmod{M_{19}}$, $r_{11} \equiv 409010 \pmod{M_{19}}$, $r_{12} \equiv 208425 \pmod{M_{19}}$, $r_{13} \equiv 132664 \pmod{M_{19}}$, $r_{14} \equiv 470878 \pmod{M_{19}}$, $r_{15} \equiv 399999 \pmod{M_{19}}$, $r_{16} \equiv 439061 \pmod{M_{19}}$, $r_{17} \equiv 523263$ (mod)M_{19}, $r_{18} \equiv 0 \pmod{M_{19}}$; thus, M_{19} is a prime.
9.11

```
              0 0 0 0 1 1 1
            × 0 0 0 0 1 1 1
            ---------------
              0 0 0 0 1 1 1
            0 0 0 1 1 1 0
          0 0 1 1 1 0 0
        0 0 0 0 0 0 0
      0 0 0 0 0 0 0
    0 0 0 0 0 0 0
 +0 0 0 0 0 0 0
```

since $2^{7+i} \equiv 2^i \pmod{M_7}$ → exploit wrap-around

⑦	⑥	⑤	④	③	②	①
0	0	0	0	1	1	1
0	0	0	1	1	1	0
0	0	1	1	1	0	0
0	0	0	0	0	0	0
0	0	0	0	0	0	0
0	0	0	0	0	0	0
0	0	0	0	0	0	0

⑦	⑥	⑤	④	③	②	①
0	0	0	0	1	1	1
0	0	0	1	1	1	0
0	0	1	1	1	0	0
0	0	0	0	0	0	0
0	0	0	0	1	1	0
0	0	0	1	1	0	0
0	0	0	0	0	0	0

working → with pawns

⑦	⑥	⑤	④	③	②	①
				♙	♙	♙
			♙	♙	♙	
		♙	♙	♙		

We can apply rule (iii) repeatedly and move all the pawns up.

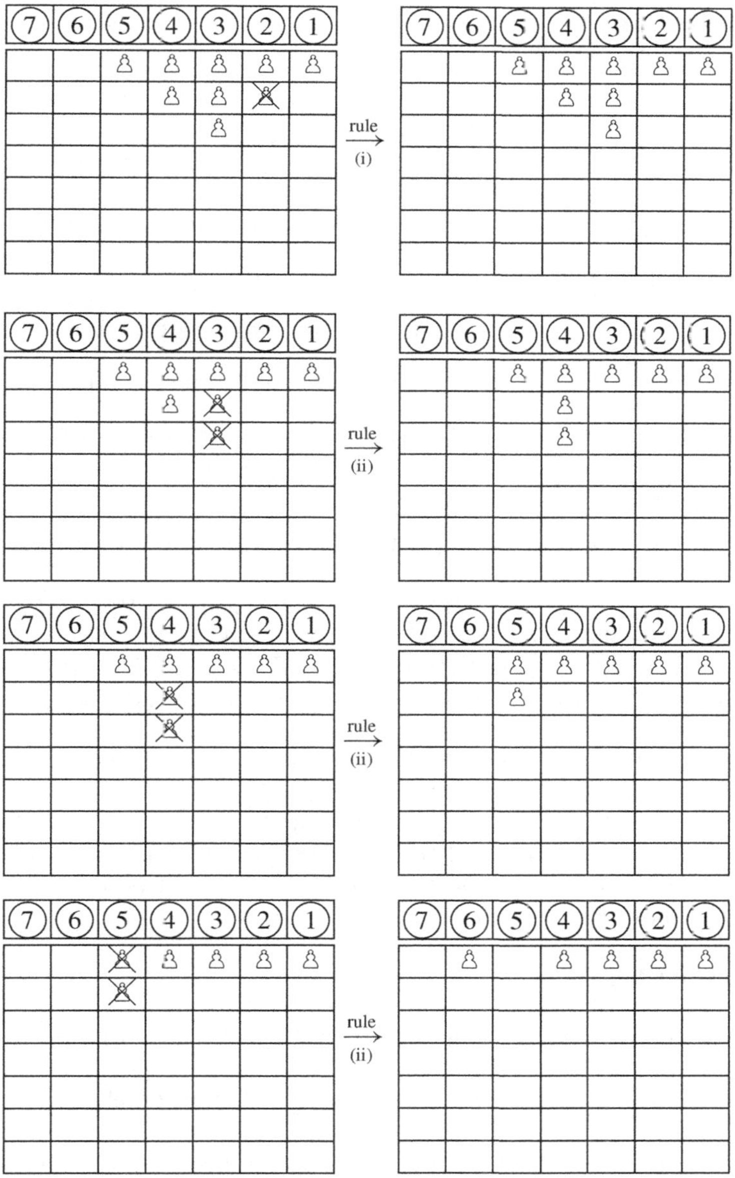

$2^5 + 2^3 + 2^2 + 2^1 + 2^0 = 47.$

Problems of Chap. 10

10.1 (a)

n	U_n	V_n	n	U_n	V_n	n	U_n	V_n
0	0	2	5	31	33	10	1023	1025
1	1	3	6	63	65	11	2047	2049
2	3	5	7	127	129	12	4095	4097
3	7	9	8	255	257	13	8191	8193
4	15	17	9	511	513	14	16383	16385

10.2 (a) $U_{n+m} = U_{2+3} = U_5 = 31$ and $U_m U_{n+1} - QU_n U_{m-1} = U_3 U_3 - 2U_2 U_2 = 7 \cdot 7 - 2 \cdot 3 \cdot 3 = 31$. $V_{n+m} = V_{2+3} = V_5 = 33$ and $V_m U_{n+1} - QU_n V_{m-1} = V_3 U_3 - 2U_2 V_2 = 9 \cdot 7 - 2 \cdot 3 \cdot 5 = 33$.
10.3 (a) $2U_{m+n} = 2U_{3+2} = 2U_5 = 2 \cdot 31 = 62$ and $U_m V_n + V_m U_n = U_3 V_2 + V_3 U_2 = 7 \cdot 5 + 9 \cdot 3 = 62$. $2V_{m+n} = 2V_{3+1} = 2V_5 = 2 \cdot 33 = 66$ and $V_m V_n + DU_m U_n = V_3 V_2 + (P^2 - 4Q)U_3 U_2 = 9 \cdot 5 - (3^2 - 4 \cdot 2) \cdot 7 \cdot 2 = 45 + 21 = 66$.
10.4 (a) $D = P^2 - 4Q = 3^2 = 4 \cdot 2 = 1$, $\epsilon(p) = (D|p) = (1|5) = 1$, $U_{p-\epsilon(p)} = U_{5-1} = U_4 = 15$ and $5 \mid 15$.
10.5 (a) $D = P^2 - 4Q = 3^2 = 4 \cdot 2 = 1$, $\epsilon(p) = (D|p) = (1|5) = 1$, $V_{p-\epsilon(p)} = V_{5-1} = V_4 = 17 \equiv 2 \equiv 2Q^0 \equiv 2Q^{(1-\epsilon(p))/2} \pmod 5$.
10.6 $r_1 = 4$, $r_2 \equiv 14 \pmod{M_{13}}$, $r_3 \equiv 194 \pmod{M_{13}}$, $r_4 \equiv 4870 \pmod{M_{13}}$, $r_5 \equiv 3953 \pmod{M_{13}}$, $r_6 \equiv 5970 \pmod{M_{13}}$, $r_7 \equiv 1857 \pmod{M_{13}}$, $r_8 \equiv 36 \pmod{M_{13}}$, $r_9 \equiv 1294 \pmod{M_{13}}$, $r_{10} \equiv 3470 \pmod{M_{13}}$, $r_{11} \equiv 128 \pmod{M_{13}}$, $r_{12} \equiv 0 \pmod{M_{13}}$; thus, M_{13} is a prime.

Problems of Chap. 11

11.1 (a) $N = 7$, $(N-1)! + 1 = 6! + 1 = 720 + 1 = 721 = 7 \cdot 103$; thus, since $N \mid (N-1)! + 1$, $N = 7$ is a prime.
11.3 (a) $\frac{x}{\log x} = \frac{20}{\log 20} \approx 6.676164\ldots$, $1.25506 \cdot \frac{x}{\log x} = 1.25506 \frac{20}{\log 20} \approx 8.37899\ldots$ and $\pi(20) = 8$.
11.4 (a) We can immediately see $N = 1891$ is not divisible by 2, 3 or 5. Since $\sqrt{N} \approx 43.4856\ldots$; the divisors of interest are

$30 \cdot 0 + 7 = 7$	$30 \cdot 0 + 19 = 19$	$30 \cdot 1 + 7 = 37$
$30 \cdot 0 + 11 = 11$	$30 \cdot 0 + 23 = 23$	$30 \cdot 1 + 11 = 41$
$30 \cdot 0 + 13 = 13$	$30 \cdot 0 + 29 = 29$	$30 \cdot 1 + 13 = 43$
$30 \cdot 0 + 17 = 17$	$30 \cdot 1 + 1 = 31$	

Upon performing trial divisions, we see $31 \mid 1891$ and $1891 = 31 \cdot 61$.
11.5 (a) $11 \cdot 7 \cdot 5 - 3 \cdot 2 = 379$.
11.6 (a) $N - 1 = 31 - 1 = 30 = 2 \cdot 3 \cdot 5$, $3^{30} \equiv 1 \pmod{31}$, $3^{15} \equiv 30 \pmod{31}$, $3^{10} \equiv 25 \pmod{31}$ and $3^6 \equiv 16 \pmod{31}$; thus, N is a prime.
11.7 (a) $F_2 = 2^{2^2} + 1 = 17$ and $3^{(F_2-1)/2} = 3^{(17-1)/2} = 3^8 = 6561 \equiv -1 \pmod{F_2}$; thus, $F_2 = 17$ is a prime.

Problems of Chap. 12

12.1 (a) 1117.
12.2 (b) 2237.
12.3 (b) 101.
12.7 (a) $n = 4$ yields $(4+1)^3 - 4^3 = 125 - 64 = 61$ and 61 is a two-digit prime.

Problems of Chap. 13

13.1 (c)

n	U_n	n	U_n	n	U_n
0	0	5	9	10	369
1	1	6	32	11	529
2	1	7	−7	12	−896
3	−4	8	−119	13	−2575
4	−7	9	−68	14	1169

13.2 (a) $U_0 = 0$, $U_1 = 1$, $U_2 = P_1 + 2R$, $U_3 = P_1^2 + RP_1 - P_2 - 3R^2$, $U_{-1} = -U_1/R^2 = -1/R^2$, $U_{-2} = -U_2/R^4 = -(P_1 + 2R)/R^4$; thus,

$$
\begin{aligned}
U_4 &= P_1U_3 - (P_2 + 3R^2)U_2 + (P_3 + 2R^2P_1)U_1 - R^2(P_2 + 3R^2)U_0 \\
&\quad + R^4P_1U_{-1} - R^6U_{-2}, \\
&= P_1(P_1^2 + RP_1 - P_2 - 3R^2) - (P_2 + 3R^2)(P_1 + 2R) + (P_3 + 2R^2P_1)(1) \\
&\quad - R^2(P_2 + 3R^2)(0) + R^4P_1(-1/R^2) - R^6(-(P_1 + 2R)/R^4).
\end{aligned}
$$

13.3 (a) $W_0 = 6$, $W_1 = P_1$, $W_2 = P_1^2 - 2P_2 - 6R^2$, $W_3 = P_1^3 + 3RP_1^2 - 3R^2P_1 - 12R^3 - 3P_1P_2 - 6RP_2$, $W_{-1} = W_1/R^2 = P_1/R^2$, $W_{-2} = W_2/R^4 = (P_1^2 - 2P_2 - 6R^2)/R^4$; thus,

$$
\begin{aligned}
W_4 &= P_1W_3 - (P_2 + 3R^2)W_2 + (P_3 + 2R^2P_1)W_1 - R^2(P_2 + 3R^2)W_0 \\
&\quad + R^4P_1W_{-1} - R^6W_{-2}, \\
&= P_1(P_1^3 + 3RP_1^2 - 3R^2P_1 - 12R^3 - 3P_1P_2 - 6RP_2) \\
&\quad - (P_2 + 3R^2)(P_1^2 - 2P_2 - 6R^2) + (P_3 + 2R^2P_1)(P_1) \\
&\quad - R^2(P_2 + 3R^2)(6) + R^4P_1(P_1/R^2) - R^6((P_1^2 - 2P_2 - 6R^2)/R^4).
\end{aligned}
$$

Problems of Chap. 14

14.1 (a)

$$
\begin{aligned}
&789 = 6 \cdot 123 + 51 \to \frac{789}{123} = 6 + \frac{51}{123} \to \frac{789}{123} = 6 + \frac{1}{\frac{123}{51}} \\
&123 = 2 \cdot 51 + 21 \to \frac{123}{51} = 2 + \frac{21}{51} \to \frac{123}{51} = 2 + \frac{1}{\frac{51}{21}} \\
&51 = 2 \cdot 21 + 9 \to \frac{51}{21} = 2 + \frac{9}{21} \to \frac{51}{21} = 2 + \frac{1}{\frac{21}{9}} \\
&21 = 2 \cdot 9 + 3 \to \frac{21}{9} = 2 + \frac{3}{9} \to 2 + \frac{1}{3};
\end{aligned}
$$

thus,

$$
\frac{789}{123} = 6 + \cfrac{1}{\cfrac{123}{51}} = 6 + \cfrac{1}{2 + \cfrac{1}{\cfrac{51}{21}}} = 6 + \cfrac{1}{2 + \cfrac{1}{2 + \cfrac{1}{\cfrac{21}{9}}}} = 6 + \cfrac{1}{2 + \cfrac{1}{2 + \cfrac{1}{2 + \cfrac{1}{3}}}}
$$

or $\dfrac{789}{123} = [6; 2, 2, 2, 3]$.

14.2 (a)

$$[4; 3, 7, 5] = 4 + \cfrac{1}{3 + \cfrac{1}{7 + \cfrac{1}{5}}} = 4 + \cfrac{1}{3 + \cfrac{1}{\cfrac{36}{5}}} = 4 + \cfrac{1}{3 + \cfrac{5}{36}} = 4 + \cfrac{1}{\cfrac{113}{36}} = 4 + \frac{36}{113} = \frac{488}{113}.$$

14.3 (a) $[a_0; a_1, a_2, a_3] = [4; 3, 7, 5]$

$$\begin{aligned} A_0 &= a_0 A_{-1} + A_{-2} = 4 \cdot 1 + 0 = 4 & B_0 &= a_0 B_{-1} + B_{-2} = 4 \cdot 0 - 1 = 1 \\ A_1 &= a_1 A_0 + A_{-1} = 3 \cdot 4 + 1 = 13 & B_1 &= a_1 B_0 + B_{-1} = 3 \cdot 1 + 0 = 3 \\ A_2 &= a_2 A_1 + A_0 = 7 \cdot 13 + 4 = 95 & B_2 &= a_2 B_1 + B_0 = 7 \cdot 3 + 1 = 22 \\ A_3 &= a_3 A_2 + A_1 = 5 \cdot 95 + 13 = 488 & B_3 &= a_3 B_2 + B_1 = 5 \cdot 22 + 3 = 113; \end{aligned}$$

thus, $[4; 3, 7, 5] = \dfrac{A_3}{B_3} = \dfrac{488}{113}$.

14.4 (a) For $143x + 77y = 55$, let $a = 143$, $b = 77$, $c = 55$; thus, $\bar{a} = \frac{143}{11} = 13$, $\bar{b} = \frac{77}{11} = 7$, $\bar{c} = \frac{55}{11}$. We have

$$\frac{\bar{a}}{\bar{b}} = \frac{a}{b} = [1; 1, 6] = \frac{A_2}{B_2} = \frac{A_n}{B_n} \quad \text{(we see } n = 2\text{)}.$$

Hence, $\frac{A_{n-1}}{B_{n-1}} = \frac{A_1}{B_1} = \frac{2}{1}$ and

$$\begin{aligned} x_0 &= (-1)^{n-1} \bar{c} B_{n-1} = (-1)^{2-1} \cdot 5 \cdot (1) = -5, \\ y_0 &= (-1)^n \bar{c} A_{n-1} = (-1)^2 \cdot 5 \cdot 2 = 10. \end{aligned}$$

Therefore all solutions are the pairs

$$\begin{aligned} x &= x_0 + \bar{b}t = -5 + 7t, \\ y &= y_0 - \bar{a}t = 10 - 13t. \end{aligned}$$

14.5 (a) Note $\sqrt{p} = \sqrt{137} \approx 11.7046$ and

$$\frac{c}{p} = \frac{37}{137} = [0; 3, 1, 2, 2, 1, 3] = [a_0; a_1, a_2, a_3, a_4, a_5, a_6].$$

Also, $\frac{A_3}{B_3} = \frac{3}{11}$ and $\frac{A_4}{B_4} = \frac{7}{26}$; thus, for $k = 3$ we have $B_k < \sqrt{p} < B_{k+1}$ and therefore

$$B_k^2 + (pA_k - cB_k)^2 = p.$$

Making the appropriate substitutions we see $11^2 + 4^2 = 137$.
14.6 (b) Note $\alpha_0 = \sqrt{83} = \frac{0+\sqrt{83}}{1}$; thus, $P_0 = 0, Q_0 = 1, D = 83$ and $a_0 = \lfloor\sqrt{83}\rfloor = 9$. We will now repeatedly use

$$P_k = a_{k-1}Q_{k-1} - P_{k-1}, \quad Q_k = \frac{D - P_k^2}{Q_{k-1}}.$$

For $k = 1$, we have $P_1 = a_0Q_0 - P_0 = 9 \cdot 1 - 0 = 9$, $Q_1 = \frac{D-P_1^2}{Q_0} = \frac{83-9^2}{1} = 2$, $\alpha_1 = \frac{P_1+\sqrt{D}}{Q_1} = \frac{9+\sqrt{83}}{2}$, $a_1 = \left\lfloor \frac{9+\sqrt{83}}{2} \right\rfloor = 9$. Similarly, for $k = 2$, we have $P_2 = a_1Q_1 - P_1 = 9 \cdot 2 - 9 = 9$, $Q_2 = \frac{D-P_2^2}{Q_1} = \frac{83-9^2}{2} = 1, \alpha_2 = \frac{P_2+\sqrt{D}}{Q_2} = \frac{9+\sqrt{83}}{1}, a_2 = \left\lfloor \frac{9+\sqrt{83}}{1} \right\rfloor = 18$. Also, for $k = 3$, $P_3 = a_2Q_2 - P_2 = 18 \cdot 1 - 9 = 9$, $Q_3 = \frac{D-P_3^2}{Q_2} = \frac{83-9^2}{1} = 2, \alpha_3 = \frac{P_3+\sqrt{D}}{Q_3} = \frac{9+\sqrt{83}}{2} = \alpha_1$. Thus, $\sqrt{83} = [9; \overline{9, 18}]$.

Problems of Chap. 15

15.1 (a) $\lfloor\sqrt{N}\rfloor = \lfloor\sqrt{390629}\rfloor = 625$. For $a = 625 + 1 \to x = 626^2 - 390629 = 1247$, which cannot be a square by modular exclusion. Next, $a = 625 + 2 \to x = 627^2 - 390629 = 2500 = 50^2$. We have $N = a^2 - b^2$ for $a = 627$ and $b = 50$; thus, $N = (a - b)(a + b) = (627 - 50)(627 + 50) = 577 \cdot 677$.
15.2 We begin by setting $w = \lfloor\sqrt{N}\rfloor = \lfloor\sqrt{54253}\rfloor = 232$; also, since $N = w^2 + r$, we have $r = 429$. The first congruence $s^2 \equiv N \pmod{p^k}$ with $s = 8$ and $p^k = 3^5$ yields $m = w + (s + yp^k) = 232 + (8 + 1 \cdot 3^5) = 483$ and $t = m(2w - m) + r = 483(2 \cdot 232 - 483) + 429 = -8748$. This is easily factored as we know $3^5 \mid t$; thus, $t = -2^2 \cdot 3^7$. Also, $N = (w - m)^2 + t \to (w - m)^2 \equiv -t \pmod{N} \to 251^2 \equiv 2^2 \cdot 3^7 \pmod{N}$. Similarly, using the second congruence with $s = 60$ and $p^k = 37^2$, we have $m = 1661$ and $t = -1987788 = -2^2 \cdot 3 \cdot 11^2 \cdot 37^2$ and $(w - m)^2 \equiv -t \pmod{N} \to 1429^2 \equiv 2^2 \cdot 3 \cdot 11^2 \cdot 37^2 \pmod{N}$. We then have

$$(251 \cdot 1429)^2 \equiv (2^2 \cdot 3^4 \cdot 11 \cdot 37)^2 \pmod{N}.$$

Thus, $a = 358679, b = 131868$ and $\gcd(a - b, N) = 239$; hence, $N = 227 \cdot 239$.
15.4 (a) We set

$$M = \prod_{\text{primes } q \leq B} q^{\lfloor \log_q B \rfloor} = 2^2 \cdot 3 \cdot 5 = 60$$

and solve $\gcd(a^M - 1, N) = \gcd(2^{60} - 1, 899) = 31$; hence, $N = 29 \cdot 31$.
15.5 (a) We have $x_1 = x_0^2 + 1 = 5^2 + 1 = 26$, $x_2 = x_2^2 + 1 = 26^2 + 1 = 677 \equiv 150 \pmod{N}$, $y_1 = x_2 - x_1 = 150 - 26 = 124$ and $\gcd(y_1, N) = \gcd(124, 527) = 31$; thus, $N = 527 = 17 \cdot 31$.
15.6 (a) Set $\alpha_0 = \sqrt{N} = \sqrt{9167}$. We have $a_0 = \lfloor\alpha_0\rfloor = \lfloor\sqrt{9167}\rfloor = 95$, $\alpha_1 = \frac{1}{\alpha_0 - a_0} = \frac{1}{\sqrt{9167}-95} = \frac{\sqrt{9167}+95}{142} = \frac{\sqrt{N}+P_1}{Q_1}$ and $\frac{A_0}{B_0} = 95$. Also, $A_{n-1} \equiv$

$(-1)^n Q_n \pmod N) \to A_0^2 \equiv (-1)^1 \cdot 142 \pmod N \to 95^2 \equiv -2 \cdot 71 \pmod N$. Next, $a_1 = \lfloor \alpha_1 \rfloor = \left\lfloor \frac{\sqrt{9167}+95}{142} \right\rfloor = 1$, $\alpha_2 = \frac{1}{\alpha_1 - a_1} = \frac{1}{\frac{\sqrt{9167}+95}{142} - 1} = \frac{\sqrt{9167}+47}{49} = \frac{\sqrt{N}+P_2}{Q_2}$ and $\frac{A_1}{B_1} = 95 + 1/1 = 96$. Now, $A_{n-1} \equiv (-1)^n Q_n \pmod N \to 96^2 \equiv 7^2 \pmod N$. Thus, $\gcd(96-7, 9167) = 89$; hence, $N = 9167 = 89 \cdot 101$.
15.7 From the cases $i = 3$ and 7, we have

$$\begin{aligned} 311^2 \cdot 319^2 &\equiv 2^4 \cdot 3 \cdot 5 \cdot 7 \cdot 2^6 \cdot 3 \cdot 5 \cdot 7 \pmod N \\ \to (311 \cdot 319)^2 &\equiv (2^5 \cdot 3 \cdot 5 \cdot 7)^2 \pmod N \\ \to 99209^2 &\equiv 3360^2 \pmod N. \end{aligned}$$

We have $\gcd(99209 - 3360, 95041) = 101$; thus $N = 95041 = 101 \cdot 941$.

Problems of Chap. 17

17.1 (a) We begin by setting $\alpha_0 = \sqrt{17}$ and noting $D = 17$, $P_0 = 0$, $Q_0 = 1$, $Q_{-1} = 17$ and $\lfloor \sqrt{17} \rfloor = 4$. Also $P_0 + \lfloor \sqrt{D} \rfloor = 0 + 4 = 4 \cdot 1 + 0 = a_0 \cdot Q_0 + R_0$, i.e. $a_0 = 4$ and $R_0 = 0$. Using the given identities, we have $P_1 = -\lfloor \sqrt{D} \rfloor - R_0 = 4 - 0 = 4$ and $Q_1 = Q_{-1} - a_0(P_1 - P_0) = 17 - 4(4-0) = 1$. Now we see $P_1 + \lfloor \sqrt{D} \rfloor = 4 + 4 = 8 = 8 \cdot 1 + 0 = a_1 \cdot Q_1 + R_1$; we have $a_1 = 8$ and $R_1 = 0$. Again, using the identities, $P_2 = \lfloor \sqrt{D} \rfloor - R_1 = 4 - 4 = 0$ and $Q_2 = Q_0 - a_1(P_2 - P_1) = 1 - 8(4-4) = 1$. Since $P_1 = P_2$ and $Q_1 = Q_2$, we have $\sqrt{17} = [4; \overline{8}]$.
17.2 (a) We begin by setting $\alpha_0 = \frac{2+\sqrt{17}}{13}$; thus, $D = 17$, $P_0 = 2$, $Q_0 = 13$, $Q_{-1} = \frac{D-P_0^2}{Q_0} = \frac{17-2^2}{13} = 1$ and $\left\lfloor \sqrt{17} \right\rfloor = 4$. Also $P_0 + \lfloor \sqrt{D} \rfloor = 2 + 4 = 0 \cdot 13 + 4 = a_0 \cdot Q_0 + R_0$, i.e. $a_0 = 0$ and $R_0 = 4$. Using the given identities, we have $P_1 = +\lfloor \sqrt{D} \rfloor - R_0 = 4 - 6 = -2$ and $Q_1 = Q_{-1} - a_0(P_1 - P_0) = 1 - 0(-2-2) = 1$. Now we see $P_1 + \lfloor \sqrt{D} \rfloor = -2 + 4 = 2 = 2 \cdot 1 + 0 = a_1 \cdot Q_1 + R_1$; we have $a_1 = 2$ and $R_1 = 0$. Again, using the identities, $P_2 = \lfloor \sqrt{D} \rfloor - R_1 = 4 - 0 = 4$ and $Q_2 = Q_0 - a_1(P_2 - P_1) = 13 - 2(4 - (-2)) = 1$. Next, $P_2 + \lfloor \sqrt{D} \rfloor = 4 + 4 = 8 = 8 \cdot 1 + 0 = a_2 \cdot Q_2 + R_2$; we have $a_2 = 8$ and $R_2 = 0$. Again, using the identities, $P_3 = \lfloor \sqrt{D} \rfloor - R_2 = 4 - 0 = 4$ and $Q_3 = Q_1 - a_2(P_3 - P_2) = 1 - 8(4-4) = 1$. Since $P_2 = P_3$ and $Q_2 = Q_3$, we have $\frac{2+\sqrt{17}}{13} = [0; 2\overline{8}]$.

Problems of Chap. 18

18.1 (a) Note $p_3 = 5$, $k = 5$, $n = 2$. We have $e_1 = \lfloor 5/2 \rfloor + \lfloor 5/2^2 \rfloor = 3$, $e_2 = \lfloor 5/3 \rfloor = 1$; thus, $P(2,5) = 2^3 \cdot 3^1 = 24$. Hence,

$$F(2,5) = \frac{5!}{P(2,5)} + \frac{P(2,5)}{(5-1)!} - \left\lfloor \frac{(5-1)!}{P(2,5)} \right\rfloor = \frac{120}{24} + \frac{24}{24} - \left\lfloor \frac{24}{24} \right\rfloor = 5 + 1 - 1 = 5 = k.$$

18.2 (a) Note $p_3 4$, $k = 4$, $n = 2$. We have $e_1 = \lfloor 4/2 \rfloor + \lfloor 4/2^2 \rfloor = 3$, $e_2 = \lfloor 4/3 \rfloor = 1$; thus, $P(2,4) = 2^3 \cdot 3^1 = 24$. Hence,

$$F(2,4)=\frac{4!}{P(2,4)}+\frac{P(2,4)}{(4-1)!}-\left\lfloor\frac{(4-1)!}{P(2,4)}\right\rfloor=\frac{24}{24}+\frac{24}{6}-\left\lfloor\frac{6}{24}\right\rfloor=1+4-0=5=k+1.$$

18.3 (a)

$$\begin{aligned}p_1&=1+\sum_{i=1}^{2^1}\left\lfloor\left(\frac{1}{\sum_{j=1}^{i}\left\lfloor\left(\cos\frac{(j-1)!+1}{j}\pi\right)^2\right\rfloor}\right)^{1/1}\right\rfloor\\&=1+\left\lfloor\frac{1}{\sum_{j=1}^{1}\left\lfloor\left(\cos\frac{(j-1)!+1}{j}\pi\right)^2\right\rfloor}\right\rfloor+\left\lfloor\frac{1}{\sum_{j=1}^{2}\left\lfloor\left(\cos\frac{(j-1)!+1}{j}\pi\right)^2\right\rfloor}\right\rfloor\\&=1+\left\lfloor\frac{1}{\left\lfloor\left(\cos\frac{0!+1}{1}\pi\right)^2\right\rfloor}\right\rfloor+\left\lfloor\frac{1}{\left\lfloor\left(\cos\frac{0!+1}{1}\pi\right)^2\right\rfloor+\left\lfloor\left(\cos\frac{1!+1}{2}\pi\right)^2\right\rfloor}\right\rfloor\\&=1+\left\lfloor\frac{1}{\lfloor(1)^2\rfloor}\right\rfloor+\left\lfloor\frac{1}{\lfloor(1)^2\rfloor+\lfloor(-1)^2\rfloor}\right\rfloor=1+\lfloor 1\rfloor+\lfloor 1/2\rfloor=1+1+0=2.\end{aligned}$$

18.3 (a)

$$\begin{aligned}f(2)&=\left\lfloor\frac{2!\bmod(2+1)}{2}\right\rfloor(2-1)+2=\left\lfloor\frac{2\bmod(3)}{2}\right\rfloor(1)+2\\&=\lfloor 1\rfloor\cdot 1+2=3=2+1\end{aligned}$$

Problems of Chap. 19

19.1 (a) In this case the period $p=6$; since this is even, $(t,u)=(A_{p-1},B_{p-1})=(A_5,B_5)=(295,28)$.

Problems of Chap. 20

20.1 (a) We have $ax^2+bxy+cy^2+dx+ey+f=x^2-4y^2+2x+4=0$; thus,

$$\begin{aligned}D&=b^2-4ac=0^2-4\cdot 1\cdot(-4)=16,\\E&=bd-2ae=0\cdot 2-2\cdot 1\cdot 0=0,\\F&=d^2-4af=2^2-4\cdot 1\cdot 4=-12,\\N&=E^2-DF=0^2-16(-12)=192=2^6\cdot 3.\end{aligned}$$

Note $X^2-DY^2=N\to X^2-16Y^2=192\to(X-4Y)(X+4Y)=2^6\cdot 3$. Hence, $X+4Y$ must belong to

$$k \in K = \{1, 2, 3, 4, 6, 8, 12, 16, 24, 32, 48, 64, 96, 192\}.$$

Upon inspection we see that there are possible solutions $X = \pm 16$, $Y = \pm 2$, when $k = 8$. We will now recover x, y, from the above using $X = Dy + E = 16y$ and $Y = 2ax + by + d = 2x + 2$:

$$X = 16 \to y = 1, \quad X = -16 \to y = -1, \quad Y = 2 \to x = 0, \quad Y = -2 \to x = -2;$$

thus, the possible integer solutions are $(0, 1)$, $(0, -1)$, $(-2, 1)$, $(-2, -1)$.

20.2 (b)

$$(k - 1, k + 1, 4k, 16k^3 - 4k) = (10 - 1, 10 + 1, 4 \cdot 10, 16 \cdot 10^3 - 4 \cdot 10) = (9, 11, 40, 15960).$$

20.3 (b)

$$\begin{aligned}&(F_{2k}, F_{2k+2}, F_{2k+4}, 4F_{2k+1}F_{2k+2}F_{2k+3}) = (F_6, F_8, F_{10}, 4F_7F_8F_9)\\ &= (8, 21, 55, 4 \cdot 13 \cdot 21 \cdot 34) = (8, 21, 55, 37128).\end{aligned}$$

20.4 (b) Since $X^3 + Y^3 = (X + Y)(X^2 - XY + Y^2)$, we see that $X + Y$ must be a divisor of $k = 1241 = 17 \cdot 73$. Thus, $X + Y$ must be in the set $\{1, 17, 73, 1241\}$. It is easily seen that if $X + Y = 73$ or 1241, then $X^3 + Y^3$ is much larger than 1241. Also, $X + Y = 1$ is also not helpful. Hence, $X + Y = 17$. Without loss of generality, if we insist $X \leq Y$, then

$$(X, Y) \in \{(0, 17), (1, 16), (2, 15), (3, 14), (4, 13), (5, 12), (6, 11), (7, 10), (8, 9).$$

Upon checking each of these, we see that $(8, 9)$ is our desired pair.

20.5 (a)

$$x = m^2 - n^2 = 3^2 - 2^2 = 5, \quad y = 2mn = 2 \cdot 3 \cdot 2 = 12, \quad z = m^2 + n^2 = 3^2 + 2^2 = 13.$$

Index

E. L. F. Roettger and H. C. Williams, *The Enchantment of Numbers*, CMS/CAIMS Books in Mathematics 16, https://doi.org/10.1007/978-3-031-76568-1

The manufacturer's authorised representative in the EU is Springer Nature Customer Service Centre GmbH, Europaplatz 3, 69115 Heidelberg, Germany. If you have any concerns regarding our products, please contact ProductSafety@springernature.com

Printed and bound by CPI Group (UK) Ltd, Croydon, CR0 4YY
30/09/2026
02225570-0012